Empathy Pathways

Andeline dos Santos

Empathy Pathways

A View from Music Therapy

Andeline dos Santos
School of the Arts
University of Pretoria
Pretoria, South Africa

ISBN 978-3-031-08555-0 ISBN 978-3-031-08556-7 (eBook)
https://doi.org/10.1007/978-3-031-08556-7

Cover credit: Hanna Tsiarleyeva/Alamy Stock Photo

This Palgrave Macmillan imprint is published by the registered company Springer Nature Switzerland AG
The registered company address is: Gewerbestrasse 11, 6330 Cham, Switzerland

Acknowledgements

Thank you, Miguel, Emilio, Judah, and Khaya for coming along on this adventure with me. As Team dos Santos, we've all learned about empathy together and there's no one I'd rather do that with than you. Carol Lotter, Lisa Shippey, Charlotte James, and Sunelle Fouché, you live and breathe the wisest kinds of empathy. I appreciate your patience in letting me bounce all these ideas off you. They took shape through our conversations and through your care. Duncan Reyburn, without you, I probably wouldn't have had the courage to take the leap and write a book. Thank you for your sharp mind and kind friendship. Mercédès Pavlicevic, I wish you were here so I could share this with you. Thank you for lighting the fire of curiosity in me. To every one of my Music Therapy Master's students: you have contributed significantly to who I am, how I think, and my love for music therapy. Thank you for the rich journey of learning together. A final "thank you" to all the music therapists across the world, to the social development practitioners and community musicians, and the teenagers and children who attended the empathy workshops that helped me refine the ideas that are presented in this book. I am grateful.

Contents

1

An Introduction to Empathies

Curiosity is oxygen for empathy. Without it, empathy struggles to ignite and battles to endure. Ironically, we can lose our sense of curiosity towards empathy itself. The study of empathy has been turbulent in fields such as philosophy, aesthetics, psychology, and neuroscience because of remarkable and ongoing disagreement about the concept's definition (Van Baaren et al., 2009). While empathy is a core pillar of music therapy, we, however, have left it almost entirely unquestioned. As a field, we have not actively taken part with curiosity in debates about what the concept means, how we make meaning with it, and the implications of those meanings.

Many descriptions of empathy revolve around sharing in and understanding another person's emotions. One separate person gains access to the emotional world of another. An entire worldview holds up this idea. It is individualistic and affirms the possibility of access to other people's "inner world." Can we really see inside another, though? And are we discrete, separate selves? Is that your worldview? Even if it is, would you be happy stating that it's the only valid worldview? If our answer to either of these questions is "no," then what happens to the concept of empathy? Do we abandon it, or do we invite it to expand?

A. dos Santos, *Empathy Pathways*,
https://doi.org/10.1007/978-3-031-08556-7_1

Empathy may look, sound, and behave differently in diverse contexts. Music therapy itself is a diverse field. Some have argued that music therapy should be so sensitive to the context that it resists "one-size-fits-all-anywhere" (Ansdell, 2003, p. 31) approaches. If this is the case, we would need to acknowledge a pluriversality (Mbembe, 2015) of practices, concepts, theories, and research methods. As empathy is a central component of our work (Bruscia, 2001), our understanding of the concept needs to be flexible enough so that it is fit for purpose within music therapy's advance as a responsive practice. When we notice that we play in a pluriverse, new possibilities open up for all of us. As Robin Dunford (2017) argued in *Towards a decolonial global ethics*, a certain value, way of living, policy, or practice is compatible with a pluriverse "if it allows other worlds to survive and thrive, and incompatible if it inevitably involves the destruction of other life-worlds" (p. 390). By definition, empathy invites us to acknowledge that there are other ways of being in the world (Clohesy, 2013).

Empathy has different meanings for different clients (Bachelor, 1988). Who are our thinking tools for? Our satisfaction or our clients' best interests? Our task is to stand behind our clients, look over their shoulders and ask them what they see. We may feel passionately invested in our orientation of choice but caring for the client always supersedes how much we care about our frameworks. My goal in this book is not to present an argument for the "rightness" of one view of empathy. The ideas in these pages are thinking tools that we can use flexibly.

We will explore four pathways of empathy, two of which are situated in a constituent approach (that prioritises discrete individuals who then enter into relationships with one another) and two of which are located in relational approaches (that acknowledge the foundational reality of relationships themselves). The philosophical positions underpinning these empathy pathways are rarely present in a "pure" state. I've attempted to separate central themes to offer a map, but I recognise overlaps in the terrain. I invite us to move within this empathy map, try out different positions, and notice what each allows us to see, hear, feel, think, and do and how each one blocks or limits us. While this book offers immersion into rich theoretical worlds, I hope that after reading it, you will be motivated towards more empathic practice, with critical, accessible, and actionable ideas for doing so. At the end of this book, you will find an Appendix within which reflection sheets are included for each empathy pathway. These invite you to explore your own and your clients' empathic processes.

Let us question the habitual. When we are habituated, we stop questioning, and we stop feeling questioned. This, as Perec (1973) wrote, is a form of anaesthesia. Instead, "what we need to question is bricks, concrete, glass, our table manners, our utensils, our tools, the way we spend our time, our rhythms. To question that which seems to have ceased forever to astonish us" (p. 205). Let's allow empathy to astonish us once again.

Empathy is an Expansive Concept

The roots of the English word "empathy" lie in the arts. The term originated from the German *Einfühlung*, meaning "feeling into" (Coplan & Goldie, 2011), which Robert Vischer first used in 1873 in the context of aesthetics. Soon after, Theodor Lipps (1903) employed *Einfühlung* to describe how people encounter aesthetic objects and how they can know the mental states of others. Edward Titchener translated the idea of *Einfühlung* into "empathy" in 1909. Interestingly, the English word "empathy" is derived from the Greek word εμπάθεια, which means malevolence or malice. When brought into English, it assumed the opposite meaning (we refer to this as a "false friend"). In Greek, the word "sympathy" relates more closely to what we now refer to when we mean empathy ("syn" meaning with or together and "pathos" meaning suffering, passion, emotion and feeling) (Stavropoulos, 2001).

Empathy was then moulded in Euro-American thought through phenomenology and hermeneutics, clinical psychology, social and developmental psychology, and neuroscience research. Instead of being refined and clarified, however, meanings of the term expanded in multiple directions. Up to 52 definitions of empathy have been listed by Nowak (2011) and 43 by Cuff et al. (2016). This diversity continues to proliferate (Hall & Schwartz, 2019). While little consensus has emerged in the literature about what counts as empathy, most definitions include some form of shared feeling, emotion, or affect as an essential component (Cuff et al., 2016).

Within Euro-American empirical research of the last century, empathy has frequently been addressed through two distinct but interrelated avenues. Affective empathy entails an emotional response to another's emotions, while cognitive empathy involves understanding another's emotions. As Zurek and Scheithauer (2017) carefully explained, although often used in overlapping ways, cognitive empathy differs from broad perspective-taking (which includes attempting to gain access to another's thoughts, beliefs, opinions, expectations, political perspectives, and so on). Cognitive empathy refers only to the cognitive understanding of another's emotional states. Affective and cognitive empathy appear to have different developmental paths (Singer, 2006), phylogenetic trajectories (Preston & de Waal, 2002), and related neurological systems (Decety & Jackson, 2004).

Primatologist Frans de Waal (2009) contended that empathy is an evolved human response, but one that is shaped by social and cultural cues. The meaning of and responses to empathy and overlapping phenomena vary widely cross-culturally (Murray et al., 2017). Until we better understand empathic experiences, expressions, and understandings in diverse cultural contexts, economic and political conditions, and social situations, definitions of empathy remain limited, imprecise, and biased. While it's not helpful to stretch out the concept so widely that it becomes unrecognisable and unusable for theory, research, or practice, it's essential to consider various "things called empathy" (Batson, 2009, p. 3) as they unfold in specific cultural contexts.

Although many people groups across the world refer to types of social knowing resembling the English constructions of empathy, very few use identical concepts. Hollan and Throop (2011) brought together anthropological accounts of empathy in the Pacific region, for example, to respond to the dearth of research on empathy across cultural contexts. While there doesn't seem to be a local word equivalent to "empathy" in this geographic area, there are significant overlaps in the use of terms equivalent to the English words "compassion," "love," "pity," and "concern" (which may, in part, be due to transculturation through Christianisation and colonisation). Murray et al. (2017) conducted another anthropological exploration of empathy, focused on relatedness and care among rural Mapuche women in Chile. They explored how the Spanish notion of carinõ (affection) functions similarly to how empathy is considered in other societies.

In his exploration of empathy among the Tzotzil Maya of Chiapas, Mexico, Groark (2008) engaged with the notion of social opacity, the widely held belief that it's nearly impossible, yet very helpful, to know another's inner state accurately. As a result, "empathic in-sight" among the Maya is an ambivalent, complex, and often paradoxical balancing act. In Israel, Kaneh-Shalit (2017) articulated a type of "emotion-free empathy" (p. 98) as part of a notion of care within an egalitarian ethos of solidarity. Omoiyari is a Japanese term whose meaning, according to Shimizu (2000), includes compassion (nino), sincerity (makoto), indulging another, and having the other's goodwill at heart (amae).

Wholeness and interrelatedness were emphasised in *Sacred Tree*, a book written by Bopp et al. (1989), a native American inter-tribal group as part of the Four Worlds Development Project. The authors highlighted how it's "possible to understand something only if we can understand how it is connected to everything else" (p. 26). According to Ly (2014), seeing yourself as a part of "the whole" is one of the most natural ways to become empathic. Dorlando (2011) researched how empathy may be part of dynamic resilience processes within the relational worldview of Native American adults. Communal empathy emerged in her data as a dynamic and relational process of collectively shared emotions and compassionate action towards the good of the community.

Steven Edwards (2010), a South African psychologist, noted that human beings cannot help but feel with each other because individually unique existences are always radically social and intersubjective. In both the Zulu and Xhosa languages, this empathic process is referred to through the phrase *umuntu ngumuntu ngabantu*, which means "a person is a person through others" (p. 323).

This very limited "mind-map" of empathy and closely related concepts within diverse sociocultural spaces give a glimpse into an expanse of meanings. As music therapists, we have primarily built our work on conventional, limited, Western views of empathy, grounded in individualistic beliefs. As Cameroonian philosopher Mbembe (2015) has highlighted, the Eurocentric canon has tended to attribute truth only to Western ways of knowledge production. This tradition has become hegemonic and disregards other knowledge traditions. According to O'Hara (1997), a psychologist in the United States, the Western notion of empathy as a "royal road to understanding" (p. 2) another individual has vastly underplayed and even obscured relational realities.

I write as a White South African, saturated in Western thinking through a privileged upbringing in an African country. It is impossible for me to speak from within Indigenous Knowledge Systems with deeply relational understandings of empathy and empathy-like constructs. I have no desire to appropriate cultural ideas or claim insight that is not my own. My socialisation, training, knowledge, experiences, and questions regarding empathy are incomplete and I write with a humble and uncomfortable acknowledgement of that. I do not draw on varied references in this book with the intention of paying lip service to superficial diversity. I seek to listen. I offer this book as part of a much broader conversation that is rich, polyphonic, sonorous, and offers varied healing pathways. I also write as a music therapist (and human being) who is aware that I still have a long journey ahead of me as I continue to develop my own empathic skills. We are learning together.

A Focus on Emotions

If empathy is a process that involves emotions at its core, then we need to unpack what we mean by "emotions." Yet again, though, we've bumped up against a word that offers no consensus regarding its meaning (Mulligan & Scherer, 2012). The term "emotion" is derived from the Latin *emovere*, meaning "to stir up," "to agitate," or "to move" (Plutchik, 1994). Emotions are explained in the literature as both states and processes (Johnson, 2009). Salzen (2001) reviewed nearly 100 theories of emotion, grouping them into five types according to whether the focus lay on adaptive observable behavioural responses, response feedback related to bodily states, cognitive appraisal, neural systems, or conflict theories where emotions are understood as interruptions to one's thought and conduct.

An argument for distinguishing between basic emotions and more complex culturally variable emotions has been presented (although authors such as Barrett [2017] have critiqued this). Basic emotions are described as being universal, innate, and physiologically discrete (Niedenthal et al., 2017). While there is some disagreement, basic emotions are frequently considered to include joy, anger, distress, fear, surprise, and disgust (Evans, 2001).

A "feeling" is often defined as "the subjective experience of an emotion or mood" (Juslin & Sloboda, 2010, p. 10), or the way our minds and bodies undergo an emotion. Feelings involve experiences of the energy associated with emotion and the movement related to that energy. The "feelingful" component of an emotional experience is particularly relevant for exploring emotional expression through music. Emotions occur over time, with variations in form and flow, similar to the unfolding of music. Due to the fact that the terms "emotions" and "feelings" are commonly used interchangeably, however, I have not sharply distinguished them in this book.

Within music therapy, emotion is always a relevant consideration (Aigen, 2005). Emotions are expressed, revealed, and explored within the musical therapeutic relationship where both therapist and client are agents in the process. Through musicking, a client may engage in an emotional experience, release emotional energy, represent their inner emotional world, and gain insight into identifying their emotions. Warrenburg (2020) explained how, within music-related literature, four main theories of emotion have predominated: basic emotions theory (emotions are functional states that are innate and universal), appraisal theory (we interpret our internal and external states in culturally variable ways which then elicit emotional responses), psychological construction theory (our brains create emotions within specific situations), and social construction theory (emotions are products of culture). Along each of the four empathy pathways in this book, we'll closely explore how different understandings of emotion dovetail specifically with each conceptualisation of empathy.

Why Dive into an In-Depth Investigation of Empathy?

A more in-depth understanding of empathy is vital for music therapists and allied professionals for many reasons. There is now robust and consistent evidence linking social connections and health. Loneliness rivals cigarette smoking, high blood pressure, lack of physical activity, and obesity as a significant health risk. Meaningful human connection is not a luxury; it's vital for survival. Empathy, particularly its expression through compassionate acts, is an antidote for loneliness (Trzeciak et al., 2019).

More than techniques, theories, or therapeutic orientations, it is the quality of the therapeutic relationship that appears to lead to positive therapeutic outcomes (Lambert & Barley, 2001; Swanick, 2019), and a central factor of this relationship is empathy, particularly client-perceived empathy (Elliott et al., 2011). In therapy, empathy is crucial for establishing authentic rapport and fostering a therapeutic alliance (Corey, 2005; Maltsberger, 2011). According to Bruscia (2001), empathy is the essential condition for developing musical rapport and engaging a client in therapeutic interactions in improvisational music therapy.

Through the empathy of the music therapist, a client can experience the therapeutic relationship as safe (Short, 2017), secure (Ruud, 2003), supportive (Kwan, 2010), trustworthy (Lindvang & Frederiksen, 1999; Pedersen, 1997), attractive (Janus et al., 2020), meaningful and beneficial (Valentino, 2006), and as a space of mutual growth (Rolvsjord, 2004). When contained by an empathic therapist, clients may be more able to accept and take ownership of their emotions (Ruud, 2003) and engage at a deeper level (So, 2017). The value of therapeutic relationships gained through empathy maybe even more significant for clients who are limited in physical and/or verbal means of expression (Kim, 2013). Mental illness can bring profound disconnection and a sense of being less than human. In this context, an empathic therapist can offer validation and opportunities for self-acceptance and belonging (Bennett, 2001).

A music therapist's ability to engage in self-empathy is also essential. Self-empathy is a protective factor for therapists against stress and towards enhancing well-being (Hojat, 2016; Nelson et al., 2018). Self-empathy improves our capacity to be empathic towards others and being empathic towards others increases our ability for self-empathy (Riess, 2018). When clients experience greater empathy for other members in a group therapeutic context, this can lead to them experiencing enhanced empathy towards themselves (Sherman, 2014).

Clients take part in music therapy for a wide range of reasons, many of which intersect with empathy challenges. Children with attention deficit hyperactivity disorder (ADHD), for example, tend to have difficulties empathising with others (Gillberg, 2007). Problems with empathy are also over-represented in individuals with Tourette's syndrome (Comings, 1990) and obsessive–compulsive disorder (OCD) with comorbid ADHD and/or tics (Thomsen, 2000). The critical distinguishing features of autism spectrum disorder (ASD) lie in the interpersonal domain, particularly in relation to empathy (Gu et al., 2015; Hobson, 2007). Some research has concluded that individuals with intellectual disabilities score lower on empathy than those without intellectual disabilities (Eyuboglu et al., 2018; Langdon et al., 2011). Children with Down's Syndrome may show concern and offer comfort to others in distress. Still, they may find empathy to be more challenging in the context of abstract and hypothetical situations (Kasari et al., 2003).

Since the earliest attempts to describe schizophrenia, difficulties in engaging with others have been recognised as key (Lee, 2007). Schizophrenia has been characterised as a disorder of representation of mental states (Frith, 1992) and intuitive attunement (Stanghellini, 2000). Persons with schizophrenia tend to display deficits in both affective and cognitive empathy (Benedetti et al., 2009).

Deficits in social functioning are strongly associated with depression (Hirschfeld et al., 2000). Empathy is frequently understood as requiring the ability to regulate one's emotional responses and flexibly shift between one's own and others' perspectives. Potential challenges in the inhibitory controls that moderate these functions may play a role in the relationship between depression and lowered empathy. Loss of emotional responsiveness to others' feelings can reduce the desire to engage in social interaction, leading to isolation and intensification of depression (Singer & Lamm, 2009). In terms of eating disorders, a significant portion of people suffering from anorexia nervosa has been found to present with empathy difficulties (Kerr-Gaffney et al., 2019).

Lower levels of affective empathy have been associated with developmental vulnerability for substance use disorders (Winters et al., 2020). Empathy can be reduced through chronic cocaine use (Quednow, 2017; Wei et al., 2020). Individuals with stimulant polysubstance use showed lower affective empathy in a study by Kroll et al. (2018). The severity of alcohol use disorder and cannabis use disorder has been associated with impairment in face processing, which appears to lead to deficits in empathy (Leiker et al., 2019).

The prototypical disorder associated with dysfunction in empathy is psychopathy (Blair, 2007; Hare, 1991). Studies have found that individuals with psychopathy may still experience empathy but are particularly skilled at switching their empathy off at will (Meffert et al., 2013). A person with psychopathy is also typically highly skilled at mimicking others' emotions (a phenomenon referred to as callous empathy) and may use their empathic skills to take pleasure in another's suffering (Aucoin & Kreitzberg, 2018). A lack of empathy is also viewed as the central dysfunctional trait of narcissistic personality disorder (NPD) (Jankowiak-Siuda & Zajkowski, 2013).

A decrease in empathy is one of the first central symptoms in the behavioural variant of frontotemporal dementia. This influences empathic concern (Baez et al., 2014), the ability to infer another's mental state, and share in another's emotional experiences (Oliver et al., 2014). Alzheimer's disease is associated with significant deficits in cognitive empathy (Demichelis et al., 2020). Many patients with traumatic brain injury experience a reduction in or loss of the ability to empathise. Even minor head injuries can disrupt aspects of social cognition (Wood & Williams, 2008).

Researchers have reached contradictory findings regarding the relationship between trauma and empathy. Couette et al. (2020) found that both affective and cognitive empathy were disturbed in individuals with posttraumatic stress disorder (PTSD). However, Mazza et al. (2015) concluded that only affective empathy was disturbed as a result of PTSD and that cognitive empathy remained intact. In contrast to studies that show impaired empathy, Greenberg et al. (2018) found that adults who reported experiencing childhood trauma had an increased ability to take and understand another's perspective, indicating how adversity can spur posttraumatic growth. Chaitin and Steinberg (2008) examined the experiences of people who have endured massive and severe trauma or who are the descendants of social trauma victims, for example, within the Palestinian-Israeli conflict. They explored whether such individuals could relate to others' suffering and develop empathy towards them and concluded that this may be very difficult. Many individuals who have suffered such extreme trauma personally or through their family history found it very hard to connect to the suffering of others who continue to be defined as "the enemy." Past horrors continue to haunt the victims and their children, re-inscribing notions of us-versus-them, a monolithic social identity that is entangled in negative interdependence of the other's identity, a collective identity founded in victimhood, distrust of others, displaced aggression, and perceptions of the world as a dangerous place. Chaitin and Steinberg argued that, while empathy is challenging in such circumstances, sometimes victimisation can still increase sensitivity towards others' pain. This can occur as one develops relationships with those labelled as the "enemy," especially when the steps of safety, acknowledgement, and reconnection have all taken place.

The relationship between aggression and empathy has been widely researched. Children with lower levels of empathy tend to exhibit increased levels of aggression (Spinrad & Eisenberg, 2009; Zhou et al., 2002). This has also been observed with adolescents (Berger et al., 2015), particularly in relation to low affective empathy (Jolliffe & Farrington, 2006).

Music therapists focus not only on individuals and small groups, but also on broader relational and societal dynamics (Elefant, 2010; Fouché & Torrance, 2005), bearing in mind that these layers notably intertwine. Intergroup conflict fuelled by cultural differences (Valentino, 2006), political disputes and inequality can still be understood at least in part as a struggle with empathy (Dovidio et al., 2010). While disputes over tangible resources often drive disagreements, emotions play a central role in all conflict instigation, maintenance, and resolution (Halperin, 2015).

If empathy speaks to how we may go about knowing the emotional experiences of others, then much (if not all) music therapy research also necessarily needs to be an empathic pursuit. Empathy hasn't been a cornerstone of academic cultures that cling to distance and objectivity and pressurise researchers to churn out findings at speed. Empathy requires responsive intersubjectivity and a slower rhythm (Bresler, 2008). If we hope that participants/co-researchers will share their emotional worlds with us, we must engage with careful, empathic consideration (Bresler, 2013; Taylor & Statler, 2014).

Much of the literature on the value of empathy draws on the concept in a way that falls within what I'm calling "insightful empathy." As this book unfolds, we'll see why travelling along all four empathy pathways can be beneficial for work and play within the music therapy field.

Four Pathways

I propose that there are four empathy pathways. As we explore each one, we'll examine what *stance* we assume, what we become *aware* of, what *emotion* processes are at play, how *meanings* are engaged with, and what our *response* may be. We could think of these as stepping stones along the paths.

As Schertz (2007) highlighted, the issue of the ontology of the subject precedes any understanding of the nature and characteristics of empathy. Any notion of empathy rests on how we think about the self and how the self connects with others. The first two pathways of empathy that we'll examine meander through the terrain of a constituent (also called a substantivist or substantialist) ontology. Here, entities are ontologically primary, and relations are subordinate. Relations do not have ontological standing; relationships are products of selves (Wildman, 2010). You and I would be seen as separate, discrete individuals, contained within ourselves, who, through coming into contact with one another, then form a relationship.

Humanism is underpinned by a constituent ontology, where individual elements are considered ontologically primary and relations are secondary (Wildman, 2010). Self-subsistent entities involve themselves in dynamic flow with one another (Emirbayer, 1997). An approach that places the client as an individual human being in the centre of the therapeutic process is supported by humanistic values of individual subjectivity and personhood (Abrams, 2010; Ansdell, 2014), agency, self-determination, dignity and autonomy (Abrams, 2018; Ruud, 2010), identity (Echard, 2019), authenticity (Yehuda, 2013), potential and self-actualisation (Abrams, 2018; Dunlap, 2017), and care for the individual (Ruud, 2010). Through music therapy grounded in this orientation, clients are invited to express their selfhood, particularly through music (Epp, 2007; Nebelung & Stensæth, 2018; Nordoff & Robbins, 2004; Pavlicevic, 1997) and access their inner wisdom (Aigen, 2015).

While philosophers such as Aristotle and Descartes considered a person to be an atomistic, sovereign self with a deep interior world (Atay, 2018), the initial emphasis of humanism, however, was not on attempting to understand another person's inner experience through privileged access to it. Humans were thought to understand one another through outward appearances. As humanism developed, and through the modern psychological concepts of "the unconscious" and the "inner voice," for example, the focus shifted back to humans' inner lives (Ansdell & Stige, 2018).

If human beings are separate units (who then form relationships with one another), a key question emerges regarding whether, and if so, how we can know one another's inner worlds. Empathy rests firmly on this question. Debates have raged over hundreds of years regarding whether separate human beings can truly know each other's inner worlds. Insightful empathy, examined in Part I of this book, operates upon the belief that "sight" "into" another's emotional states and experiences is possible. I define insightful empathy as *purposefully sharing and understanding another's emotions.* The conventional descriptions of affective and cognitive empathy found in literature depend necessarily on the capacity to grasp another's inner world and both fall under insightful empathy.

Translational empathy, explored in Part II, rests upon the belief that knowing another's inner world is not always possible and attempts to do so are not always appropriate. I define translational empathy as *a quality of presence in which a sense of withness is generated through a situated and productive process of emotion translation moves of expression and response.*

It is perhaps most helpful to consider insightful and translational empathy along a continuum rather than as a binary. I'm not arguing that complete knowledge of another is entirely and always possible; neither am I contending that the other is altogether a stranger. In reality, we may choose to build our understanding on a position of greater or lesser knowledge. We may find ourselves in situations where the person we want to empathise with presents themselves as more or less knowable.

Many authors in music therapy have shifted their interests from exploring how isolated individuals impact each other to examining ecologies, systems, networks, and relationships. Music therapy is now frequently articulated as a situated relational encounter in which all participants create and engage in therapeutic musicking. All involved are reciprocally formed through their joint engagement (Ansdell & Stige, 2015; Pavlicevic & Impey, 2013; Rolvsjord, 2006). I suggest that individualistic understandings of empathy are philosophically and practically incongruent, or at least insufficient, within this approach. Focusing on the relational "between" rather than the inner subjective worlds of participants calls for different understandings of empathy.

The last two pathways of empathy are, therefore, situated within relational ontologies. Empathising assemblages are first explored within an approach that considers relationships as ontologically primary. This form of empathy is fed by ideas from relational constructionism, posthumanism and indigenous worldviews. I define the process of an empathising assemblage in Part III as *territorialising an assemblage as mutually affectively response-able.*

Lastly, we'll consider relational empathy in Part IV. This empathy pathway is grounded in an argument against dualism and for the ontologically co-primal existence of constituents and relationships. I define relational empathy as *situated awareness of emotion co-storying.*

All accounts are partial, and this means that we can celebrate multiple points of view across the continuum of a phenomenon (Ellingson, 2011; Reddekop, 2014). These points of view do not need to be framed in terms of rigid dichotomies. We can approach them through the kind of diffraction that Haraway (1992) and Barad (2007, 2014) use to explore difference productively, in relation, and as we wonder how and for whom differences may matter. Water bottles packed? Let's embark on our first empathy pathway.

References

Abrams, B. (2010). Evidence-based music therapy practice: An integral understanding. *Journal of Music Therapy, 47*(4), 351–379.

Abrams, B. (2018). Understanding humanistic dimensions of music therapy: Editorial introduction. *Music Therapy Perspectives, 36*(2), 139–143.

Aigen, K. (2005). *Being in music: Foundations of Nordoff-Robbins music therapy.* Barcelona Publishers.

Aigen, K. (2015). A critique of evidence-based practice in music therapy. *Music Therapy Perspectives, 33*(1), 12–24.

Ansdell, G. (2003). Community music therapy: Big British balloon or future international trend? In *Community, relationship and spirit: Continuing the dialogue and debate* (pp. 1–13). BSMT Publications.

Ansdell, G. (2014). *How music helps in music therapy and everyday life.* Ashgate.

Ansdell, G., & Stige, B. (2015). Community music therapy. In J. Edwards (Ed.), *Oxford handbook of music therapy* (pp. 595–621). Oxford University Press.

Ansdell, G., & Stige, B. (2018). Can music therapy still be humanist? *Music Therapy Perspectives, 36*(2), 175–182.

Atay, A. (2018, March). *A relational ontology for peace* [Keynote lecture]. International Conference on Cultural Studies, Istanbul.

Aucoin, E., & Kreitzberg, E. (2018). Empathy leads to death: Why empathy is an adversary of capital defendants. *Santa Clara Law Review, 58*, 99–136.

Bachelor, A. (1988). How clients perceive therapist empathy: A content analysis of "received" empathy. *Psychotherapy, 25*, 227–240.

Baez, S., Manes, F., Huepe, D., Torralva, T., Fiorentino, N., Richter, F., Huepe-Artigas, D., Ferrari, J., Montanes, P., Reyes, P., Matallana, D., Vigliecca, N., Decety, J., & Ibanez, A. (2014). Primary empathy deficits in frontotemporal dementia. *Frontiers in Aging Neuroscience, 6*, article 264.

Barad, K. (2007). *Meeting the universe halfway: Quantum physics and the entanglement of matter and meaning.* Duke University Press.

Barad, K. (2014). Diffracting diffraction: Cutting together-apart. *Parallax, 20*, 168–187.

Barrett, L. F. (2017). *How emotions are made: The secret life of the brain.* Houghton Mifflin Harcourt.

Batson, C. (2009). These things called empathy: Eight related but distinct phenomena. In J. Decety & W. Ickes (Eds.), *The social neuroscience of empathy* (pp. 3–16). The MIT Press.

Benedetti, F., Bernasconi, A., Bosia, M., Cavallaro, R., Dallaspezia, S., Falini, A., Poletti, S., Radaelli, D., Riccaboni, R., Scotti, G., & Smeralido, E. (2009). Functional and structural brain correlates of theory of mind and empathy deficits in schizophrenia. *Schizophrenia Research, 114*, 154–160.

Bennett, M. (2001). *The empathic healer: An endangered species.* Academic Press.

Berger, C., Batanova, M., & Cance, J. D. (2015). Aggressive and prosocial? Examining latent profiles of behavior, social status, machiavellianism, and empathy. *Journal of Youth and Adolescence, 44*, 2230–2244.

Blair, R. (2007). Empathic dysfunction in psychopathic individuals. In T. Farrow & P. Woodruff (Eds.), *Empathy in mental illness* (pp. 3–16). Cambridge University Press.

Bopp, J., Bopp, M., Brown, L., Lane, P., & Morris, P. (1989). *Sacred tree.* Four Worlds International Institute.

Bresler, L. (2008). The music lesson. In J. G. Knowles & A. L. Cole (Eds.), *Handbook of the arts in qualitative research* (pp. 225–237). Sage.

Bresler, L. (2013). Cultivating empathic understanding in research and teaching. In B. White & T. Costantino (Eds.), *Aesthetics, empathy and education* (pp. 9–28). Peter Lang.

Bruscia, K. E. (2001). A qualitative approach to analyzing client improvisations. *Music Therapy Perspectives, 19*(1), 7–21.

Chaitin, J., & Steinberg, S. (2008). "You should know better": Expressions of empathy and disregard among victims of massive social trauma. *Journal of Aggression, Maltreatment & Trauma, 17*(2), 197–226.

Clohesy, A. (2013). *Politics of empathy: Ethics, solidarity, recognition.* Routledge.

Comings, D. E. (1990). *Tourette syndrome and human behaviour.* Hope Press.

Coplan, A., & Goldie, P. (2011). Introduction. In A. Coplan & P. Goldie (Eds.), *Empathy: Philosophical and psychological perspectives* (pp. ix–xlvii). Oxford University Press.

Corey, G. (2005). *Theory and practice of counselling and psychotherapy* (7th ed.). Brooks/Cole—Thomson Learning.

Couette, M., Mouchabac, S., Bourla, A., Nuss, P., & Ferreri, F. (2020). Social cognition in post-traumatic stress disorder: A systematic review. *British Journal of Clinical Psychology, 59*(2), 117–138.
Cuff, B., Brown, S., Taylor, L., & Howat, D. (2016). Empathy: A review of the concept. *Emotion Review, 8*(2), 144–153.
de Waal, F. (2009). *The age of empathy*. Harmony Books.
Decety, J., & Jackson, P. L. (2004). The functional architecture of human empathy. *Behavioral and Cognitive Neuroscience Review, 3*(2), 406–412.
Demichelis, O. P., Coundouris, S. P., Grainger, S. A., & Henry, J. D. (2020). Empathy and theory of mind in Alzheimer's disease: A meta-analysis. *Journal of the International Neuropsychological Society, 26*(10), 963–977.
Dorlando, H. (2011). *Communal empathy in native American older adults* [Doctoral thesis, University of Montana].
Dovidio, J., Johnson, J., Gaertner, S., Pearson, A., Saguy, T., & Ashburn-Nardo, L. (2010). Empathy and intergroup relations. In M. Mikulincer & P. Shaver (Eds.), *Prosocial motives, emotions, and behavior* (pp. 393–408). American Psychological Association.
Dunford, R. (2017). Toward a decolonial global ethics. *Journal of Global Ethics, 13*(3), 380–397.
Dunlap, A. L. (2017). *Women with addictions' experience in music therapy* [Master's thesis, College of Fine Arts of Ohio University]. https://etd.ohiolink.edu/
Echard, A. (2019). Making sense of self: An autoethnographic study of identity formation for adolescents in music therapy. *Music Therapy Perspectives, 37*(2), 141–150.
Edwards, S. (2010). A Rogerian perspective on empathic patterns in Southern African healing. *Journal of Psychology in Africa, 20*(2), 321–326.
Elefant, C. (2010). Musical inclusion, intergroup relations, and community development. In B. Stige, G. Ansdell, C. Elefant, & M. Pavlicevic (Eds.), *Where music helps: Community music therapy in action and reflection* (pp. 75–92). Ashgate.
Ellingson, L. (2011). *Engaging crystallisation in qualitative research*. Sage.
Elliott, R., Bohart, A., Watson, J., & Greenberg, L. (2011). Empathy. In J. Norcross (Ed.), *Psychotherapy relationships that work: Evidence-based responsiveness* (2nd ed., pp. 132–152). Oxford University Press.
Emirbayer, M. (1997). Manifesto for a relational sociology. *American Journal of Sociology, 103*(2), 281–317.
Epp, E. (2007). Locating the autonomous voice: Self-expression in music-centered music therapy. *Voices: A World Forum for Music Therapy, 7*(1).

Evans, D. (2001). *Emotion: A very short introduction.* Oxford University Press.
Eyuboglu, D., Bolat, N., & Eyuboglu, M. (2018). Empathy and theory of mind abilities of children with specific learning disorder (SLD). *Psychiatry and Clinical Psychopharmacology, 28*(2), 136–141.
Fouché, S., & Torrance, K. (2005). Lose yourself in the music, the moment, Yo! Music therapy with an adolescent group involved in gangsterism. *Voices: A World Forum for Music Therapy, 5*(3).
Frith, C. D. (1992). *The cognitive neuropsychology of schizophrenia.* Lawrence Erlbaum.
Gillberg, C. (2007). Non-autism childhood empathy disorders. In T. Farrow & P. Woodruff (Eds.), *Empathy in mental illness* (pp. 111–125). Cambridge University Press.
Greenberg, D., Baron-Cohen, S., Rosenberg, N., Fonagy, P., & Rentfrow, P. (2018). Elevated empathy in adults following childhood trauma. *PLoS ONE, 13*(10), e0203886.
Groark, K. (2008). Social opacity and the dynamics of empathic in-sight among the Tzotzil Maya of Chiapas, Mexico. *Ethos, 36*(4), 427–448.
Gu, X., Eilam-Stock, T., Zhou, T., Anagnostou, E., Kolevzon, A., Soorya, L., Hof, P., Friston, K., & Fan, J. (2015). Autonomic and brain responses associated with empathy deficits in Autism Spectrum Disorder. *Human Brain Mapping, 36*, 3323–3338.
Hall, J., & Schwartz, R. (2019). Empathy present and future. *The Journal of Social Psychology, 159*(3), 225–243.
Halperin, E. (2015). *Emotions in conflict: Inhibitors and facilitators of peace-making*. Routledge.
Haraway, D. (1992). The promises of monsters: A regenerative politics for inappropriate/d others. In L. Grossberg, C. Nelson, & P. Treichler (Eds.), *Cultural studies* (pp. 295–337). Routledge.
Hare, R. D. (1991). *The Hare psychopathy checklist—Revised*. Multi-Health Systems.
Hirschfeld, R., Montgomery, S. A., Keller, M. B., Kasper, S., Schatzberg, A. F., Möller, H.-J., et al. (2000). Social functioning in depression: A review. *Journal of Clinical Psychiatry, 61*, 268–275.
Hobson, P. (2007). Empathy and autism. In T. Farrow & P. Woodruff (Eds.), *Empathy in mental illness* (pp. 126–141). Cambridge University Press.
Hojat, M. (2016). *Empathy in health professions education and patient care.* Springer.

Hollan, D., & Throop, J. (2011). The anthropology of empathy: Introduction. In D. Hollan & J. Throop (Eds.), *The anthropology of empathy: Experiencing the lives of others in Pacific societies* (pp. 1–24). Berghahn Books.

Jankowiak-Siuda, K., & Zajkowski, W. (2013). A neural model of mechanisms of empathy deficits in narcissism. *Medical Science Monitor, 19*, 934–941.

Janus, S. I., Vink, A. C., Ridder, H. M., Geretsegger, M., Stige, B., Gold, C., & Zuidema, S. U. (2020). Developing consensus description of group music therapy characteristics for persons with dementia. *Nordic Journal of Music Therapy*, 1–17.

Johnson, G. (2009). Theories of emotion. In J. Fieser & B. Dowden (Eds.), *The internet encyclopedia of philosophy*. https://iep.utm.edu/emotion/

Jolliffe, D., & Farrington, D. P. (2006). Examining the relationship between low empathy and bullying. *Aggressive Behavior: Official Journal of the International Society for Research on Aggression, 32*(6), 540–550.

Juslin, P., & Sloboda, J. (2010). *Handbook of music and emotion: Theory, research, and applications*. Oxford University Press.

Kaneh-Shalit, T. (2017). The goal is not to cheer you up: Empathetic care in Israeli life coaching. *Ethos, 45*(1), 98–115.

Kasari, C., Freeman, S. F., & Bass, W. (2003). Empathy and response to distress in children with Down syndrome. *Journal of Child Psychology and Psychiatry, 44*(3), 424–431.

Kerr-Gaffney, J., Harrison, A., & Tchanturia, K. (2019). Cognitive and affective empathy in eating disorders: A systematic review and meta-analysis. *Frontiers in Psychiatry, 10*, 102.

Kim, S. (2013). *Multimodal quantification of interpersonal physiological synchrony between non-verbal individuals with severe disabilities and their caregivers during music therapy* [Doctoral dissertation, University of Toronto].

Kroll, S. L., Wunderli, M. D., Vonmoos, M., Hulka, L. M., Preller, K. H., Bosch, O. G., Baumgartner, M. R., & Quednow, B. B. (2018). Socio-cognitive functioning in stimulant polysubstance users. *Drug and Alcohol Dependence, 190*, 94–103.

Kwan, M. (2010). Music therapists' experiences with adults in pain: Implications for clinical practice. *Qualitative Inquiries in Music Therapy, 5*, 43.

Lambert, M. J., & Barley, D. E. (2001). Research summary on the therapeutic relationship and psychotherapy outcome. *Psychotherapy: Theory, Research, Practice, Training, 38*(4), 357.

Langdon, P. E., Murphy, G. H., Clare, I. C. H., Stevenson, T., & Palmer, E. J. (2011). Relationships among moral reasoning, empathy and distorted cognitions in men with intellectual disabilities and a history of criminal offending. *American Association on Intellectual and Developmental Disabilities, 116*(6), 438–456.

Lee, K. (2007). Empathy deficits in schizophrenia. In T. Farrow & P. Woodruff (Eds.), *Empathy in mental illness* (pp. 17–32). Cambridge University Press.

Leiker, E. K., Meffert, H., Thornton, L. C., Taylor, B. K., Aloi, J., Abdel-Rahim, H., Shah N., Tyler P. M., White S. F., Filbey F., Pope K., Do M. D., & Blair, R. J. R. (2019). Alcohol use disorder and cannabis use disorder symptomatology in adolescents are differentially related to dysfunction in brain regions supporting face processing. *Psychiatry Research: Neuroimaging, 292*, 62–71.

Lindvang, C., & Frederiksen, B. (1999). Suitability for music therapy: Evaluating music therapy as an indicated treatment in psychiatry. *Nordic Journal of Music Therapy, 8*(1), 47–57.

Lipps, T. (1903/1931). Empathy, inward imitation, and sense feelings. In E. F. Carritt (Ed.), *Philosophies of beauty: From socrates to Robert Bridges being the sources of aesthetic theory* (pp. 252–258). Clarendon Press.

Ly, R. (2014). *Beyond strategies: Infusing empathy and indigenous approaches in the elementary classroom* [Masters dissertation, University of Toronto].

Maltsberger, J. (2011). Empathy and the historical context, or how we learned to listen to patients. In K. Michel & D. Jobes (Eds.), *Building a therapeutic alliance with the suicidal patient* (pp. 29–48). American Psychological Association.

Mazza, M., Tempesta, D., Pino, M. C., Nigri, A., Catalucci, A., Guadagni, V., Iaria, G., & Ferrara, M. (2015). Neural activity related to cognitive and emotional empathy in post-traumatic stress disorder. *Behavioural Brain Research, 282*, 37–45.

Mbembe, A. (2015, April 23). *Decolonizing knowledge and the question of the archive.* Presentation at the University of the Witwatersrand, Johannesburg. https://wiser.wits.ac.za/system/files/Achille%20Mbembe%20-%20Decolonizing%20Knowledge%20and%2020the%20Question%20of%20the%20Archive.pdf

Meffert, H., Gazzola, V., Den Boer, J. A., Bartels, A. A., & Keysers, C. (2013). Reduced spontaneous but relatively normal deliberate vicarious representations in psychopathy. *Brain, 136*(8), 2550–2562.

Mulligan, K., & Scherer, K. (2012). Toward a working definition of emotion. *Emotion Review, 4*(4), 345–357.

Murray, M., Bowen, S., Verdugo, M., & Holtmannspötter, J. (2017). Care and relatedness among rural Mapuche women: Issues of carino and empathy. *Ethos, 45*(3), 367–385.

Nebelung, I., & Stensæth, K. (2018). Humanistic music therapy in child welfare. *Voices: A World Forum for Music Therapy, 18*(4).

Nelson, J., Hall, B., Anderson, J., Birtles, C., & Hemming, L. (2018). Self-Compassion as self-care: A simple and effective tool for counselor educators and counseling students. *Journal of Creativity in Mental Health, 13*(1), 121–133.

Niedenthal, P., Rychlowska, M., & Wood, A. (2017). Feelings and contexts: Socioecological influences on the nonverbal expression of emotion. *Current Opinion in Psychology, 17*, 170–175.

Nordoff, P., & Robbins, C. (1971/2004). *Therapy in music for handicapped children.* Barcelona Publishers.

Nowak, A. (2011). *Introducing a pedagogy of empathic action as informed by social entrepreneurs.* McGill University.

O'Hara, M. (1997). Relational empathy: Beyond modernist egocentricism to postmodern holistic contextualism. In A. C. Bohart & L. S. Greenberg (Eds.), *Empathy reconsidered* (pp. 295–319). American Psychological Association.

Oliver, L., Mitchell, D., Dziobek, I., MacKinley, J., Coleman, K., Rankin, K., & Finger, E. (2014). Parsing cognitive and emotional empathy deficits for negative and positive stimuli in frontotemporal dementia. *Neuropsychologia, 67*, 14–26.

Pavlicevic, M. (1997). *Music therapy in context.* Jessica Kingsley.

Pavlicevic, M., & Impey, A. (2013). Deep listening: Towards an imaginative reframing of health and well-being practices in international development. *Arts & Health, 5*(3), 238–252.

Pedersen, I. (1997). The Music Therapist's listening perspectives as source of information in improvised musical duets with grown-up: Psychiatric patients, suffering from Schizophrenia. *Nordic Journal of Music Therapy, 6*(2), 98–111.

Perec, G. (1973). *Species of spaces and other pieces.* Penguin.

Plutchik, R. (1994). *The psychology and biology of emotion.* Harper-Collins.

Preston, S. D., & de Waal, F. B. M. (2002). Empathy: Its ultimate and proximate bases. *Behavioral and Brain Sciences, 25*, 1–72.

Quednow, B. B. (2017). Social cognition and interaction in stimulant use disorders. *Current Opinion in Behavioral Sciences, 13*, 55–62.

Reddekop, J. (2014). *Thinking across worlds: Indigenous thought, relational ontology, and the politics of nature; Or, if only Nietzsche could meet a Yachaj* [Doctoral dissertation, University of Western Ontario]. Electronic Thesis and Dissertation Repository. 2082.

Riess, H. (2018). *The empathy effect*. Sounds True.

Rolvsjord, R. (2004). Therapy as empowerment. *Nordic Journal of Music Therapy, 13*(2), 99–111.

Rolvsjord, R. (2006). Whose power of music? *British Journal of Music Therapy, 20*(1), 5–12.

Ruud, E. (2003). "Burning scripts" self psychology, affect consciousness, script theory and the BMGIM. *Nordic Journal of Music Therapy, 12*(2), 115–123.

Ruud, E. (2010). *Music therapy: A perspective from the humanities*. Barcelona Publishers.

Salzen, E. (2001). A century of emotion theories—Proliferation without progress? *History and Philosophy of Psychology, 3*, 56–75.

Schertz, M. (2007). Empathy as intersubjectivity: Resolving Hume and Smith's divide. *Studies in Philosophy and Education, 26*(2), 165–178.

Sherman, N. (2014). Recovering lost goodness: Shame, guilt, and self-empathy. *Psychoanalytic Psychology, 31*(2), 217–235.

Shimizu, H. (2000). Japanese cultural psychology and empathic understanding: Implications for academic and cultural psychology. *Ethos, 28*(2), 224–247.

Short, H. (2017). It feels like Armageddon: Identification with a female personality-disordered offender at a time of cultural, political and personal attack. *Nordic Journal of Music Therapy, 26*(3), 272–285.

Singer, T. (2006). The neuronal basis and ontogeny of empathy and mind reading: Review of literature and implications for future research. *Neuroscience & Biobehavioral Reviews, 30*(6), 855–863.

Singer, T., & Lamm, C. (2009). The social neuroscience of empathy. *Annals of the New York Academy of Sciences, 1156*(1), 81–96.

So, H. (2017). US-trained music therapists from East Asian countries found personal therapy during training helpful but when cultural disconnects occur these can be problematic: A qualitative phenomenological study. *The Arts in Psychotherapy, 55*, 54–63.

Spinrad, T., & Eisenberg, N. (2009). Empathy, prosocial behaviour, and positive development in schools. In R. Gilman, E. Huebner, & M. Furlong (Eds.), *Handbook of positive psychology in schools* (pp. 119–129). Routledge.

Stanghellini, G. (2000). Vulnerability to schizophrenia and lack of common sense. *Schizophrenia Bulletin, 26*(4), 775–787.

Stavropoulos, D. N. (2001). *Oxford Greek-English dictionary* (12th ed.). Oxford University Press.

Swanick, R. (2019). What are the factors of effective therapy? Encouraging a positive experience for families in music therapy. *Approaches: An Interdisciplinary Journal of Music Therapy.* https://approaches.gr/wp-content/uploads/2019/11/Approaches_FirstView_a20191109-swanick-1.pdf

Taylor, S., & Statler, M. (2014). Material matters: Increasing emotional engagement in learning. *Journal of Management Education, 38*(4), 586–607.

Thomsen, P. (2000). Obsessions: The impact and treatment of obsessive-compulsive disorder in children and adolescents. *Journal of Psychopharmacology, 14*(2 Suppl 1), S31–S37.

Trzeciak, S., Mazzarelli, A., & Booker, C. (2019). *Compassionomics: The revolutionary scientific evidence that caring makes a difference* (pp. 287–319). Studer Group.

Valentino, R. (2006). Attitudes towards cross-cultural empathy in music therapy. *Music Therapy Perspectives, 24*, 108–114.

Van Baaren, R., Decety, J., Dijksterhuis, A., van der Leij, A., & Leeuwen, M. (2009). Being imitated: Consequences of nonconsciously showing empathy. In J. Decety & W. Ickes (Eds.), *The social neuroscience of empathy* (pp. 31–42). The MIT Press.

Warrenburg, L. (2020). Comparing musical and psychological emotion theories. *Psychomusicology: Music, Mind, and Brain, 30*(1), 1.

Wei, L., Wu, G. R., Bi, M., & Baeken, C. (2020). Effective connectivity predicts cognitive empathy in cocaine addiction: a spectral dynamic causal modeling study. *Brain Imaging and Behavior,* 1–9.

Wildman, W. J. (2010). An introduction to relational ontology. In J. Polkinghorne (Ed.), *The Trinity and an entangled world: Relationality in physical science and theology* (pp. 55–73). Wm. B. Eerdmans Publishing.

Winters, D., Wu, W., & Fukui, S. (2020). Longitudinal effects of cognitive and affective empathy on adolescent substance use. *Substance Use & Misuse, 55*(6), 983–989.

Wood, R., & Williams, C. (2008). Inability to empathize following traumatic brain injury. *Journal of the International Neuropsychological Society, 14*, 289–296.

Yehuda, N. (2013). 'I am not at home with my client's music… I felt guilty about disliking it': On 'musical authenticity' in music therapy. *Nordic Journal of Music Therapy, 22*(2), 149–170.

Zhou, Q., Eisenberg, N., Losoya, S. H., Fabes, R. A., Reiser, M., Guthrie, I. K., et al. (2002). The relations of parental warmth and positive expressiveness to

children's empathy-related responding and social functioning: A longitudinal study. *Child Development, 73*, 893–915.
Zurek, P. P., & Scheithauer, H. (2017). Towards a more precise conceptualization of empathy: An integrative review of literature on definitions, associated functions, and developmental trajectories. *International Journal of Developmental Science, 11*(3–4), 57–68.

Part I

Insightful Empathy

Purposefully sharing and understanding another's emotions.

2

A Stance of Receptivity

Much of the foundation of music therapy is built on the premise that we can know our clients through music. Kim and Whitehead-Pleaux (2015) wrote, "it is vital that the music therapist get to know the client fully" (p. 59). We seek and experience "direct musical-emotional knowing" (Pavlicevic, 1999, p. 62) as we read human expression through musical expression (Robarts, 2006). As the therapist actively engages in the process, the client also increasingly knows the therapist. Priestley (1994) believed that the music therapist's inner experience is as transparent through improvisation as the client's experience. If one were to assume a person-centred perspective, congruence would be the key attitude. The therapist engages in a real and genuine way as an open, integrated, and authentic person (Rogers, 2007).

In a range of music therapy texts, the concept of empathy is referred to directly as knowing another. A few (of the many) examples include Jackson and Gardstrom's (2012) definition of empathy as understanding and appreciating the client's perspective and position. Næss and Ruud (2007) referred to empathy as seeking to read another person's intentions, taking part in their emotions, and experiencing something they are experiencing. Préfontaine (2006) discussed empathy as receiving the

A. dos Santos, *Empathy Pathways*,
https://doi.org/10.1007/978-3-031-08556-7_2

client, their energy level, personal tempo, and affective environment, responding to this musically to offer resonance to their affective and musical experiences. Gardstrom and Hiller (2010) described empathy as the capacity to be aware of, sensitive to, and understand the feelings and thoughts of another person. Kim (2013) referred to empathy as the ability to accurately perceive how another person is feeling, see through their eyes, and adopt their frame of reference. According to Cooper (2010), music therapists try to enter their clients' worlds while improvising by internalising their experiences. Bruscia (2001) famously explained empathy as entering the client's music, engaging and following their process until we can sense how their body is creating the sounds, feeling what their sounds are expressing, and understanding the ideas they are employing to organise the sounds. Bruscia described living in the client's melody or rhythm and following it along its path.

Texts that focus on the client's social and cultural situatedness still emphasise the therapist's intention to gain deep knowledge of the client's inner world. In *Attitudes towards cross-cultural empathy in music therapy*, Valentino (2006) described how music therapists rely on various cues from a client to gain an empathic understanding of their internal reality and experience. Valentino highlighted that this reality develops over time and is influenced by the values and beliefs of one's social and familial context. In Grimmer and Schwantes' (2018) article on cross-cultural music therapy, they refer to empathic listening that "goes beyond the basic definition to a deep level of understanding" (p. 30), enabling the therapist to meet their clients regardless of communication barriers. As we seek to deeply know our clients through insightful empathy, we're purposefully trying to share in and understand their emotions. The first part of this process entails assuming a stance of receptivity.

Positioning Ourselves to Empathise

Humans decide whether to place themselves in situations where empathy can come into play. We do this depending on whether we want to feel the emotions we expect to encounter (Gross, 1998), our attitudes towards the emotions we think will be present (Markovitch et al., 2017), and our beliefs about how well we can control our own emotions (Rovenpor & Isbell, 2018). As therapists, we choose to repeatedly place ourselves in situations where empathy is likely to play a role in the interpersonal encounter(s) that will emerge. These situations offer invitations to participate in another person's subjectivity, and we respond by turning towards this person and allowing ourselves to be led into their emotional world.

As we embark on our journey into insightful empathy, rather than focusing on assuming "a position" that can lead to empathic engagement, we are purposefully concentrating on the idea of "positioning." Empathy is a dynamic process related to the moment-by-moment flow between people. As music therapists, we recognise this flow and encourage the "stream" to keep flowing. For example, during an improvisation, we employ particular techniques such as mirroring, imitating, meeting, and matching (Wigram, 2004) to encourage the ongoing development of reciprocal musicking, through which empathy can unfold.

Self-Awareness

Empathy necessitates turning towards the person I'm encountering, shifting away from focusing on myself, my plans, theories, and expectations, and making this person the centre of my experience. Paradoxically, one of the ways I can ensure that I'm focusing on the other person and not myself is to be aware of what's going on within me, what I bring to this encounter and how I'm responding as the interaction proceeds so that these aspects do not cloud my perspective of the other person.

Gadamer (2007) argued that the problem of relating does not lie in a struggle to understand the other, but in a struggle to understand ourselves. Brene Brown (2008) defined empathy as "the skill or ability to tap into our own experiences in order to connect with an experience someone is relating to us" (p. 33). This entails self-awareness, which involves the capacity to be the object of one's own attention (Eckroth-Bucher, 2010). Self-awareness is a state of being where one consciously notices one's own (potentially multifaceted, multi-layered, and complex) multi-sensorial inner dialogue, including emotions, thoughts, imagery, somatosensory experiences, behaviours, beliefs, and attitudes, which are formed through developmental, social, and cultural processes (Pieterse et al., 2013). Self-awareness is dynamic, including constantly shifting levels of awareness (Rochat, 2003) and an ongoing, evolving process of self-discovery (Eckroth-Bucher, 2010). Tasha Eurich (2017) explained that self-awareness has two dimensions. Internal self-awareness refers to seeing yourself clearly. It has to do with an inward understanding of your reactions, values, passions, patterns, aspirations, and impact on others. External self-awareness refers to understanding yourself "from the outside in" (p. 23), in other words, understanding how other people perceive you. Eurich found that there is no relationship between these two categories. For example, a person may have accurate internal self-awareness but weak external self-awareness.

Self-awareness can lead to enhanced empathic encounters with others (Haley et al., 2017; Horwitz, 2018; Moran, 2017; Rasheed et al., 2019; Waite & McKinney, 2016). To know one's client well, one first needs to know oneself well (Lee, 2016). Individuals who show greater interoceptive sensitivity (who can perceive changes within sensations in their bodies) appear to show higher affective and cognitive empathy towards others (Grynberg & Pollatos, 2015). Emotional self-awareness is also a prerequisite for empathy (Haley et al., 2017; Trentini et al., 2021). Emotional self-awareness is an attentional process that involves monitoring, distinguishing, and communicating emotional states, and identifying relationships between emotions and circumstances (Rieffe et al., 2008).

As therapists, we must pay attention to our own emotions and take our "emotional temperature" (Knapp et al., 2017, p. 3) before, during, and after sessions to identify the emotions expressed by clients and our reactions to these emotions. Welcoming and experiencing our own range of complex feelings assists us in recognising feelings in others (Cochran & Cochran, 2015; Tangen, 2017). Therapists who monitor their emotions appear to have better client outcomes (Hayes et al., 2011). We know this isn't always easy. Therapists may sometimes experience the desire to cope by shutting off their own painful feelings that arise from the difficult stories they hear. This can cause emotional blindness towards the self, hindering self-awareness and leading to illness, fatigue, burnout, and isolation (Warren & Nash, 2019).

Self-doubt, anxiety, and prejudice can focus our attention inwards, detracting from our therapeutic presence with others and preventing empathy from taking place. As we become more aware of these dynamics, we open ourselves up to becoming more empathic (Davis, 1990). A journey of self-awareness invariably includes travelling into some dark and difficult places within ourselves. Aveline (1990) wrote, "what therapists can bear to hear in themselves, they can hear in their patients" (p. 333).

Self-awareness can be consciously developed (Rasheed et al., 2019). Healthcare practitioners can enhance their self-awareness through, for example, seeking feedback (Rasheed, 2015), supervision (Pieterse et al., 2013), by reflective practices (Jack & Smith, 2007) like journaling (Williams et al., 2009), through expressive arts (Warren & Nash, 2019), and working on becoming conscious of their repressed feelings (So, 2017). Moran (2017) explained how important it is for music therapists to develop self-awareness through mindfulness and personal reflection, especially using processes that include creative modalities.

Motivation, Ability, and Capacity

Our motivation to empathise with others has evolutionary benefits. Empathy can foster the adaptive social behaviours that are crucial for human survival, such as helping and cooperating (Weisz & Zaki, 2018). Empathy enhances parental caregiving behaviours, which are essential for infants' survival. (Interestingly, parents with musical training are better able to discriminate infant distress (Parsons et al., 2014).

Individual differences in the ability to empathise seem to result from a complex relationship between biological predisposition and environmental factors (Knafo et al., 2008). Parental empathy is associated positively with children's secure attachment (Stern et al., 2015). In turn, more securely attached children appear to display a greater ability to empathise (Panfile & Laible, 2012). However, children's behaviours can also impact parents' abilities to empathise. For example, Psychogiou et al. (2008) found that behaviours linked to child ADHD and child psychopathy decreased parents' empathy.

While growing up in an empathic environment is likely to enhance our empathy, a range of other experiences may also improve this ability. Empathy can be invoked, awakened, nurtured, taught (Lockwood, 2016), refined, increased (Gerdes & Segal, 2011) and practised (Jeffrey & Downie, 2016). Therapeutic competencies can be developed that feed into one's empathic ability, such as listening (Préfontaine, 2006), perspective-taking, building rapport, and engaging with a client spontaneously, warmly, and playfully (Winter, 2013). Stress management enhances empathic ability (Crenshaw et al., 2019), as does effective communication techniques (Stepien & Baernstein, 2006; Valentino, 2006), diverse musical knowledge and skills (Behrens, 2012), clinical improvisation (Préfontaine, 2006), mirroring (McGarry & Russo, 2011), reflexive use of cultural schemas, and broad understandings of historical, social, and political contexts (Valentino, 2006). Several studies with music therapy students have explored the potential of experiential learning and role-play to increase empathy for clients (Jackson & Gardstrom, 2012; Murphy, 2007; Schmid & Rolvsjord, 2020; Winter, 2013). Personal therapy can also improve one's empathic abilities (Fox & McKinney, 2016; So, 2017).

The ability to empathise doesn't always equate to the motivation to empathise (Carpenter et al., 2016). The more important a relationship is to us, the more effort we will put into trying to empathise (Redmond, 2018). The evolutionary benefits of empathy appear to play out predominantly within in-groups (Hackel et al., 2017), referred to as intergroup empathy bias (Hasson et al., 2018). Humans express more empathy when the target is familiar (Preston & de Waal, 2002), relationally close (Porat et al., 2016), and in their cultural group (Galinsky et al., 2005). We are more empathic towards those who look like us (Han, 2018), are kind to us (Singer et al., 2006), and are not in competition with us (Bruneau et al., 2017). Negative attribution and stigma seem to act as catalysts to decrease empathy. Once stigmas form, they can resist being loosened (Breithaupt, 2019). A sense of kinship with a client can drive a therapist's empathy due to perceived shared characteristics (dos Santos & Brown, 2021; Valdesolo & DeSteno, 2011).

Other interpersonal and situational variables (and how we appraise them) also modulate empathy (de Vignemont & Singer, 2006). For example, we are more likely to empathise when we're in a positive mood (Nelson, 2009). Our empathy tends to decrease when we perceive the other person as behaving unfairly (Singer et al., 2006). We tend not to express empathy towards the suffering of people we consider morally in the wrong and deserving of punishment. Empathy also reduces when the observer knows that the pain the other person is experiencing is for their good (such as to cure them) (Lamm et al., 2007). The intensity of a person's emotional display modulates empathic responses from others, as does the empathiser's past experiences with similar situations (Hein & Singer, 2008). At times, the suffering of another may be so severe that we do not wish to imagine what their experience is like (Frie, 2010). It's essential for therapists to bear in mind that humans tend to be more prepared to experience empathy for a person they perceive to have the potential to change or heal or be on the path to recovery already (Breithaupt, 2019). This is a natural bias that we too can fall into.

Being in a position of power can lower our empathy. People with higher empathic abilities may be drawn to become therapists. Still, the structure of therapy contains stubborn built-in problems of control and power, regardless of the good intentions of the practitioner. Clients reach the therapy room hoping and expecting that the therapist will be able to help through their wisdom, understanding and expertise, and they, therefore, hand over a degree of power (Totton, 2017). Compared to low power individuals, those with high power tend to have reduced interpersonal sensitivity (Pril, 2017), are less capable of emotion recognition (Kraus et al., 2010) and perspective-taking (Galinsky et al., 2005), have reduced mirroring (Hogeveen et al., 2013), poorer empathic accuracy (Galinsky et al., 2008; Kraus et al., 2010), are more prone to stereotyping (Fiske, 1993; Goodwin et al., 2000), more likely to objectify others (Gruenfeld et al., 2008), and show less altruistic behaviour (Piff et al., 2012).

Besides prosocial motivations for empathy (Eisenberg et al., 2015), a large body of research has shown that antisocial motivations can underpin empathy (this is sometimes called "tactical empathy"). These can include the desire to manipulate (O'Connell, 2018), to be cunning

(Carr, 2000), to seduce, deceive or be violent (Bubandt & Willerslev, 2015), or to be on the "winning side" during negotiations (Galinsky et al., 2008). The most realistic assessment is probably that humans are prosocially motivated to empathise in certain situations and towards certain people and antisocially motivated to empathise in other situations and towards other people (O'Connell, 2018).

From within sociology, Ruiz-Junco (2017) articulated the notion of "empathy frames." These are discursive structures guided by moral claims, and we usually use them without conscious awareness. One component of empathy frames is empathy rules, which are socially learned expectations about empathy that a person has internalised and that vary across social contexts. To begin empathising, one must have identified the situation as "empathy-deserving." Setting norms at the start of a group music therapy process, for example, can help produce the situation as one where empathy can be invited and allowed to flourish. Empathy also entails identifying "rightful empathisers" according to a social hierarchy (for example, some cultures identify women as more empathic than men). In group sessions, the unspoken assumption may be that the music therapist is the empathiser. Clients may need to be intentionally invited to take hold of the role of empathiser towards others in the group. Doing empathy also involves defining appropriate "empathy recipients" (those who are "deserving" of receiving empathy). Music therapy is a space that constructs all participants as worthy of empathy.

One may have the ability and motivation to empathise (and to do so prosocially), but one's *capacity* to empathise varies. We are not constantly empathising with others (de Vignemont, 2006). In day-to-day life, we witness many contradictory emotions from people around us. If we were empathising with all of them all the time, we would be in a state of emotional turmoil and would struggle to have room for our own emotions. Our capacity to empathise can depend on the time we have at our disposal, our cognitive and emotional resources, and our flexibility (Dewall et al., 2008; Gleichgerrcht & Decety, 2013). Our level of maturity (Dileo, 2021), ego strength (Germer et al., 2013), and self-esteem (Winter, 2013) also influence our capacity to experience and express empathy towards others. In addition, the more confident we are in our ability to assess others' emotional experiences, the more motivated we are to empathise (Moskowitz, 2005). When our energy becomes depleted, our willingness to help others diminishes (particularly others who are less familiar to us) (Stewart-Williams, 2007). Burnout can affect music therapists, negatively impacting the care they can provide to clients (Clements-Cortes, 2013). If therapists continue to engage in therapeutic work while experiencing burnout, they can risk reducing their capacity for empathy (Simionato et al., 2019; Wilkinson et al., 2017). Empathy itself can, however, also be energising and enhance our well-being (Doré et al., 2017).

Paying Attention

Empathy requires our full and purposeful attention as we turn to the other (Fleischacker, 2019; Haviland, 2014). In the words of Noddings (2010), "In many, perhaps most, situations, we listen or observe receptively, and then we feel empathy; that is, attention precedes empathy" (p. 9). We don't simply pay attention to what we already expect the situation to contain (which would be an exercise in confirmation bias). We also need to wonder what is not fitting our expectations of this situation and pay attention to that.

Effortful control has been positively associated with empathy (Eisenberg et al., 2007; Valiente et al., 2004; Zorza et al., 2019). In particular, the voluntary control of the allocation of attention feeds our capacity to empathise (Eisenberg & Eggum, 2009). If we want to empathise, the best place to start is to shift our attention to focus on the other person's emotions (Zaki, 2014).

The ability to control attention strategically during sessions is a crucial skill for therapists (Ivey et al., 2017). As Bruscia (1998) explained, music therapists consistently need to centre, expand, and shift their focus of attention to navigate the flow of encounters in a music therapy session. Paying mindful attention is particularly important. This involves moving one's focus from "doing" to "being," which can enhance nonjudgemental acceptance of events and feelings in sessions and increase one's capacity to remain focused on the present moment and engage empathically (Greason & Cashwell, 2009).

A Non-Judgemental Attitude

Empathic positioning includes a committed posture of non-judgement. Within phenomenology, "bracketing" refers to the suspension of previous beliefs, knowledge, presuppositions, and prejudices while attempting to attend to a person, event, or object (Edwards, 2012; Forinash & Gonzalez, 1989). Bracketing demands critical concentration and deep reflection, curiosity and sustained naïveté to experience the world, the situation, and the person one is encountering freshly (Finlay, 2011). However, it would be unrealistic to believe that one can entirely discard one's presuppositions. In bracketing, one attempts to see around them at least temporarily, hold them "in abeyance" (Finlay, 2014, p. 122), or stand above them and be consciously aware of and more fully present to the phenomenon that one beholds (Finlay, 2009). For example, in a study I conducted on aggression and empathy with teenagers in group music therapy, I needed to see around preconceived ideas I held about "who aggressive teenagers were" and "what aggressive teenagers do." Only then could I begin to explore the adaptive features of aggression, the multifaceted understandings that these teenagers created, and how aggression could act as a sign of creativity, intelligence, and skill, rather than (only) social deficits, incompetence, and psychopathology (dos Santos, 2020). This opened a richer window for me into their lived experience.

Hinojosa (2021) conducted a music therapy project with high school students who openly identified as LGBTQ+. She explained how music afforded a safe space for exploration and expression, compared to the censorship and confinement that the participants experienced in their daily lives. In sessions, they could explore their identities and sexualities through non-judgemental and creative exploratory techniques to create a common connection and build a sense of belonging through music. Music therapy can be a space where clients feel they can bring their "uncensored self" (Lipson, 2019). Unconditional positive regard entails the therapist's absolute respect for their client, their acceptance of them precisely as they are, and their trust in their potential to grow when in nurturing conditions (Hall, 2009).

Through a trusting and unconditionally accepting relationship, clients may feel more able to disclose their struggles, feelings of shame (Gardstrom & Hiller, 2010), spiritual pain (Dimaio, 2010), and forbidden feelings (Bullard, 2011). When this occurs in group therapy contexts, empathy can be fostered as members engage in non-judgemental positioning toward each other (Gardstrom & Hiller, 2010). Through this, they grow in their ability to see others in the group truly for who they are rather than as projective screens (Ahonen & Mongillo Desideri, 2014).

The Courage to Engage in Playful Risk-Taking

Being a music therapist requires being fully present within playful and spontaneous interactions that demand courage and risk-taking (Dileo, 2021; Winter, 2013). We don't know and are not attempting to control where the interaction will lead. In *How to be a master therapist*, Kottler and Carlson (2014) acknowledged that there are tremendous risks involved in being a therapist. We are affected by the pain and suffering that we can hold, and at a certain point, we may begin to shut down vulnerable parts of ourselves. Opening ourselves to others often leads to change within ourselves. As Rogers and Farson (1957) concluded, it requires significant inner bravery and security and to risk one's self in understanding another.

References

Ahonen, H., & Mongillo Desideri, A. (2014). Heroines' journey-emerging story by refugee women during group analytic music therapy. *Voices: A World Forum for Music Therapy, 14*(1). https://doi.org/10.15845/voices.v14i1.686

Aveline, M. (1990). The training and supervision of individual therapists. In W. Dryden (Ed.), *Individual therapy: A handbook* (pp. 434–466). Open University Press.

Behrens, G. A. (2012). Use of traditional and nontraditional instruments with traumatized children in Bethlehem, West Bank. *Music Therapy Perspectives, 30*(2), 196–202.

Breithaupt, F. (2019). *The dark sides of empathy*. Cornell University Press.

Brown, B. (2008). *I thought it was just me (but it isn't): Making the journey from "what will people think?" to " I am enough"*. Avery.

Bruneau, E. G., Cikara, M., & Saxe, R. (2017). Parochial empathy predicts reduced altruism and the endorsement of passive harm. *Social Psychological and Personality Science, 8*, 934–942.

Bruscia, K. (1998). *The dynamics of music psychotherapy*.

Bruscia, K. E. (2001). A qualitative approach to analyzing client improvisations. *Music Therapy Perspectives, 19*(1), 7–21.

Bubandt, N., & Willerslev, R. (2015). The dark side of empathy: Mimesis, deception, and the magic of alterity. *Comparative Studies in Society and History, 57*(1), 5–34.

Bullard, E. (2011). Music therapy as an intervention for inpatient treatment of suicidal ideation. *Qualitative Inquiries in Music Therapy, 6*, 75–121.

Carpenter, J. M., Green, M. C., & Vacharkulksemsuk, T. (2016). Beyond perspective-taking: Mind-reading motivation. *Motivation and Emotion, 40*, 358–374.

Carr, D. (2000). Emotional intelligence, PSE and self-esteem: A cautionary note. *Pastoral Care in Education, 18*(3), 27–33.

Clements-Cortés, A. (2013). Burnout in music therapists: Work, individual, and social factors. *Music Therapy Perspectives, 31*(2), 166–174.

Cochran, J. L., & Cochran, N. H. (2015). *The heart of counseling* (2nd ed.). Routledge.

Cooper, M. (2010). Clinical-musical responses of Nordoff-Robbins music therapists: The process of clinical improvisation. *Qualitative Inquiries in Music Therapy, 5*, 86–115.

Crenshaw, A. O., Leo, K., & Baucom, B. R. (2019). The effect of stress on empathic accuracy in romantic couples. *Journal of Family Psychology, 33*(3), 327.

Davis, C. M. (1990). What is empathy, and can empathy be taught? *Physical Therapy, 70*(11), 707–711.

de Vignemont, F. (2006). When do we empathize? In G. Bock & J. Goode (Eds.), *Empathy and fairness* (pp. 181–196). Wiley.

de Vignemont, F., & Singer, T. (2006). The emphatic brain: How, when, and why? *Trends in the Cognitive Sciences, 10*, 435–441.

Dewall, C., Baumeister, R., Gailliot, M., & Maner, J. (2008). Depletion makes the heart grow less helpful: Helping as a function of self-regulatory energy and genetic relatedness. *Personality & Social Psychology Bulletin, 34*, 1653–1662.

Dileo, C. (2021). *Ethical thinking in music therapy* (2nd ed.). Jeffrey Books.
Dimaio, L. (2010). Music therapy entrainment: A humanistic music therapist's perspective of using music therapy entrainment with hospice clients experiencing pain. *Music Therapy Perspectives, 28*(2), 106–115.
Doré, B. P., Morris, R. R., Burr, D. A., Picard, R. W., & Ochsner, K. N. (2017). Helping others regulate emotion predicts increased regulation of one's own emotions and decreased symptoms of depression. *Personality and Social Psychology Bulletin, 43*(5), 729–739.
dos Santos, A. (2020). The usefulness of aggression as explored by becoming-teenagers in group music therapy. *Nordic Journal of Music Therapy, 29*(2), 150–173. https://doi.org/10.1080/08098131.2019.1649712
dos Santos, A., & Brown, T. (2021). Music therapists' empathic experiences of shared and differing orientations to religion and spirituality in the client-therapist relationship. *The Arts in Psychotherapy, 74*, 101786.
Eckroth-Bucher, M. (2010). Self-awareness: A review and analysis of a basic nursing concept. *Advances in Nursing Science, 33*(4), 297–309.
Edwards, J. (2012). We need to talk about epistemology: Orientations, meaning, and interpretation within music therapy research. *Journal of Music Therapy, 49*(4), 372–394.
Eisenberg, N., & Eggum, N. D. (2009). Empathic responding: Sympathy and personal distress. In J. Decety & W. Ickes (Eds.), *The social neuroscience of empathy* (pp. 71–83). MIT Press.
Eisenberg, N., Michalik, N., Spinrad, T. L., Hofer, C., Kupfer, A., Valiente, C., Liew, J., Cumberland, A., & Reiser, M. (2007). The relations of effortful control and impulsivity to children's sympathy: A longitudinal study. *Cognitive Development, 22*(4), 544–567.
Eisenberg, N., Spinrad, T. L., & Knafo-Noam, A. (2015). Prosocial development. In L. Liben & U. Muller (Eds.), *Handbook of child psychology and developmental science* (pp. 1–47). Wiley.
Eurich, T. (2017). *Insight*. Crown Business.
Finlay, L. (2009). Debating phenomenological research methods. *Phenomenology and Practice, 3*(1), 6–25.
Finlay, L. (2011). *Phenomenology for therapists.* Wiley-Blackwell.
Finlay, L. (2014). Engaging phenomenological analysis. *Qualitative Research in Psychology, 11*, 121–141.
Fiske, S. T. (1993). Controlling other people: The impact of power on stereotyping. *American Psychologist, 48*(6), 621–628.
Fleischacker, S. (2019). *Being me being you: Adam Smith and empathy.* University of Chicago Press.

Forinash, M., & Gonzalez, D. (1989). A phenomenological perspective of music therapy. *Music Therapy, 8*(1), 35–46.

Fox, E. I., & McKinney, C. H. (2016). The Bonny method of guided imagery and music for music therapy interns. *Music Therapy Perspectives, 34*(1), 90–98.

Frie, R. (2010). Compassion, dialogue, and context: On understanding the other. *International Journal of Psychoanalytic Self Psychology, 5*, 451–466.

Galinsky, A., Ku, G., & Wang, C. (2005). Perspective-taking and self-other overlap. *Group Processes & Intergroup Relations, 8*(2), 109–124.

Galinsky, A., Maddux, W., Gilin, D., & White, J. (2008). Why it pays to get inside the head of your opponent. *Psychological Science, 19*(4), 378–384.

Gadamer, H. (2007). Letter exchange with Karl Löwith on being and time. In T. Kisiel & T. Sheehan (Eds.), *Becoming Heidegger: On the trail of his occasional writings, 1910–1927* (pp. 289–303). Northwestern University Press.

Gardstrom, S., & Hiller, J. (2010). Song discussion as music psychotherapy. *Music Therapy Perspectives, 28*(2), 147–156.

Gerdes, K. E., & Segal, E. (2011). Importance of empathy for social work practice: Integrating new science. *Social Work, 56*(2), 141–148.

Germer, C., Siegel, R., & Fulton, P. (2013). *Mindfulness and psychotherapy*. Guilford press.

Gleichgerrcht, E., & Decety, J. (2013). Empathy in clinical practice: How individual dispositions, gender, and experience moderate empathic concern, burnout, and emotional distress in physicians. *PLoS ONE, 8*(4), e61526.

Goodwin, S. A., Gubin, A., Fiske, S. T., & Yzerbyt, V. Y. (2000). Power can bias impression processes: Stereotyping subordinates by default and by design. *Group Processes & Intergroup Relations, 3*(3), 227–256.

Greason, P., & Cashwell, C. (2009). Mindfulness and counseling self-efficacy: The mediating role of attention and empathy. *Counselor Education and Supervision, 49*(1), 2–19.

Grimmer, M. S., & Schwantes, M. (2018). Cross-cultural music therapy: Reflections of American music therapists working internationally. *The Arts in Psychotherapy, 61*, 21–32.

Gross, J. (1998). The emerging field of emotion regulation: An integrative review. *Review of General Psychology, 2*, 271–299.

Gruenfeld, D., Inesi, M., Magee, J., & Galinsky, A. (2008). Power and the objectification of social targets. *Journal of Personality and Social Psychology, 95*(1), 111–127.

Grynberg, D., & Pollatos, O. (2015). Perceiving one's body shapes empathy. *Physiology & Behavior, 140*, 54–60.
Hackel, L., Zaki, J., & Van Bavel, J. (2017). Social identity shapes social valuation: Evidence from prosocial behavior and vicarious reward. *Social Cognitive and Affective Neuroscience, 12*(8), 1219–1228.
Haley, B., Heo, S., Wright, P., Barone, C., Rao Rettiganti, M., & Anders, M. (2017). Relationships among active listening, self-awareness, empathy, and patient-centered care in associate and baccalaureate degree nursing students. *NursingPlus Open, 3*, 11–16.
Hall, S. (2009). *Anger, rage and relationship*. Routledge.
Han, S. (2018). Neurocognitive basis of racial ingroup bias in empathy. *Trends in Cognitive Sciences, 22*(5), 400–421.
Hasson, Y., Tamir, M., Brahms, K. S., Cohrs, J. C., & Halperin, E. (2018). Are liberals and conservatives equally motivated to feel empathy toward others? *Personality and Social Psychology Bulletin, 44*(10), 1449–1459.
Haviland, J. (2014). *Exploring empathy in music therapy* [Masters Dissertation, Molloy College].
Hayes, J. A., Gelso, C. J., & Hummel, A. M. (2011). Managing countertransference. *Psychotherapy, 48*, 88–97.
Hein, G., & Singer, T. (2008). I feel how you feel but not always: The empathic brain and its modulation. *Current Opinion Neurobiology, 18*(2), 153–158.
Hinojosa, K. (2021). *Building community and finding identity: Queer sounds, an inclusive school based music therapy group* [Doctoral dissertation, Saint Mary's College of California].
Hogeveen, J., Inzlicht, M., & Obhi, S. S. (2013). Power changes how the brain responds to others. *Journal of Experimental Psychology. General, 142*, 1–9.
Horwitz, E. B. (2018). Humanizing the working environment in health care through music and movement. In L. O. Bonde & T. Theorell (Eds.), *Music and public health: A Nordic perspective* (pp. 187–199). Springer.
Ivey, A., Daniels, T., Zalaquett, C., & Ivey, M. (2017). Neuroscience of attention: Empathy and counselling skills. In T. Field, L. Jones, & L. Russell-Chapin (Eds.), *Neurocounseling: Brian-based clinical approaches* (pp. 83–100). American Counseling Association.
Jack, K., & Smith, A. (2007). Promoting self-awareness in nurses to improve nursing practice. *Nursing Standard, 21*(32), 47–52.
Jackson, N., & Gardstrom, S. (2012). Undergraduate music therapy students' experiences as clients in short-term group music therapy. *Music Therapy Perspectives, 30*(1), 65–82.

Jeffrey, D., & Downie, R. (2016). Empathy-can it be taught? *Journal of the Royal College of Physicians of Edinburgh, 46*(2), 107–112.

Kim, S. (2013). *Multimodal quantification of interpersonal physiological synchrony between non-verbal individuals with severe disabilities and their caregivers during music therapy* [Doctoral dissertation, University of Toronto].

Kim, S., & Whitehead-Pleaux, A. (2015). Music therapy and cultural diversity. In B. Wheeler (Ed.), *Music therapy handbook* (pp. 51–63). Guilford Press.

Knafo, A., Zahn-Waxler, C., van Hulle, C., Robinson, J. L., & Rhee, S. H. (2008). The developmental origins of a disposition toward empathy: Genetic and environmental contributions. *Emotion, 8*, 737–752.

Knapp, S., Gottlieb, M. C., & Handelsman, M. M. (2017). Self-awareness questions for effective psychotherapists: Helping good psychotherapists become even better. *Practice Innovations, 2*(4), 163.

Kottler, J., & Carlson, J. (2014). *On being a master therapist*. Wiley.

Kraus, M. W., Cote, S., & Keltner, D. (2010). Social class, contextualism, and empathic accuracy. *Psychological Science, 21*(11), 1716–1723.

Lamm, C., Batson, C. D., & Decety, J. (2007). The neural substrate of human empathy: Effects of perspective-taking and cognitive appraisal. *Journal of Cognitive Neuroscience, 19*(1), 42–58.

Lee, J. H. (2016). A qualitative inquiry of the lived experiences of music therapists who have survived cancer who are working with medical and hospice patients. *Frontiers in Psychology, 7*, 1840.

Lipson, J. (2019). Seeking the uncensored self. In B. MacWilliam, B. Harris, D. Trottier, & K. Long (Eds.), *Creative arts therapies and the LGBTQ community: Theory and practice* (pp. 171–184). Jessica Kingsley.

Lockwood, P. L. (2016). The anatomy of empathy: Vicarious experience and disorders of social cognition. *Behavioural Brain Research, 311*, 255–266.

Markovitch, N., Netzer, L., & Tamir, M. (2017). What you like is what you try to get: Attitudes toward emotions and situation selection. *Emotion, 17*, 728–739.

McGarry, L. M., & Russo, F. A. (2011). Mirroring in dance/movement therapy: Potential mechanisms behind empathy enhancement. *The Arts in Psychotherapy, 38*(3), 178–184.

Moran, D. (2017). *Mindfulness and the music therapist: An approach to self-care* [Doctoral dissertation, Concordia University].

Moskowitz, G. (2005). *Social cognition: Understanding self and others*. Guilford Press.

Murphy, K. (2007). Experiential learning in music therapy: Faculty and student perspectives. *Qualitative Inquiries in Music Therapy, 3*, 31–61.

Næss, T., & Ruud, E. (2007). Audible gestures: From clinical improvisation to community music therapy. *Nordic Journal of Music Therapy, 16*(2), 160–171.

Nelson, D. (2009). Feeling good and open-minded: The impact of positive affect on cross cultural empathic responding. *The Journal of Positive Psychology, 4*(1), 53–63.

Noddings, N. (2010). Complexity in caring and empathy. *Abstracta, 6*(2), 6–12.

O'Connell, J. E. (2018). *Beyond prosocial motivations to empathize* [Doctoral dissertation, University of Ontario Institute of Technology].

Panfile, T. M., & Laible, D. J. (2012). Attachment security and child's empathy: The mediating role of emotion regulation. *Merrill-Palmer Quarterly, 58*(1), 1–21.

Parsons, C. E., Young, K. S., Jegindø, E. M. E., Vuust, P., Stein, A., & Kringelbach, M. L. (2014). Music training and empathy positively impact adults' sensitivity to infant distress. *Frontiers in Psychology, 5*, 1440.

Pavlicevic, M. (1999). Thoughts, words and deeds: Harmonies and counterpoints in music therapy theory: A response to Elaine Streeter's 'finding a balance between psychological thinking and musical awareness in music therapy theory: A psychoanalytic perspective. *British Journal of Music Therapy, 13*(2), 59–62.

Pieterse, A., Lee, M., Ritmeester, A., & Collins, N. (2013). Towards a model of self-awareness development for counselling and psychotherapy training. *Counselling Psychology Quarterly, 26*(2), 190–207.

Piff, P., Stancato, D., Cote, S., Mendoza-Denton, R., & Keltner, D. (2012). Higher social class predicts increased unethical behavior. *Proceedings of the National Academy of Sciences of the United States of America, 109*(11), 4086–4091.

Porat, R., Halperin, E., & Tamir, M. (2016). What we want is what we get: Group-based emotional preferences and conflict resolution. *Journal of Personality and Social Psychology, 110*(2), 167.

Préfontaine, J. (2006). On becoming a music therapist. *Voices: A World Forum for Music Therapy, 6*(2).

Preston, S. D., & de Waal, F. B. M. (2002). Empathy: Its ultimate and proximate bases. *Behavioral and Brain Sciences, 25*, 1–72.

Priestley, M. (1994) *Essays on analytical music therapy.* Barcelona Publishers

Pril, D. (2017). *Toward a better understanding of the interpersonal effects of power: Power decreases interpersonal sensitivity, but not toward people within the power relationship* [Doctoral dissertation, Universität zu Köln].

Psychogiou, L., Daley, D., Thompson, M. J., & Sonuga-Barke, E. J. (2008). Parenting empathy: Associations with dimensions of parent and child psychopathology. *British Journal of Developmental Psychology, 26*(2), 221–232.

Rasheed, S. (2015). Self-awareness as a therapeutic tool for nurse/ client relationship. *International Journal of Caring Sciences, 8*(1), 211–216.

Rasheed, S., Younas, A., & Sundus, A. (2019). Self-awareness in nursing: A scoping review. *Journal of Clinical Nursing, 28*(5–6), 762–774.

Redmond, M. (2018). *Social decentering: A theory of other-orientation encompassing empathy and perspective taking*. De Bruyter.

Rieffe, C., Oosterveld, P., Miers, A. C., Meerum Terwogt, M., & Ly, V. (2008). Emotion awareness and internalising symptoms in children and adolescents: The emotion awareness questionnaire revised. *Personality and Individual Differences, 45*(8), 756–761.

Robarts, J. (2006). Music therapy with sexually abused children. *Clinical Child Psychology and Psychiatry, 11*(2), 249–269.

Rochat, P. (2003). Five levels of self-awareness as they unfold early in life. *Consciousness and Cognition, 12*(4), 717–731.

Rogers, C. (2007). The necessary and sufficient conditions of therapeutic personality change. *Psychotherapy, 44*(3), 240–248.

Rogers, C., & Farson, R. (1957). *Active listening*. https://wholebeinginstitute.com/wp-content/uploads/Rogers_Farson_Active-Listening.pdf

Rovenpor, D. R., & Isbell, L. M. (2018). Do emotional control beliefs lead people to approach positive or negative situations? Two competing effects of control beliefs on emotional situation selection. *Emotion, 18*, 313–331.

Ruiz-Junco, N. (2017). Advancing the sociology of empathy: A proposal. *Symbolic Interaction, 40*(3), 414–435.

Schmid, W., & Rolvsjord, R. (2020). Becoming a reflexive practitioner: Exploring music therapy students' learning experiences with participatory role-play in a Norwegian context. *Scandinavian Journal of Educational Research*, 1–13.

Simionato, G., Simpson, S., & Reid, C. (2019). Burnout as an ethical issue in psychotherapy. *Psychotherapy, 56*(4), 470–482.

Singer, T., Seymour, B., O'Doherty, J. P., Stephan, K. E., Dolan, R. J., & Frith, C. D. (2006). Empathic neural responses are modulated by the perceived fairness of others. *Nature, 439*(7075), 466–469.

So, H. (2017). US-trained music therapists from East Asian countries found personal therapy during training helpful but when cultural disconnects

occur these can be problematic: A qualitative phenomenological study. *The Arts in Psychotherapy, 55*, 54–63.
Stepien, K. A., & Baernstein, A. (2006). Educating for empathy. *Journal of General Internal Medicine, 21*(5), 524–530.
Stern, J. A., Borelli, J. L., & Smiley, P. A. (2015). Assessing parental empathy: A role for empathy in child attachment. *Attachment & Human Development, 17*(1), 1–22.
Stewart-Williams, S. (2007). Altruism among kin vs. nonkin: Effects of cost of help and reciprocal exchange. *Evolution and Human Behavior, 28*, 193–198.
Tangen, J. L. (2017). Attending to nuanced emotions: Fostering supervisees' emotional awareness and complexity. *Counselor Education and Supervision, 56*, 65–78.
Totton, N. (2017). Power in the therapeutic relationship. In R. Tweedy (Ed.), *The political self* (pp. 29–42). Taylor & Francis.
Trentini, C., Tambelli, R., Maiorani, S., & Lauriola, M. (2021). Gender differences in empathy during adolescence: Does emotional self-awareness matter? *Psychological Reports*, 0033294120976631.
Valdesolo, P., & DeSteno, D. (2011). Synchrony and the social tuning of compassion. *Emotion, 11*(2), 262–266.
Valentino, R. (2006). Attitudes towards cross-cultural empathy in music therapy. *Music Therapy Perspectives, 24*, 108–114.
Valiente, C., Eisenberg, N., Fabes, R. A., Shepard, S. A., Cumberland, A., & Losoya, S. H. (2004). Prediction of children's empathy-related responding from their effortful control and parents' expressivity. *Developmental Psychology, 40*, 911–926.
Waite, R., & McKinney, N. S. (2016). Capital we must develop: Emotional competence educating pre-licensure nursing students. *Nursing Education Perspectives, 37*(2), 101–103.
Warren, J. A., & Nash, A. (2019). Using expressive arts in online education to identify feelings. *Journal of Creativity in Mental Health, 14*(1), 94–104.
Weisz, E., & Zaki, J. (2018). Motivated empathy: A social neuroscience perspective. *Current Opinion in Psychology, 24*, 67–71.
Wigram, T. (2004). *Improvisation: Methods and techniques for music therapy clinicians, educators and students.* Jessica Kingsley.
Wilkinson, H., Whittington, R., Perry, L., & Eames, C. (2017). Examining the relationship between burnout and empathy in healthcare professionals: A systematic review. *Burnout Research, 6*, 18–29.

Williams, G., Gerardi, M., Gill, S., Soucy, M., & Taliaferro, D. (2009). Reflective journaling: Innovative strategy for self-awareness for graduate nursing students. *International Journal for Human Caring, 13*(3), 36–43.

Winter, P. (2013). *Effects of experiential music therapy education on student's reported empathy and self-esteem: A mixed methods study* [Doctoral dissertation, Temple University].

Zaki, J. (2014). Empathy: A motivated account. *Psychological Bulletin, 140*(6), 1608.

Zorza, J. P., Marino, J., & Acosta Mesas, A. (2019). Predictive influence of executive functions, effortful control, empathy, and social behavior on the academic performance in early adolescents. *The Journal of Early Adolescence, 39*(2), 253–279.

3

Awareness Through Multifaceted Perceiving

Once we have assumed a stance of receptivity, we need to pay attention to the other person and their emotions in multifaceted ways. We perceive their emotions in the context of who they are as a whole person, a musical being, an embodied being, a being with conscious and unconscious dimensions, and as a being who is situated in a particular context. We engage in finely attuned listening, retaining a clear sense of the distinction between self and other.

Perceiving Emotions

Emotion has been defined in countless ways, as explored in Chapter 1. Along our current pathway of insightful empathy, certain definitions of emotion are particularly helpful, as their theoretical underpinnings support this way of thinking about and practising empathy. Denzin (2009) defined emotions as situated, lived, believed-in, and temporal experiences that run through a person's body and radiate through their stream of consciousness. In the process of living an emotion, one is

A. dos Santos, *Empathy Pathways*,
https://doi.org/10.1007/978-3-031-08556-7_3

plunged into the reality created by the emotional experience. Juslin (2016), a distinguished researcher of emotions in relation to musical experience, framed emotions as intense but brief responses to potentially important occurrences in the internal or external environment that entail cognitive appraisal, subjective feelings, physiological responses, expressions, and action tendencies. These facets tend to be synchronised during an emotional episode. Moment-by-moment changes in the quality, intensity, and complexity of emotion can occur, which can be described in relation to valence (whether an emotion is pleasant or unpleasant) and arousal (the degree of calmness or agitation). The notion of emotions being short-lived has been challenged. While indeed an emotion may be fleeting, it could last for days depending on the circumstance and type of emotion. One might also experience an emotion from a past event when ruminating or when the current context makes an event salient in one's mind (Levine et al., 2012; Verduyn et al., 2011).

A central feature of emotions is that they are directed towards a particular event or object (Lench & Carpenter, 2018). In other words, they are "about" something. In functional evolutionary accounts, emotions are understood to serve the purpose of resolving the problems that elicited them (Lench & Carpenter, 2018). Bericat (2016) summarised emotions as constituting the bodily manifestation of the importance that an event in the natural or social world has for a subject. Emotion is a bodily consciousness that signals and indicates this importance, regulating a specific subject's relationships with the world. In its most basic expression, this involves three elements: (a) the assessment/appraisal; (b) of an event in the world; that's (c) made by an individual. As we guide our attention to the other person's emotional experience, we are, therefore, seeking to notice and track this lived, temporal, situated, cognitive, the physiological experience they are having, which is pitched at a particular level of valence and arousal, is about something in their environment, and that provokes certain action tendencies.

We perceive emotion through musical experiences in multiple ways, often simultaneously. Schiavio et al. (2017) described the "where" problem as having to do with how emotion may be perceived as "located in," "expressed by," "attributed to," or "possessed by" the music itself, while the "how" problem refers to the way that music may act as a stimulus to generate emotions inside listeners. These two orientations are not necessarily mutually exclusive. Hiller (2015) summarised key theories that attempt to explain the relationship between music and emotion, especially regarding how this plays out in music therapy. Somatic theory concentrates on felt experiences of emotion through changes within the body. A client may create musical representations of their emotions by musically imitating what they are experiencing in their body. Expressive code theory explores how a performer can communicate emotions to an audience by using the voice-like qualities of an instrument. According to contour theory, musical sounds convey emotions through the similarities between features in the music and observable body movements of human beings (tensing muscles, wringing hands, and so on). A client's downward melodic motion on a xylophone or flat thud on a drum may mirror slumped shoulders, a bowed head, and slow body movements. Expression theory claims that the emotions expressed in a piece belong to the composer or performer. If we hear a person shouting, we may infer that they are angry. Similarly, we recognise the composer's or performer's emotions through their musical expressions. Gesture theory highlights how we recognise communicative intent through particular gestures. If I wave to you, you may recognise that I am greeting you. When listening to music, we may perceive the shape or pattern of sounds as communicative gestures.

Perceiving Another as a Musical Being

The therapeutic mechanisms in music-centred music therapy are understood to be located in the experiences, processes, and structure of the music itself (Aigen, 2014). This is not only music that is "external" to the participant. Besides being composed of "blood and bone," Aldridge (1989) argued that people are made of patterns, rhythms, and melodic contours. In clinical improvisation, music therapists listen to the person in the music and the person *as* the music. We read the client's music as containing relational and emotional material (Pavlicevic, 1997). Physiological problems can be conceptualised as instances of being "out of tune" or "out of time." Through the isomorphism between biological and musical forms, music therapists understand a client's musical expressions as a portrait of their identity. Ansdell (2014) noted that this "music-as-person" metaphor is complex because our experience of being in relationship to music could entail feeling as if we are "in" the music and/or feeling as if the music is "in" us. Even in the context of analytic music therapy, where verbal interactions play a more significant role, Mary Priestley (1975) stated that music therapists "listen for the words in the music and the music in the words" (p. 250).

Perceiving Embodied Resonance

60–65% of interpersonal communication occurs nonverbally (Foley & Gentile, 2010). Nonverbal behaviours include proxemics (how relationships and behaviours change depending on the distance between two people), kinesics (how bodies move, including posture, gestures, facial expressions, touching behaviours, and self-soothing behaviours), and paralanguage (features of speech such as rhythm, prosody, tempo, volume, tone, and pitch) (Knapp & Hall, 2010; Shea, 1998).

Embodied resonance is a fundamental component of meaningful communication. An empathic dialogue can come about with any client, regardless of developmental level, verbal capacity, level of confusion, or psychosis, as long as the therapist can tune into the client's embodied modes of expression (Dekeyser et al., 2009). First and foremost, shared experiences entail a "meeting of bodies" rather than a meeting of minds. Even at the very start of human life, an experiencing subject has perceptual access to their mother's body (Ciaunica, 2019). When we come into contact with others, we unconsciously and subtly mimic their gestures, facial expressions, posture, and vocal pitch. When we can't mimic others, we find it harder to figure out what they're feeling. In a fascinating study, people who were injected with Botox for wrinkle reduction and experienced mild paralysis of their facial muscles were less accurate in their perceptions of others' emotions. The researchers concluded that this was because they were now less accurate in mirroring others' facial expressions, so they couldn't simulate the other person's emotions within themselves (Baumeister et al., 2016). As therapists, we use our bodies as a kind of radar. We read our bodies for clues to what our clients may feel (Murphy, 2021).

Simulation theory explains how we gain a sense of the mind of another through physiological resonance or mirroring (Gallese & Goldman, 1998). In the 1990s, Rizzolatti and his colleagues were studying an area labelled F5 in the premotor cortex of macaques. They'd already gained a significant understanding of the neural mechanisms of motor control of the monkey's hands when, one day, a member of the team reached for an object and heard a notification from the computer attached to the electrodes implanted in the brain of a macaque who happened to be sitting quietly watching him. This notification signalled activity in the F5 area. While the monkey's hands were completely still, cells related to the movement of the muscles of its hands—soon to be called mirror neurons—were firing as it watched someone else move their hands (Rizzolatti et al., 1996). Evidence showing the existence of the Mirror Neuron System (MNS) in humans has accumulated since this discovery (Gallese, 2009).

When I observe an action and this activates the very motor circuits used when I execute that action, this may allow me to grasp the other person's intention (Colle et al., 2008). Some mirror neurons have also been observed to respond to whether an action is seen, heard and seen, or merely heard (Keysers et al., 2003; Kohler et al., 2002). Mirror function isn't only activated in relation to motor acts but is also at play during emotional reactions (Aragona et al., 2013; de Vignemont & Singer, 2006), perception of pain (Botvinick et al., 2005; Ebisch et al., 2008), and auditory experiences, such as listening to music (Overy & Molnar-Szakacs, 2009). Gallese (2009) concluded that empathy should be understood as the result of a natural tendency to first experience interpersonal relations at an implicit level of intercorporeity, in other words, through the mutual resonance of meaningful sensory-motor behaviours. When encountering another person, we are aware of their similarity to us because we embody it through our MNS.

In the realm of phenomenology, Husserl (1977) argued that our own and others' bodies (as something alive, not merely material objects) are the essential instruments that lead to our ability to share experiences. For Merleau-Ponty (1962), bodily reciprocity formed the basis for empathy. The mind is not a separate substance that interacts mysteriously with a mechanical body, as a Cartesian understanding would suggest. Consciousness is embodied. There can be no perception without awareness of an acting body (Gallese, 2009).

Through awareness of my own body, I can recognise another as a lived body expressive of emotions, thoughts, and intentions (Walsh, 2014). Stein (1989) referred to the empathic experiencing of another's body by its own field of sensations as "sensual empathy." Just as we experience our own sensing, living bodies, we perceive others as also having sensing, living bodies. The behaviour of another becomes meaningful to me as I experience it as linked to my own situated, lived experience of the same behaviour. We are involved sympathetically, sensually, and imaginatively with the movements of those around us because what they are doing resembles how we imagine we would move in that circumstance and for that purpose (Trevarthen & Fresquez, 2015).

Physiological synchrony refers to the spontaneous correlation of autonomic activity between people as they interact socially (Feldman, 2007). Physiological synchrony can lead to empathy, and empathy can enhance physiological synchrony (Ebisch et al., 2012; Marci et al., 2007; Papp et al., 2009; Stephens et al., 2010). Kim (2013) built on this body of literature to examine physiological synchrony between severely disabled nonverbal individuals and their parents and between the same individuals and their music therapists. She measured synchrony in terms of electrodermal activity (EDA), heart rate, and cortisol levels for all dyads and neural activity for the client–parent dyads. Kim observed a higher level of physiological synchrony in the client–parent than in the client–therapist dyads. However, the level of synchrony increased with respect to EDA and cortisol in both groups. By the latter stages of the study, the synchrony of the client–therapist dyad exceeded that of the child–parent dyad. She argued that physiological synchrony is apparent in the presence of disability and that this can facilitate deeper levels of understanding of individuals with disabilities.

Bruscia (2001) explained how he enters his client's music, following and engaging with their process until he can sense how their body makes and organises the sounds. This is how he "lives" in and follows the client's rhythm and melody. In De Backer's (2008) description of music therapy with clients experiencing psychosis, he explained how the therapist listens to the client's body and attempts to come into resonance with the client's affect through their own body. Similarly, Pedersen (1997) described how music therapists engage in "extrasensitive," embodied, empathic listening by resonating with clients who have, for example, schizophrenia.

Intentional embodied mirroring is a central feature of improvisational music therapy. Therapeutic encounters can become healing as therapists resonate with and offer modified expressions of embodied emotions (Meekums, 2012). Collaborative music-making offers exquisite opportunities for motor resonance, imitation, and entrainment (Rabinowitch et al., 2012). The term "entrainment" refers to spatiotemporal coordination, usually in response to rhythmic signals. Components of entrainment include shared rhythmical periodicity in the brain and body (Goebl & Palmer, 2009; Phillips-Silver et al., 2010) and synchrony or affective concordance, generated through mutual emotion sharing between individuals (Phillips-Silver & Keller, 2012). We can begin entrainment simply by breathing in time with a client, which starts to generate resonance (Dimaio, 2010). We then develop the empathic space further by sharing music experiences, synchronised actions, and emotions (Winter, 2013). When there is close temporal sensorimotor synchronisation, we experience groove and affiliation (Janata et al., 2012).

Listening Perspectives and Practices

Hearing entails engaging in a physiological process of noise awareness. Listening is significantly more complex and multidimensional. It requires attending, interpreting, and responding affectively to verbal and nonverbal messages. This is a cognitive, emotional, and behavioural process (Jones, 2011). Alice Duer Miller (cited in Miller, 2018) wrote that "Listening is not merely not talking…[I]t means taking a vigorous, human interest in what is being told us. You can listen like a blank wall or like a splendid auditorium where every sound comes back fuller and richer" (p. 26).

If we want to empathise, we have to listen. In therapeutic literature, multiple forms of listening are discussed. Rogers (1951), for example, developed the practice of active listening. This requires active emotional commitment (consciously restraining personal needs and wants) and active responsiveness to the other's emotions, thoughts, and experiences (Walker, 1997). One listens for and responds to the total meaning of the other's expression, including verbal and nonverbal content and feelings. One genuinely respects the worth of the person one is listening to while valuing and trusting their inner wisdom and capacity for self-direction.

Double listening, an approach drawn from narrative therapy, involves listening to how a client's identity is constituted through their problem-saturated stories (and, I would suggest, their problem-saturated ways of improvising, for example, "stuckness"). It also involves listening to alternative stories (and alternative ways of playing) that can lead to preferred ways of being. Alternative stories speak of other possibilities and hold the potential for more freedom and agency (Guilfoyle, 2015). In many ways, this notion of an alternative story is congruent with Nordoff and Robbin's (2007) idea of the music child. As music therapists, we recognise and welcome our clients' healthy, whole parts that sickness, disorder, and struggle cannot annihilate. We support the musicality, creativity, vibrancy, resilience, playfulness, and strength of the music child (although I've experienced how sometimes this part of the client appears in other ways too, for example, as a music warrior or a music elder). We "thicken the narrative" of the preferred story as we notice it, meet it musically, and encourage its growth.

Paul Nordoff and Clive Robbins brought acute, active listening to music therapy. By attuning carefully to each child they worked with, Nordoff and Robbins could draw them "into a shared musical world where they felt more at home yet were also increasingly involved with other people and the broader world of their musical culture" (Ansdell, 2014, p. 46). Ansdell (1995) identified three main listening components within Creative Music Therapy. The music therapist aims to listen to the purely musical domain as a musician, to the emotional and physical qualities of the client through the music and in other domains, and the quality of the relationship in the music. De Backer (2008) added embodied listening to this list. Music therapists listen to the body of the client through their own body. The therapist attempts to resonate through their body with the client's affect. Music therapists also, importantly, listen to silence (Sutton, 2002).

Authors such as Arnason (2003) have offered detailed frameworks that can guide music therapists' listening (to the music, from memory, by observing, feeling and thinking clinically, and for significance, imagery and the intangible). Several music therapists have also adapted Ferrara's (1984) approach to listening to clinical improvisations (Amir, 1990; Arnason, 2003; Bonde, 1997; dos Santos, 2019a, 2019b; Forinash & Gonzalez, 1989; Skewes, 2001; Trondalen, 2003). This involves focusing on a specific form of listening at a time and attempting to bracket other levels of meaning.

Based on his work on empathy, Bruscia (2001) formulated three listening positions (as experiential musical spaces for listeners to locate themselves) that music therapists can use when analysing client improvisations. The first is an empathic listening position where one attempts to listen to the improvisation from the client's perspective and identify with the client's unfolding experiences. The second is a complementary listening position where one listens from another individual's perspective who took part in the improvisation (for example, the therapist or another group member). Third, in a reactive listening position, one takes up an "audience" role, reflecting on one's own personal responses. Importantly, as one listens to a client, one also listens to how they listen (Black et al., 2017).

Perceiving Unconscious Dynamics

Insightful empathy isn't locked into one therapeutic orientation. If we were to approach insightful empathy within psychodynamically oriented music therapy, for example, there would be a need to explore the role that awareness of unconscious emotional dynamics may play. Psychoanalytic theory—object relations and self-psychology in particular—strongly emphasise the role of empathy in the therapeutic alliance (Duan & Hill,

1996; Eagle & Wolitzky, 1997; Feller & Cottone, 2003). Austin (1996) described psychodynamic music therapy as a creative process that uses music and words within the relationship between client and therapist to facilitate dialogue between conscious and unconscious material. Music may give voice to that which is still inaudible by providing a language for the symbolic expression of unconscious content and intrapsychic processes. Music can function as a bridge for unconscious material to cross over into consciousness to be integrated. From a psychoanalytic perspective, focusing one's awareness within insightful empathy would include perceiving deeper meaning underneath what the client is saying and not saying, playing and not playing, and in the affective signals that they are often unknowingly transmitting (Basch, 1988).

Being available to a client as an empathically attuned companion often involves reflecting on transference-countertransference reactions (Austin, 2001). Some authors clearly distinguish between empathy and countertransference. For example, Siebert et al. (2007) argued that the central distinction is that empathy entails the transfer of emotions, while countertransference is a reaction to the client's experiences. Within music therapy literature, authors have varied in their descriptions of the direction and quality of the relationship(s) between empathy and countertransference. According to Priestley (1994), we can conceptualise three categories of countertransference: "countertransference" as per the classical Freudian conceptualisation; "c-countertransference," which refers to the process whereby a therapist identifies with a client's introject; and "e-countertransference," which entails the therapist's empathic resonance with the client's unconscious or becoming-conscious emotions. The therapist's awareness of e-countertransference develops through emotional awareness, attention to somatic sensations, and musical interactions, particularly improvisation. When experiencing negative countertransference responses, we can temper natural reactions such as defensiveness, frustration, fear, judgement, or anger by drawing on empathy to consciously shift out of our experiential world into the client's world.

Scheiby (2005) described how empathising could lead to powerful countertransference reactions. As an example of this, Short (2017b) explained how one of her clients in mental health services experiencing homelessness and psychosis expressed destructive and violent lyrics that seemed incongruent with his verbal presentation. She interpreted her feelings of shock as a countertransference response, which offered a window into his emotional experience and compelled her to empathise with him. Pavlicevic (2002) noted a potentially problematic relationship between empathy and countertransference in work with a client who seemed vacant. She experienced him as not being "there" to like or dislike, which led to her experiencing lower levels of empathy.

In Short's (2017a) work with Karly, a young woman diagnosed with a personality disorder in a secure unit for women in the United Kingdom, Karly seemed to experience safety with Short, who experienced a strong countertransference response to nurture her. Short remembered her own intense memories of adolescence and empathised with Karly's position, but this left her struggling to separate her sense of Karly's experience from her own memories of rage. When Karly would state that she just wanted someone to love her, Short found that this mirrored her feelings of desiring to be valued and have her work recognised. Personal psychotherapy assisted her in navigating these inner conflicts. Musical improvisation offered a safe way for Karly and Short to explore their relationship and for Karly to engage in interactive, affective communication.

Stern (2009) urged therapists to stay with what unfolds in sessions, especially the affects that are in the background or destabilising the therapeutic relationship. When therapists simply and freely allow all of this to affect them, guide their reverie, and inform their interventions, their capacity for empathy remains active. Similarly, Scheiby (2005) described openness to musical countertransference as entailing letting music unfold in sessions naturally rather than obsessively focusing on an awareness of potential musical countertransference itself.

Considering empathy within psychodynamic work entails not only analysis of countertransference with the intention of offering an interpretation. Austrian-born psychoanalyst Heinz Kohut (1968, 1971) was a follower of Freud's model of psychoanalysis until he had a pivotal experience with a client, Miss F, who insisted that Kohut offer her almost perfect attunement, confirmation, and mirroring rather than interpretation and analysis of defences (Feller & Cottone, 2003). Kohut began to develop what became Self Psychology as he listened to this patient in an empathic manner, claiming that understanding patients is "unthinkable" without empathy (Kohut, 1977, p. 306). Crucially, Kohut highlighted how empathy places the analyst's attention on what it is like to be the patient rather than on what it is like to be the target of the patient's demands and wishes (Rowe & Isaac, 2004). For Kohut, empathy was a perceptual and experience-near mode of observation (Schwaber, 1984) and functioned as a data-gathering tool in psychoanalysis (Siegel, 1996). He described how corrective emotional experiences could help a patient realise that, contrary to their childhood experiences, "the sustaining echo of empathic resonance is indeed available in this world" (Kohut, 1984, p. 78).

Perceiving Spiritual Dimensions

A core value in music therapy is regard for the client as a "whole" person, which includes spiritual dimensions (dos Santos & Brown, 2021; Kidwell, 2014; Potvin & Argue, 2014). The term "spirituality" seems to resist conclusive definition. It originated from the Latin word *spiritus*, meaning "breath." Broadly, spirituality refers to intangible aspects of individual and collective existence (Sperry, 2012; Tanyi, 2002). Spirituality includes multidimensional considerations of relationships with others, relations to the transcendent, a quest for meaning and truth (Aldridge, 2000), the mystery of creation, and personal transformation (LaPierre, 1994).

Musical expressions inform us about the psychological, physical, social, and spiritual experiences and worlds of those taking part. Listening for the intangible "is about the spiritual dimensions of improvisational music therapy, a sense of sacredness that is grounded in the musical-creative-human experience" (Arnason, 2003, p. 133). The dimension of spirituality can be at play when music therapists tune into the inner state of a client as this is musically expressed (Aigen, 1999). Several authors have used the term "transpersonal empathy." Edwards (2010), for example, explained transpersonal empathy as feeling into or with the spiritual world, while Gilewski (1993) described it as relying on intuitive forms of knowing.

Perceiving (in) Contexts

Attempting to empathise with contextual sensitivity requires music therapists to consider multiple interfacing layers and how emotions arise in relation to them. These layers include musical structures and form, events in the session, phases of therapy, the client's worlds outside of the therapy room, their family dynamics, diagnostic features, as well as cultural, social, political, and economic factors (Arnason, 2002, 2003; Behrens, 2012; Stige, 2016; Viega, 2018). Ansdell (2014), in collaboration with DeNora (2000, 2011), highlighted the value of a "meso-level" of analysis that concentrates on people and music within specific situations and their unique conditions.

From an ecological perspective, the empathiser must also acknowledge their own contextual positioning and potential to operate from an ethnocentric viewpoint. When I engage in insightful empathy, trusting that I can come to know the emotional world of another, I need to exercise bracketing (as discussed earlier) because I too am a contextually situated being with a particular history. I am also part of distinct "sonic ecologies."

In Brown's (2002) much-cited article *Towards a culture-centred music therapy practice*, she drew on the work of Ridley and Lingle (1996). They argued that, in multicultural contexts, "generic empathy" is insufficient. When therapists and clients have differing sociocultural experiences, their processing of information and styles of self-expression may differ. Brown also cited Hays' (1996) model called "ADRESSING," which emphasises consideration of the following when working with clients: age, disability, religion, ethnicity, social status, sexual orientation, Indigenous heritage, and gender. Brown concluded that cultural empathy is a dynamic concept experienced in emotional, cognitive, and behavioural domains. Therapists also need to know clients' attitudes towards empathy and how empathy is expressed in their culture (Thomas & Sham, 2014; Valentino, 2006).

Perceiving the Distinctions Between Self and Other

To engage in insightful empathy by noticing and tuning into clients' emotions we are required to remain aware of the distinction between self and other (Decety & Jackson, 2004). While empathy is often described as the capacity to personally experience an emotion we observe in another (Trentini et al., 2021), the key to this process is full awareness that the origin of the emotion lies in the other (Preckel et al., 2018). We perceive the other's experience as theirs, which differs from how we would experience it if it were our own (Zahavi, 2010). We're not simply projecting our feelings onto the other person (Stein, 1989). When Rogers (1959) explained empathy as sharing in another's experience as if one were that person, he crucially added that we do this "without losing the as-if condition" (p. 210). If we don't maintain this self-other distinction, we risk merging or confusing whose feelings belong to whom (Batson et al., 1997; Preston & de Waal, 2002).

Having a coherent, clear, and stable self-concept (including emotional states, personality, values, attitudes, attributes, preferences, and roles that one claims as "mine" or "me") is crucial for mature empathic responding (Krol & Bartz, 2021). When one's self-concept is challenged and is less stable (perhaps when experiencing critique during supervision as a student; when going through difficult relational challenges in one's personal life, or experiencing rejection), one can expect heightened vulnerability to experience distress in the face of another's suffering, and even wanting to withdraw from the other rather than responding in a prosocial manner. If increasing empathy is a therapeutic goal with a client, there may be a preceding need to offer opportunities for enhancing the stability of their self-concept and their capacity to distinguish between self and others. Musical improvising can foster the development of a person's sense of self and self-other distinction (Purdon, 2002).

In her book *Anger, Rage, and Relationships*, Hall (2009) distinguished between "empathic anger" and "disempowering anger." She described a therapeutic situation where a therapist may feel angry on behalf of their client or respond to their client's experience with their own unprocessed anger about a similar situation they've encountered in their life (or that other clients they see have been dealing with). There is then a blurring of self/other. A therapist's expression of anger is only therapeutic when it entails a pure response to the client's immediate experience and disclosure. Otherwise, their anger burdens the client and is not empathic.

Identifying with another person can help us understand what they're feeling. However, sometimes perspective-taking itself can create a sense of merging (Davis et al., 1996). When I represent another in my mind, this image may closely relate to how I see myself. This could be because I've applied characteristics of myself to the other person (especially if I don't know them particularly well), or I've drawn on their characteristics when viewing myself (which tends to happen when we're close to or very familiar with the other person). Confusion between self and other then takes place as boundaries weaken. Therefore, a delicate and flexible balance must be maintained between identifying and retaining separation (Hermansson, 1997). Moving both in and out of synchrony with others appears to lead to enhanced perception of self, of the other as a separate person, and the reciprocal bodily interaction between the participants, refining the sense of self-other distinction necessary for empathy (Behrens, 2012).

Intersubjectivity (The Version That Refers to Shared Agreement About an Objective Reality)

While insightful empathy requires that we acknowledge the separation between ourselves and the person we are empathising with, we are still in an intersubjective relationship with them. The term intersubjectivity is polysemic. Its multiple meanings are situated within a wide range of theoretical orientations and refer to various possible relationships between people's perspectives and experiences of each other (Gillespie & Cornish, 2010).

The German word *Intersubjektivität* was used for the first time in 1885 by Johannes Volkelt. James Ward carried it into the English language in 1896. Initially, it referred to something valid for everyone, independent of subjects who may encounter it. Husserl (1989) offered the first extensive and systematic discussion of the concept of intersubjectivity. He used the term "gegenstand," translated as "objective" (with a lowercase "o"), to refer to an object as "being an object for consciousness" (Hermberg, 2006, p. 8) or an object as it is experienced. He used "objectiv" (usually translated as Objective with a capital "O") to describe an object that is "there for everyone" (p. 9). In this sense, intersubjectivity refers to agreement: we can have a shared definition of an object (Mori & Hayashi, 2006). Objectivity is intersubjective because the perspectives of others are crucial in addition to one's own for the validity of the world to be established. Husserl (1989) used the example of a car. You might view one side while I'm looking at the other, but we assume we can make the same conclusion about this being a car because we share the same objective world. While we can't be in precisely the same position at the same time, we can hold the idea that if we were to be, we would see the same thing. This constitutes the potential for seeing the world from the other's perspective (Duranti, 2010; Husserl, 1989). We can interact meaningfully because we have a shared sense of objective reality.

Intersubjectivity and empathy mutually inform one another. Husserl (1969) explained that a shared world is an accomplishment that becomes possible through empathy, which he understood as a primordial experience of participating in the feelings and actions of another without becoming the other. This implies that empathy precedes intersubjectivity. In other words, we use empathy to reach a sense of shared understanding (Reuther, 2014). It is empathy that makes an intersubjectively Objective (with a capital "O") world possible (Hermberg, 2006). At other times, Husserl (1989) argued that empathy becomes possible because of intersubjectivity. From this view, it is within intersubjective encounters that we can experience empathy for one another (Jardine, 2014). I need to have some pre-existing sense of what it may be like for you within your situation to recognise when I've reached empathic accuracy. The similarities between us (our sharing of objective reality) enable me to transpose myself into your situation and understand your experiences from your side (Daly, 2016). As Daniel Stern (2005) explained, people wouldn't share their experiences with a therapist if they did not assume the therapist would be able to see the mental landscape they were describing.

Music therapists (Birnbaum, 2014; Pavlicevic, 1997; Robarts, 1996, 2003, 2009; Turry, 1998, 2009, 2011) have drawn on understandings of intersubjectivity formed within developmental psychology by authors such as Stern (2005), Papoušek and Papoušek (1981), Trevarthen (1994), and Trevarthen and Malloch (2000). Through conducting microanalyses of infant-caregiver interactions, these researchers focused on how caregivers and infants know each other through mutually impactful displays of emotion and movement. This intersubjectivity occurs at a prelinguistic level, which is the same level tapped into within a musical improvisation between client and therapist (Næss & Ruud, 2007).

Building on the work of Trevarthen (1994), Pavlicevic (1997) defined intersubjectivity as "knowing and interacting with another's internal state" (p. 109). Suppose an infant excitedly bashes bells against a table, making a delightfully harsh jingle-clash, and their mother laughs brightly. In that case, the infant may realise that the form of the mother's vocalisation relates to their arm movements. This is a cross-modal exchange. The infant feels that the mother shares their feeling. Trondalen (2019) summarised intersubjectivity as the experience of knowing that you know that I know and feeling that you feel that I feel. She argued that "it is about being together and also having the feeling of being together" (p. 2). Two inner worlds meet within an intersubjective interaction frame (Trondalen, 2016). Stern (2005) distinguished between asymmetrical intersubjectivity, where one party feels that they know what the other is experiencing, but that person doesn't feel known, and two-way intersubjectivity, where both parties are within a similar experiential landscape and mutually validate each other's knowledge.

References

Aigen, K. (1999). The true nature of music-centred music therapy theory. *British Journal of Music Therapy, 13*(2), 77–82.

Aigen, K. (2014). Music-centered dimensions of Nordoff-Robbins music therapy. *Music Therapy Perspectives, 32*(1), 18–29.

Aldridge, D. (1989). A phenomenological comparison of the organization of music and the self. *Arts in Psychotherapy, 16*(2), 91–97.

Aldridge, D. (2000). *Spirituality, healing and medicine: Return to the silence.* Jessica Kingsley.

Amir, D. (1990). A song is born: Discovering meaning in improvised songs through a phenomenological analysis of two music therapy sessions with a traumatic spinal cord injured young adult. *Music Therapy, 9*(1), 62–81.

Ansdell, G. (1995). *Music for life.* Jessica Kingsley.

Ansdell, G. (2014). *How music helps in music therapy and everyday life.* Ashgate.

Aragona, M., Kotzalidis, G., & Puzella, A. (2013). The many faces of empathy: Between phenomenology and neuroscience. *Archives of Psychiatry and Psychotherapy, 4*, 5–12.

Arnason, C. (2002). An eclectic approach to the analysis of improvisations in music therapy sessions. *Music Therapy Perspectives, 20*(1), 4–12.

Arnason, C. (2003). Music therapists' listening perspectives in improvisational music therapy. *Nordic Journal of Music Therapy, 12*(2), 124–138.

Austin, D. (1996). The role of improvised music in psychodynamic music therapy with adults. *Music Therapy, 14*(1), 29–43.

Austin, D. (2001). In search of the self: The use of vocal holding techniques with adults traumatized as children. *Music Therapy Perspectives, 19*(1), 22–30.

Basch, M. F. (1988). *Understanding psychotherapy: The science behind the art.* Basic Books.

Batson, C., Early, S., & Salvarini, G. (1997). Perspective taking: Imagining how another feels versus imagining how you would feel. *Personality and Social Psychology Bulletin, 23*, 751–758.

Baumeister, J. C., Papa, G., & Foroni, F. (2016). Deeper than skin deep: The effect of botulinum toxin-A on emotion processing. *Toxicon, 118*, 86–90.

Behrens, G. A. (2012). Use of traditional and nontraditional instruments with traumatized children in Bethlehem, West Bank. *Music Therapy Perspectives, 30*(2), 196–202.

Bericat, E. (2016). The sociology of emotions: Four decades of progress. *Current Sociology, 64*(3), 491–513.

Birnbaum, J. (2014). Intersubjectivity and Nordoff-Robbins music therapy. *Music Therapy Perspectives, 32*(1), 30–37.

Black, S., & Zimmermann, C., & Rodin, G. (2017). Comfort, connection and music: Experiences of inter-active listening on a palliative care unit. *Music & Medicine, 9*(4), 227–233.

Bonde, L. O. (1997). Music analysis and image potentials in classical music. *Nordic Journal of Music Therapy, 7*(2), 121–128.

Botvinick, M., Jha, A. P., Bylsma, L. M., Fabian, S. A., Solomon, P. E., & Prkachin, K. M. (2005). Viewing facial expressions of pain engages cortical areas involved in the direct experience of pain. *NeuroImage, 25*, 315–319.

Brown, J. M. (2002). Towards a culturally centered music therapy practice. *Voices: A World Forum for Music Therapy, 2*(1).

Bruscia, K. E. (2001). A qualitative approach to analyzing client improvisations. *Music Therapy Perspectives, 19*(1), 7–21.

Ciaunica, A. (2019). The 'meeting of bodies': Empathy and basic forms of shared experience. *Topoi, 38*, 185–195.

Colle, L., Becchio, C., & Bara, B. G. (2008). The non-problem of the other minds: A neurodevelopmental perspective on shared intentionality. *Human Development, 51*(5–6), 336–348.

Daly, A. (2016). *Merleau-Ponty and the ethics of intersubjectivity*. Palgrave Macmillan.

Davis, M. H., Conklin, L., Smith, A., & Luce, C. (1996). Effect of perspective taking on the cognitive representation of persons: A merging of self and other. *Journal of Personality and Social Psychology, 70*(4), 713.

De Backer, J. (2008). Music and psychosis: A research report detailing the transition from sensorial play to musical form by psychotic patients. *Nordic Journal of Music Therapy, 17*(2), 89–104.

de Vignemont, F., & Singer, T. (2006). The emphatic brain: How, when, and why? *Trends in the Cognitive Sciences, 10*, 435–441.

Decety, J., & Jackson, P. L. (2004). The functional architecture of human empathy. *Behavioral and Cognitive Neuroscience Review, 3*(2), 406–412.

Dekeyser, M., Elliott, R., & Leijssen, M. (2009). Empathy in psychotherapy: Dialogue and embodied understanding. In J. Decety & W. Ickes (Eds.), *The social neuroscience of empathy* (pp. 113–124). The MIT Press.

DeNora, T. (2000). *Music in everyday life*. Cambridge University Press.

DeNora, T. (2011). Practical consciousness and social relation in MusEcological perspective. In D. Clarke & E. Clarke (Eds.), *Music and consciousness:*

Philosophical, psychological, and cultural perspectives (pp. 309–326). Oxford University Press.

Denzin, N. (2009). *On understanding emotion*. Transaction Publishers.

Dimaio, L. (2010). Music therapy entrainment: A humanistic music therapist's perspective of using music therapy entrainment with hospice clients experiencing pain. *Music Therapy Perspectives, 28*(2), 106–115.

dos Santos, A. (2019a). Empathy and aggression in group music therapy with teenagers: A descriptive phenomenological study. *Music Therapy Perspectives, 37*(1), 14–27.

dos Santos, A. (2019b). Group music therapy with adolescents referred for aggression. In K. McFerran, P. Derrington, & S. Saarikallio (Eds.), *Handbook of music, adolescents, and wellbeing* (pp. 15–24). Oxford University Press.

dos Santos, A., & Brown, T. (2021). Music therapists' empathic experiences of shared and differing orientations to religion and spirituality in the client-therapist relationship. *The Arts in Psychotherapy, 74*, 101786.

Duan, C., & Hill, C. E. (1996). The current state of empathy research. *Journal of Counseling Psychology, 43*(3), 261.

Duranti, A. (2010). Husserl, intersubjectivity and anthropology. *Anthropological Theory, 10*(1–2), 16–35.

Eagle, M., & Wolitzky, D. L. (1997). Empathy: A psychoanalytic perspective. In A. C. Bohart & L. S. Greenberg (Eds.), *Empathy reconsidered: New directions in psychotherapy* (pp. 217–244). American Psychological Association.

Ebisch, S. J. H., Perrucci, M. G., Ferretti, A., Gratta, C. D., Romani, G. L., Gallese, V. (2008). The sense of touch: Embodied simulation in a visuotactile mirroring mechanism for observed animate or inanimate touch. *Journal of Cognitive Neuroscience, 20*(9), 1611–1623.

Ebisch, S. J., Aureli, T., Bafunno, D., Cardone, D., Romani, G. L., & Merla, A. (2012). Mother and child in synchrony: Thermal facial imprints of autonomic contagion. *Biological Psychology, 89*(1), 123–129.

Edwards, S. (2010). A Rogerian perspective on empathic patterns in Southern African healing. *Journal of Psychology in Africa, 20*(2), 321–326.

Feldman, R. (2007). Parent-infant synchrony and the construction of shared timing; physiological precursors, developmental outcomes, and risk conditions. *Journal of Child Psychology and Psychiatry, 48*(3–4), 329–354.

Feller, C. P., & Cottone, R. R. (2003). The importance of empathy in the therapeutic alliance. *The Journal of Humanistic Counseling, Education and Development, 42*(1), 53–61.

Ferrara, L. (1984). Phenomenology as a tool for musical analysis. *The Musical Quarterly, 70*(3), 355–373.
Foley, G. N., & Gentile, J. P. (2010). Nonverbal communication in psychotherapy. *Psychiatry, 7*(6), 38.
Forinash, M., & Gonzalez, D. (1989). A phenomenological perspective of music therapy. *Music Therapy, 8*(1), 35–46.
Gallese, V. (2009). Mirror neurons, embodied simulation, and the neural basis of social identification. *Psychoanalytic Dialogues, 19*(5), 519–536.
Gallese, V., & Goldman, A. (1998). Mirror neurons and the simulation theory of mind-reading. *Trends in Cognitive Sciences, 2*(12), 493–501.
Gilewski, M. (1993). The use of transpersonal empathy with child abuse survivors. In *Proceedings from The American Psychological Association*. Toronto, Canada. https://files.eric.ed.gov/fulltext/ED371247.pdf
Gillespie, A., & Cornish, F. (2010). Intersubjectivity: Towards a dialogical analysis. *Journal for the Theory of Social Behaviour, 40*(1), 19–46.
Goebl, W., & Palmer, C. (2009). Synchronization of timing and motion among performing musicians. *Music Perception, 26*(5), 427–438.
Guilfoyle, M. (2015). Listening in narrative therapy: Double listening and empathic positioning. *South African Journal of Psychology, 45*(1), 36–49.
Hall, S. (2009). *Anger, rage and relationship*. Routledge.
Hays, P. (1996). Addressing the complexities of a culture and gender in counseling. *Journal of Counseling and Development, 74*, 332–337.
Hermansson, G. (1997). Boundaries and boundary management in counselling: The never-ending story. *British Journal of Guidance and Counselling, 25*(2), 133–146.
Hermberg, K. (2006). *Husserl's phenomenology: Knowledge, objectivity and others*. Continuum International Publishing Group.
Hiller, J. (2015). Aesthetic foundations of music therapy: Music and emotion. In B. Wheeler (Ed.), *Music therapy handbook* (pp. 29–39). The Guilford Press.
Husserl, E. (1969). *Formal and transcendental logic*. Martinus Nijhoff.
Husserl, E. (1977). *Cartesian meditations* (D. Cairns, Trans.). Kluwer Academic.
Husserl, E. (1989). *Ideas pertaining to a pure phenomenology and to a phenomenological philosophy, second book: Studies in the phenomenology of constitution* (R. Rojcewicz & A. Schuwer, Trans.). Kluwer.
Janata, P., Tomic, S. T., & Haberman, J. M. (2012). Sensorimotor coupling in music and the psychology of the groove. *Journal of Experimental Psychology: General, 141*(1), 54.

Jardine, J. (2014). Husserl and Stein on the phenomenology of empathy: Perception and explication. *Synthesis Philosophica, 58*, 273–288.

Jones, S. (2011). Supportive listening. *The International Journal of Listening, 25*(1–2), 85–103.

Juslin, P. (2016). Emotional reactions to music. In S. Hallam, I. Cross, & M. Thaut (Eds.), *The Oxford handbook of music psychology* (pp. 197–213). Oxford University Press.

Keysers, C., Kohler, E., Umiltà, M. A., Nanetti, L., Fogassi, L., & Gallese, V. (2003). Audiovisual mirror neurons and action recognition. *Experimental Brain Research, 153*(4), 628–636.

Kidwell, M. (2014). Music therapy and spirituality: How can I keep from singing? *Music Therapy Perspectives, 32*, 129–134.

Kim, S. (2013). *Multimodal quantification of interpersonal physiological synchrony between non-verbal individuals with severe disabilities and their care-givers during Music Therapy* [Doctoral dissertation, University of Toronto].

Knapp, M., & Hall, J. A. (2010). *Nonverbal communication in human interaction* (7th ed.). Cengage Learning.

Kohler, E., Keysers, C., Umilta, M. A., Fogassi, L., Gallese, V., & Rizzolatti, G. (2002). Hearing sounds, understanding actions: Action representation in mirror neurons. *Science, 297*(5582), 846–848.

Kohut, H. (1968). The psychoanalytic treatment of narcissistic personality disorders, outline of a systematic approach. *The Psychoanalytic Study of the Child, 23*, 86–113.

Kohut, H. (1971). *The analysis of the self.* International Universities Press.

Kohut, H. (1977). *The restoration of the self.* International Universities Press.

Kohut, H. (1984). *How does analysis cure?* University of Chicago Press.

Krol, S., & Bartz, J. (2021). The self and empathy: Lacking a clear and stable sense of self undermines empathy and helping behavior. *Emotion, Advance Online Publication.* https://doi.org/10.1037/emo0000943

LaPierre, L. L. (1994). A model for describing spirituality. *Journal of Religion and Health, 33*(2), 153–161.

Lench, H., & Carpenter, Z. (2018). What do emotions do for us? In H. Lench (Ed.), *The functions of emotions: When and why emotions help us* (pp. 1–8). Springer.

Levine, L. J., Lench, H. C., Kaplan, R. L., & Safer, M. A. (2012). Accuracy and artifact: Re-examining the intensity bias in affective forecasting. *Journal of Personality and Social Psychology, 103*, 584–605.

Marci, C. D., Ham, J., Moran, E., & Orr, S. P. (2007). Physiologic correlates of perceived therapist empathy and social-emotional process during psychotherapy. *The Journal of Nervous and Mental Disease, 195*(2), 103–111.

Meekums, B. (2012). Kinesthetic empathy and movement metaphor in dance movement psychotherapy. In D. Reynolds & M. Reason (Eds.), *Kinesthetic empathy in creative and cultural practices* (pp. 51–66). Intellect.

Merleau-Ponty, M. (1962). *Phenomenology of perception* (C. Smith, Trans.). Routledge & Kegan Paul.

Miller, W. (2018). *Listening well: The art of empathic understanding.* Wipf & Stock.

Mori, J., & Hayashi, M. (2006). The achievement of intersubjectivity through embodied completions: A study of interactions between first and second language speakers. *Applied Linguistics, 27*(2), 195–219.

Murphy, A. (2021). *The extended mind.* Houghton Mifflin Harcourt.

Næss, T., & Ruud, E. (2007). Audible gestures: From clinical improvisation to community music therapy. *Nordic Journal of Music Therapy, 16*(2), 160–171.

Nordoff, P., & Robbins, C. (2007). *Creative music therapy: A guide to fostering clinical musicianship* (2nd ed.). Barcelona Publishers.

Overy, K., & Molnar-Szakacs, I. (2009). Being together in time: Musical experience and the mirror neuron system. *Music Perception, 26*(5), 489–504.

Papoušek, M., & Papoušek, H. (1981). Musical elements in the infant's vocalization: Their significance for communication, cognition, and creativity. In L. P. Lipsitt & C. K. Rovee-Collier (Eds.), *Advances in infancy research* (Vol. 1, pp. 163–224). Ablex.

Papp, L., Pendry, P., & Adam, E. (2009). Mother-adolescent physiological synchrony in naturalistic settings: Within-family cortisol associations and moderators. *Journal of Family Psychology, 23*(6), 882–894.

Pavlicevic, M. (1997). *Music therapy in context.* Jessica Kingsley.

Pavlicevic, M. (2002). South Africa: Fragile rhythms and uncertain listenings: Perspectives from music therapy with South African children. In J. P. Sutton (Ed.), *Music, music therapy and trauma* (pp. 97–118). Jessica Kingsley Publishers.

Pedersen, I. (1997). The Music Therapist's listening perspectives as source of information in improvised musical duets with grown-up: Psychiatric patients, suffering from Schizophrenia. *Nordic Journal of Music Therapy, 6*(2), 98–111.

Phillips-Silver, J., Aktipis, C., & Bryant, G. (2010). The ecology of entrainment: Foundations of coordinated rhythmic movement. *Music Perception, 28*(1), 3–14.

Phillips-Silver, J., & Keller, P. (2012). Searching for roots of entrainment and joint action in early musical interactions. *Frontiers in Human Neuroscience, 6*(26), 1–11.

Potvin, N., & Argue, J. (2014). Theoretical considerations of spirit and spirituality in music therapy. *Music Therapy Perspectives, 32*(2), 118–128.

Preckel, K., Kanske, P., & Singer, T. (2018). On the interaction of social affect and cognition: Empathy, compassion and theory of mind. *Current Opinion in Behavioral Sciences, 19*, 1–6.

Preston, S. D., & de Waal, F. B. M. (2002). Empathy: Its ultimate and proximate bases. *Behavioral and Brain Sciences, 25*, 1–72.

Priestley, M. (1975). *Music therapy in action.* Constable.

Priestley, M. (1994). *Essays on analytical music therapy.* Barcelona Publishers

Purdon, C. (2002). The role of music in analytical music therapy: Music as a carrier of stories. In J. Eschen (Ed.), *Analytical music therapy* (pp. 104–114). Jessica Kingsley.

Rabinowitch, T., Cross, I., & Burnard, P. (2012). Long-term musical group interaction has a positive influence on empathy in children. *Psychology of Music, 41*(4), 484–498.

Reuther, B. (2014). Intersubjectivity, Overview. In *Encyclopedia of Critical Psychology*. Springer.

Ridley, C. R., & Lingle, D. W. (1996). Cultural empathy in multicultural counseling: A multidimensional process model. In P. B. Pedersen, J. G. Draguns, W. J. Lonner, & J. E. Trimble (Eds.), *Counseling across cultures* (pp. 21–46). Sage.

Rizzolatti, G., Fadiga, L., Gallese, V., & Fogassi, L. (1996). Premotor cortex and the recognition of motor actions. *Cognitive Brain Research, 3*(2), 131–141.

Robarts, J. (1996). Music therapy for children with autism. In C. Trevarthen, K. Aitken, D. Papoudi, & J. Robarts (Eds.), *Children with autism* (pp. 134–160). Jessica Kingsley.

Robarts, J. (2003). The healing function of improvised songs in music therapy with a child survivor of early trauma and sexual abuse. In S. Hadley (Ed.), *Psychodynamic music therapy: Case Studies* (pp. 141–182). Barcelona.

Robarts, J. (2009). Supporting the development of mindfulness and meaning: Clinical pathways in music therapy with a sexually abused child. In S.

Malloch & C. Trevarthen (Eds.), *Communicative musicality: Exploring the basis of human companionship* (pp. 377–400). Oxford University Press.
Rogers, C. (1951). *Client-centered therapy: Its current practice, implications, and theory*. Houghton Mifflin Company.
Rogers, C. (1959). A theory of therapy, personality, and interpersonal relationships as developed in the client-centered framework. In S. Koch (Ed.), *Psychology: A study of a science, Vol. 3: Formulations of the person and the social context* (pp. 184–256). McGraw Hill.
Rowe, C., & Isaac, D. (2004). *Empathic attunement: The "technique" of psychoanalytic self psychology*. Rowman & Littlefield.
Scheiby, B. (2005). An intersubjective approach to music therapy: Identification and processing of musical countertransference in a music psychotherapeutic context. *Music Therapy Perspectives, 23*(1), 8–17.
Schiavio, A., van der Schyff, D., Cespedes-Guevara, J., & Reybrouck, M. (2017). Enacting musical emotions. Sense-making, dynamic systems, and the embodied mind. *Phenomenology and the Cognitive Sciences, 16*(5), 785–809.
Schwaber, E. (1984). Empathy: A mode of analytic listening. In J. Lichtenberg, M. Bornstein, & D. Silver (Eds.), *Empathy II* (pp. 143–172). Routledge.
Shea, S. (1998). *Psychiatric interviewing: The art of understanding* (2nd ed.). Sanders.
Short, H. (2017a). It feels like Armageddon: Identification with a female personality-disordered offender at a time of cultural, political and personal attack. *Nordic Journal of Music Therapy, 26*(3), 272–285.
Short, H. (2017b). "Big up West London Crew": One man's journey within a community rap/music therapy group. *Music Therapy Perspectives, 35*(2), 151–159.
Siebert, D., Siebert, C., & Taylor-McLaughlin, A. (2007). Susceptibility to emotional contagion: Its measurement and importance to social work. *Journal of Social Service Research, 33*(3), 47–56.
Siegel, A. (1996). *Heinz Kohut and the psychology of the self*. Routledge.
Skewes, K. (2001). *The experience of group music therapy for six bereaved adolescents* [Doctoral dissertation, University of Melbourne].
Sperry, L. (2012). *Spirituality in clinical practice: Theory and practice of spiritually oriented psychotherapy* (2nd ed.). Routledge.
Stein, E. (1989). *On the problem of empathy* (W. Stein, Trans.). ICS Publishers.
Stephens, G., Silbert, L., & Hasson, U. (2010). Speaker-listener neural coupling underlies successful communication. *Proceedings of the National Academy of Sciences of the United States of America, 107*(32), 14425–14430.

Stern, D. (2005). Intersubjectivity. In E. Person, A. Cooper, & G. Gabbard (Eds.), *Textbook of psychoanalysis* (pp. 77–92). American Psychiatric Publishing.

Stern, S. (2009). The dialectic of empathy and freedom. *International Journal of Psychoanalytic Self Psychology, 4*, 132–164.

Stige, B. (2016). *Culture-centered music therapy*. Oxford University Press.

Sutton, J. (2002). The pause that follows. *Nordic Journal of Music Therapy, 11*(1), 27–38.

Tanyi, R. A. (2002). Towards clarification of the meaning of spirituality. *Journal of Advanced Nursing, 39*(5), 500–509.

Thomas, A., & Sham, F. T. Y. (2014). "Hidden rules": A duo-ethnographical approach to explore the impact of culture on clinical practice. *Australian Journal of Music Therapy, 25*, 81–91.

Trentini, C., Tambelli, R., Maiorani, S., & Lauriola, M. (2021). Gender differences in empathy during adolescence: Does emotional self-awareness matter? *Psychological Reports*, 0033294120976631.

Trevarthen, C. (1994). The self born in intersubjectivity: The psychology of an infant communicating. In U. Neisser (Ed.), *The perceived self: Ecological and interpersonal sources of self-knowledge* (pp. 121–173). Cambridge University Press.

Trevarthen, C., & Fresquez, C. (2015). Sharing human movement for well-being: Research on communication in infancy and applications in dance movement psychotherapy. *Body, Movement and Dance in Psychotherapy, 10*(4), 194–210.

Trevarthen, C., & Malloch, S. (2000). The dance of wellbeing: Defining the musical therapeutic effect. *Norwegian Journal of Music Therapy, 9*(2), 3–17.

Trondalen, G. (2003). "Self listening" in music therapy with a young woman suffering from Anorexia Nervosa. *Nordic Journal of Music Therapy, 12*(1), 3–17.

Trondalen, G. (2016). *Relational music therapy: An intersubjective perspective.* Barcelona.

Trondalen, G. (2019). Musical intersubjectivity. *The Arts in Psychotherapy, 65*, 101589.

Turry, A. (1998). Nordoff-Robbins: Transference and countertransference. In K. E. Bruscia (Ed.), *The dynamics of music psychotherapy* (pp. 161–212). Barcelona Publishers.

Turry, A. (2009). Integrating musical and psychological thinking: The relationship between music and words in clinically improvised songs. *Music and Medicine, 1*(2), 106–116.

Turry, A. (2011). *Between music and psychology.* Lambert Academic Publishing.

Valentino, R. (2006). Attitudes towards cross-cultural empathy in music therapy. *Music Therapy Perspectives, 24*, 108–114.

Verduyn, P., Van Mechelen, I., & Tuerlinckx, F. (2011). The relation between event processing and the duration of emotional experience. *Emotion, 11*, 20–28.

Viega, M. (2018). A humanistic understanding of the use of digital technology in therapeutic songwriting. *Music Therapy Perspectives, 36*(2), 152–160.

Walker, K. L. (1997). Do you ever listen?: Discovering the theoretical underpinnings of empathic listening. *International Journal of Listening, 11*(1), 127–137.

Walsh, P. (2014). Empathy, embodiment, and the unity of expression. *Topio, 33*, 215–226.

Winter, P. (2013). *Effects of experiential music therapy education on student's reported empathy and self-esteem: A mixed methods study* [Doctoral dissertation, Temple University].

Zahavi, D. (2010). Empathy, embodiment and interpersonal understanding: From Lipps to Schutz. *Inquiry, 53*(3), 285–306.

4

Sharing Emotions

Once we are open to empathising and then direct our attention towards perceiving multiple layers of the emotional experience of the person in front of us, we may also begin to share in this person's emotions. Types of emotion-sharing occur along a spectrum. We could unconsciously "catch" their emotion (which would not technically be classified as empathy, as we'll explore in a moment), or we could consciously seek to share in their emotional experience with a clear sense of the distinction between self and other. This is the attribute of "sharing" that is part of insightful empathy, but let's begin by exploring what is meant by "catching" emotions so the distinction is clear.

Emotional Contagion (Is Not Empathy)

As we saw in the previous chapter, through the Mirror Neuron System, a human will mirror another person's vocalisations, facial expressions, postures, and movements that correspond with their emotional state. Consequently, they may also start to feel this person's emotion (Hatfield et al., 1994). Our movements affect not only how others understand our

A. dos Santos, *Empathy Pathways*,
https://doi.org/10.1007/978-3-031-08556-7_4

intentions but also provide us with kinaesthetic body feedback, which helps us to perceive our own emotional states. What we think and feel influences how we move, and our expressive movement has a causal impact on our emotional states, the formation of our attitudes, and our subsequent behaviour regulation (Koch & Fuchs, 2011).

Infants mimic their caregiver's facial expressions and may cry when hearing another infant's crying. Automatic mimicry continues into adulthood (Krol & Bartz, 2021). As the philosopher Eric Hoffer (1955) famously said, "When people are free to do as they please, they usually imitate each other" (aph. 33). Decety and Jackson (2004) used the term "perception–action coupling" to refer to how people connect with others—particularly in their in-group (Gutsell & Inzlicht, 2010)—by resonating with their movements and posture as well as their emotions. When spending time with a jovial group of people, one may find their joy to be infectious, while, when spending time with a friend who is feeling mournful, one may experience a personal sense of despondency (Maibom, 2017). This process of "catching" another person's emotions is known as emotional contagion. Witnessing others in pain also activates the parts of the brain involved in the perception of physical pain (Lamm et al., 2011), although this also occurs more strongly when we're observing people we perceive to be in our in-group (Xu et al., 2009).

Some authors contend that emotional contagion is a primitive form of empathy (Maibom, 2017), but seeing as insightful empathy includes the notion of intentionality, emotional contagion is then excluded (a distinction made by Zurek and Scheithauer [2017]). One also loses the self-other distinction in emotional contagion because the emotion is experienced as being one's own (Preston & de Waal, 2002). In empathy, the focus of the experience remains on the other person. In addition, emotional contagion concerns the quality rather than the object of the emotion. I may "catch" your sadness but not be aware of what you are sad about (Zahavi, 2008). Primitive emotional contagion is relatively unintentional, automatic, uncontrollable, and largely inaccessible to conscious awareness (Hatfield et al., 1994). It's possible that emotional contagion can lead to empathy if one gains awareness of what is taking place, generating conscious intentionality (Zurek & Scheithauer, 2017).

We can experience emotional contagion when we listen to music. We may feel as if we're absorbing feelings of nervousness, joy, lightness, or longing from the piece we're listening to or playing with others, for example. Juslin and Västfjäll (2008) described music-emotional contagion as a process through which a piece of music induces emotion in a listener. The listener perceives the emotional expression of the music and then internally mimics this expression (through peripheral muscle feedback and/or through neurological activation of the relevant emotional representation), leading to an experience of that emotion.

Durkheim (1912/1964) used the term "collective effervescence" to describe the emotional intensification that takes place in social groups, particularly in interaction rituals (when people sing together in church or at a soccer game, for example, or during a group improvisation in a music therapy session). During interaction rituals, people focus their attention on the same thing. They become mutually aware of this shared focus (in other words, there is intersubjectivity), and they feel a shared emotion. When these three conditions are present, both the focus and the emotion intensify, and rhythmic entrainment draws the participants into an emotional state that feels stronger than how any of them would have experienced it individually (as if a force is carrying them along). This experience generates solidarity, belonging, and a common identity, as well as emotional energy that can be long-lasting as individuals take it away with them when they leave the event (Collins, 2004).

As music therapists, while we may have invigorating experiences of collective effervescence, we may also experience "catching" painful feelings from individuals and groups we work with. We may leave a session having absorbed some of a client's weariness, heartache, or anxiety. Emotional contagion can lead to personal distress when we become overaroused from vicariously experiencing another's suffering. Even though the origin of this is the other's distress, the result is a feeling of distress for oneself (Maibom, 2017). This could lead to prioritising easing one's own distress above assisting the other (Trentini et al., 2021) through, for example, distancing ourselves or avoiding the other person (Eisenberg, 2000). We may decide to help them, but this could be at least in part with the intention of lowering our own feelings of distress (Hornstein, 1991).

The good news, as Hoffman (1991) explained, is that alleviating one's own resonant distress is rarely a conscious intention of people who help others. Even when we relieve our own uncomfortable feelings by assisting another person, their welfare is often still enhanced. Researchers such as Cialdini et al. (1997) have found that vicarious distress can indeed lead to prosocial choices and helping others. However, it appears that the more intense the distress becomes, the more likely it is to become self-focused (López-Pérez et al., 2014; Maibom, 2017). As humans, if we feel that our discomfort will last unless we ease the other's suffering, helping tends to increase (offering ourselves a "difficult escape"). If we conclude that the discomfort will cease if we simply leave the situation, then we're more likely to do so (taking the route of an "easy escape") (Carrera et al., 2013).

The fork in the road that can lead to either emotional contagion or insightful empathy is emotional regulation, with low regulators experiencing more significant personal distress (Spinrad & Eisenberg, 2009). Emotional regulation refers to processes through which an individual pays attention to their emotions, manages the duration and intensity of emotional arousal, and transforms the nature and meaning of their feeling states when faced with distressing or stressful events and situations (Thompson, 1994). To be empathic, we need to regulate our emotions (Paden, 2018).

How do we do this? One mindfulness strategy that has been proposed draws on the acronym "RAIN" (Brach, 2003): (R)—recognise what's going on and consciously acknowledge your feelings, thoughts, and behaviours; (A)—allow these simply to be there without judging what you're thinking, feeling, or doing; (I)—draw on your curiosity to investigate what is happening inside yourself; and (N)—this is described differently by various authors, for example, as either "non-identification" (not fusing your sense of self with what you're feeling and thinking at this moment) (Ellis et al., 2018) or "nurture" (King, 2018) (treating yourself with kindness and appreciation instead of criticism). Seeing as the physical mirroring of the other leads to corresponding emotional states (Hatfield et al., 1994), one may also find it helpful when overwhelmed by emotional contagion to self-regulate by purposefully mismatching breathing rhythms and body posture to reduce the emotional resonance.

It's common for music therapists to encounter clients who feel sadness and despair. Music therapists who are highly in tune with their clients' emotions, who struggle to regulate their feelings and detach from their clients' problems may experience emotional overwhelm, which can lead to burnout (Siebert et al., 2007). Burnout appears to result more strongly from emotional contagion than empathy (in fact, empathy can protect against burnout) (Hunt et al., 2017; Wilkinson et al., 2017). Burnout itself can be contagious (Bakker et al., 2001, 2005). Negative emotions appear even more contagious than positive ones (McIntosh et al., 1994). When we "soak up" others' emotions like sponges we can become "anger-absorbed" or "frustration-absorbed," which can have a negative impact on decision-making (Petitta et al., 2017, 2019).

As music therapists, we can employ strategies to assist us in maintaining healthy psychological boundaries between our emotional world and the emotional worlds of our clients (Jacobowitz, 1992; Lee, 1996). For example, we can reflect on whether feelings belong to ourselves or to the client, improvise or draw our emotional experiences after sessions, attend personal therapy and supervision, and reflect with a co-therapist if we have one.

In light of working with clients who have experienced trauma, Levine (2010) highlighted that "trauma is not what happens to us, but what we hold inside in the absence of an empathetic witness" (p. xii). To be an empathetic witness requires one to be a separate self rather than to be fused with the emotional world of the other. In Calhoun and Tedeschi's (2006) discussion of posttraumatic growth, they referred to this witness as an expert companion. Austin (2001) explained that a highly empathic musical environment offers fertile soil for trust to grow and for the client to feel safe enough to begin to differentiate and express feelings such as rage, terror, and grief. For this to happen, the client needs to experience the music therapist and the music as resilient enough to withstand such intense emotions. Dimaio (2010) wrote how she purposefully entered her client's pain while also implementing boundaries to protect herself from her client's pain. She explained that resonance does not equate to feeling their pain as they do: "I can resonate with their pain but not be in pain myself" (p. 114).

The Nature of Affective Empathy

"Affective empathy" or "emotional empathy" entails consciously sharing in another's feelings while being aware that they belong to the other. In music therapy literature, sharing in another's emotion has also been described with terms such as "resonating" (Abbott, 2018; Dileo, 2021; Kenny, 1999; Kwan, 2010; Lindvang & Frederiksen, 1999; Pedersen, 1997; Préfontaine, 2006; Priestley, 1994; Wilkerson et al., 2017; Winter, 2013), "creating emotional connection" (Valentino, 2006), empathic "joining in" (Layman et al., 2013; Pedersen, 1997), "tuning in" (Kwan, 2010) or "being in tune" (Robarts, 2006), "emotion sharing" (Haviland, 2014; Jackson, 2010; Jones, 2018; Sena Moore, 2017), "meeting" the client emotionally or in their emotional space (Haviland, 2014; Swanick, 2019; Wigram & Elefant, 2009) and "being on the same wavelength" (Robarts, 2006).

Affective empathy is distinct from sympathy, which tends to be understood as feeling *for* rather than *with* another (Zurek & Scheithauer, 2017). An example of sympathy could be seen in the statement, "I feel sad for you that you are scared." Affective empathy can potentially lead to sympathy (Decety & Lamm, 2011).

Isomorphism means that the emotion felt by the empathiser must correspond to the emotion of the person they are empathising with. According to De Vignemont and Singer (2006), this is a necessary condition for empathy. If I'm experiencing your fear as restlessness, for example, I'm not empathising with you. The degree to which one accurately perceives the content of the other person's emotions is termed "content accuracy," and the accuracy with which one perceives the degree of pleasantness or unpleasantness of the other's emotions is referred to as "valence accuracy" (Ickes, 2009).

The idea of empathic accuracy can be traced back to Roger's (1957) description of the therapist correctly inferring the thoughts and feelings of the client moment-to-moment. Empathic accuracy improves as relationships deepen, as therapists spend more time with a client, and receive feedback from their client (Marangoni et al., 1995). Empathic accuracy is dependent on factors such as the empathiser's previous personal experiences, imagination, and simulation abilities (Coplan, 2011), available resources for sharing emotions (Cuff et al., 2016), the empathiser's ability to read their own emotions accurately (Batson, 2011), and whether they are correctly reading the situation as a whole (Elfenbein, 2014). Empathic accuracy is a complex and ongoing process because people's emotional states fluctuate, sometimes rather quickly (Hollan, 2008).

In affective empathy, the emotion I feel is more appropriate to the other's situation or state than my own (Hoffman, 2000). (Empathically sharing emotions cannot simply entail two people who happen to both be experiencing the same emotion simultaneously). Let's consider a situation where a member of your music therapy group, Tracey, makes a racially disparaging statement towards another group member, Nina. Nina feels angry with Tracey. You also feel angry with Tracey. Is this a case of feeling "with" Tracey? Perhaps both you and Nina are simply angry with Tracey. For this anger to be empathic, your emotion needs to relate to Nina's in a particular way. If you are angry because you have perceived that Nina is angry with Tracey (through her music, her facial expressions, her clenched fists) and you are feeling Nina's emotion with her, then that would be empathy. In reality, your anger may be a combination of empathic anger with Tracey because of what has happened to her and personal anger towards Tracey for how she treated Nina (and perhaps also how this caused you offence).

There are contrasting views on whether we must know what another person's emotion is about to fully empathise with them. A teenage client in my music therapy group may, for example, feel embarrassed when singing in front of the other group members. When I empathically feel embarrassment at that moment, I'm not experiencing embarrassment for myself. I'm embarrassed with her. I may be embarrassed with her because I know and understand what caused her feelings. My emotion then also has the same object as hers (the other group members who laughed at her when she sang). According to Maibom (2017), this is full empathic emotion. There are other times when we may not know or understand what has caused the specific emotion in the other person. My emotion may then have the other person's emotion as its object, and not also the object of their emotion (what their emotion is about). Some (such as Maibom) consider this to be empathy to a lesser degree.

Others, such as Bradford and Robin (2021), have contended that one could feel someone's anger with them because you have felt anger before and know the emotion, even if the situation they are in would not necessarily cause you to feel angry and even if you have no idea why they are feeling anger at all. You still know what anger feels like and can, therefore, empathise with them. This extends to situations where there is disagreement about the underlying issue, as in the event of a conflict. In this case, empathy would involve a non-judgemental response of emotion-sharing that reflects a resonance at the level of the emotion behind the issue (Chaitin & Steinberg, 2008).

Smith (2017) argued that to share in another's emotional state, the empathiser, A, must both know that the other, B, is experiencing a particular affective state, and A must know how it feels for B to be in this particular state. For this to be possible, A must either have had a personal experience and memory of this affective state to draw on or must be able to transpose themselves imaginatively into that state as long as they have been in states themselves that sufficiently match this state in a degree of similarity. Zahavi (2017) disagreed with the argument that it is necessary to have had a similar experience to share empathically in another's emotional state. From his phenomenological perspective, the object of the act of empathy is not the emotion as it is given to the other person first-personally; one is given an experience of the other person living through an emotional experience. I am not experiencing your sadness; I am experiencing you feeling your sadness. The experience of another person experiencing sadness is, by definition, different from a personal experience of sadness. This is fundamentally important because the asymmetry allows me to understand that I am experiencing another's mind (Zahavi, 2008). I am not merely imagining their sadness, however. Through seeing their facial expressions, posture, and movements and hearing their voice, I am experiencing aspects of their emotional life. Through empathy, I grasp their emotion.

Also, from a phenomenological perspective, Stein (1989) described this process in terms of stages. She gave the example of watching a trapeze artist. One begins by watching with a clear, separate sense of self (sitting on a chair in the stands). At one point, as one watches the trapeze artist flying through the air, one may be drawn into the experience and feel one's stomach lurch along with them. One then shifts back to experiencing oneself as an observer. In a music therapy session, I may begin by "reading" the client's sadness off their face and the quality of their music. The client's sadness then pulls me into a shared experience. I then "lean back out," and the client's experience faces me once again as an object (as it was in the beginning), but I now have a fuller experience of it due to having entered into the sadness with them (still as their sadness, though).

The resonant, shared experience in the middle stage that Stein (1989) referred to has been articulated similarly by Hart (1997). Early on in an encounter with a client, we may find ourselves entraining: our vocal qualities, breathing, posture, gestures, and music fall into alignment with theirs as we "link up." The client may be feeling helpless, "foggy," and lost. We, as the therapist, may also begin to experience these emotions and even the physical sensation of fogginess, dullness, and cloudiness. A psychic unity takes place as the self-other boundary *temporarily* dissolves. The client may experience difficulty articulating their feelings, but as the therapist is able to experience their feelings and then move into Stein's third stage, taking a step back and identifying the emotions, they can reflect them back to the client (verbally or, in our case as music therapists, through the music). The client may have the resulting experience of realising, "Oh yes, that is how I feel." Through the modelling of the therapist, the client is invited to tune deeply into their own experiences. This co-feeling is a form of psychological resonance. It can be likened to a violin string that is excited by the sounding of other strings tuned to the same notes. The two strings do not become one but connect at a particular frequency.

As discussed in Chapter 3, the body plays a crucial role in this process. Through embodied resonance, my body mirrors the motor movements of your body. My brain then interprets my movements as indicating that I feel a certain way, and I then have the experience of feeling that emotion myself. If I become consciously aware that this is taking place and that the emotion originated in you, I transition from emotional contagion into purposeful empathy as I intentionally share in your emotional state.

Shlien (1997), a student of Carl Rogers, referred to empathy as a whole-body response, operating on data such as sight, sound, and smell (the smell of fear, the sound of yawns, tones, cadences, and cries, the sight of sagging shoulders and blushing cheeks). Empathy includes the process of "bodily-living-in" (Rosan, 2012, p. 124), which involves a dance between subjects, somatic memory, autobiographical reverberations, and re-enactment of movements from the other's narrated drama, all of which offer visceral experiences of meaning related to the other's emotional world.

Dekeyser et al. (2009) also understood empathy as a bodily rather than conceptual process. They highlighted five metaphors that capture the experience: letting go (setting aside preformed ideas and personal issues and allowing, opening up and clearing space for the client's experiences); moving towards or into (dwelling on, joining in, stepping into, and becoming immersed in the client's experience while actively reaching out to enter their world); resonating (through body-oriented attention, opening up to sensations, thoughts, feelings, and memories that emerge, while paying close attention to the client); discovering or discerning (sorting through, finding, detecting, and differentiating); and grasping or taking hold (of what is central, critical, and most alive).

Pizer (2018) wrote of the "tension of tenderness" (p. 110). Therapists welcome, with willing and disciplined receptiveness, the multiplicity of the clients' emotions and the emotions that arise within themselves through the negotiated therapeutic relationship. We are required to sustain the struggle in ourselves tenaciously as we generously contain our clients' hopelessness, rage, dread, and anguish, showing fundamental respect and an openness to feel confusion, pain, and self-doubt in the service of our clients' journeys towards greater wellbeing. Our task is to continue moving towards the emotion even while our own "buttons" may be pushed and while we do not deny these personal emotions. It's fundamental for us as therapists to register the "imprint" of the client's state while preserving our own equilibrium and integrity. A therapist communicates to their client through containment, acceptance, and recognition: I am with you.

Sharing Emotions Through Music

In Small's (1998) words, "the act of musicking establishes in the place where it is happening a set of relationships, and it is in those relationships that the meaning of the act lies" (p. 13). Musical relationships afford social attuning and empathy in particular (Leman, 2007) through their ability to express the quality of emotion itself beyond what verbal language can accomplish (Ansdell, 1995). Music can invite participants into a joint emotional endeavour and emotional alignment (Laurence, 2008; Molnar-Szakacs, 2017). In musicking, we enter into emotional "co-experiencing" (Rosan, 2012) through a variety of techniques that invite immersion and shared intentionality, flexibility in remaining attuned throughout musical changes, and freedom for authentic creativity (Cross et al., 2012). As music therapists engage in matching, imitating, reflecting, exaggerating, enhancing, synchronising, incorporating, and mirroring (Bruscia, 1987), as well as, for example, facilitating discussions of lyrics (Clements-Cortés, 2010; Viega, 2016), we take part in the experience of the other's emotions, ranging from deep despair to exuberant joy. We enter and move around in our client's emotional worlds with them.

When improvising with clients in sessions, we may share their emotions as they express them through their music. We are also jointly taking part in creating an emotional world together through co-musicking. In terms of the first process, as we listen to a client's music, we perceive the emotion in their music and/or emotions are elicited in us as we listen. Theorists who hold a cognitivist position maintain that music does not induce genuine emotions in us as listeners. Frightening-sounding music may not elicit fear equivalent to that evoked in a genuinely threatening encounter. Alternatively, those who support an emotivist position contend that music can indeed produce specific, genuine emotions in listeners (Corrigall & Schellenberg, 2013). There appear to be several musical emotion induction mechanisms. As Juslin (2013) identified, brain stem reflexes are activated when a loud, sudden, or dissonant music feature brings about an automatic, quick, and unlearned neurological response. Rhythmic entrainment engages a process of bodily, emotional, and social synchronisation. Through evaluative conditioning, a particular aspect of the music stimulates a conditioned response. Emotion can also be induced through emotional contagion, visual imagery, and episodic memory. An emotion may be generated in a listener when a piece violates, confirms, or delays expectancy. A listener's aesthetic attitude can bring about emotional responses through how they judge certain music events as beautiful, novel, distasteful, and so on. Levinson (2006) added persona theory as an explanatory mechanism for how a piece of music may be perceived to express the emotions of a character "within" it (which may be the composer, a singer, or an imagined expressive character). Music can also effectively communicate emotional meaning by mimicking the expressive qualities of the human voice and bodily movement (Jackendoff & Lerdahl, 2006) and through culturally learned cues (Dalla Bella et al., 2001).

Support for the hypothesis that music induces genuine emotion has also come from researchers who have demonstrated that pleasurable music activates regions in the brain that usually respond to other pleasures and rewards (such as the ventral striatum, the orbitofrontal cortex, anterior insula, and parietal and somatosensory areas) (Blood & Zatorre, 2001; Menon & Levitin, 2005). Sad music has been shown to activate the amygdala, hippocampus and medial temporal lobe areas involved in negative affective states and anxiety. Fear and tension evoked by music may also increase activity in the amygdala, just as other fear-inducing stimuli do (Aubé et al., 2015; Koelsch & Skouras, 2014; Lehne et al., 2014). Vuilleumier and Trost (2015) found that this evocation of "true" emotions extends beyond basic emotions to more complex ones, such as awe, wonder, and nostalgia. However, explanatory complexities arise because the mechanisms that induce emotions are ubiquitous, shared, and intertwined (Moors & Kuppens, 2008; Scherer & Zentner, 2008). Also, areas in the brain associated with emotions have not been found to be consistent or specific (Barrett, 2017). Some researchers have argued that, rather than a cognitivist position or an emotivist position, a constructive theory is most realistic: emotions are formed through meaning-making processes (Christensen, 2018). When exploring how music may evoke, express, and create emotions, it's also crucial that we examine the context and our active, creative role in the process (Krueger, 2013; Van der Schyff & Schiavio, 2017).

References

Abbott, E. A. (2018). Subjective observation in music therapy: A study of student practicum Logs. *Music Therapy Perspectives, 36*(1), 117–126.

Ansdell, G. (1995). *Music for life.* Jessica Kingsley.

Aubé, W., Angulo-Perkins, A., Peretz, I., Concha, L., & Armony, J. L. (2015). Fear across the senses: Brain responses to music, vocalizations and facial expressions. *Social Cognitive and Affective Neuroscience, 10*(3), 399–407.

Austin, D. (2001). In search of the self: The use of vocal holding techniques with adults traumatized as children. *Music Therapy Perspectives, 19*(1), 22–30.

Bakker, A. B., Le Blanc, P. M., & Schaufeli, W. B. (2005). Burnout contagion among intensive care nurses. *Journal of Advanced Nursing, 51*(3), 276–287.

Bakker, A. B., Schaufeli, W. B., Sixma, H. J., & Bosveld, W. (2001). Burnout contagion among general practitioners. *Journal of Social and Clinical Psychology, 20*(1), 82–98.

Barrett, L. F. (2017). *How emotions are made: The secret life of the brain.* Houghton Mifflin Harcourt.

Batson, C. (2011). *Altruism in humans.* Oxford University Press.

Blood, A. J., & Zatorre, R. J. (2001). Intensely pleasurable responses to music correlate with activity in brain regions implicated in reward and emotion. *Proceedings of the National Academy of Sciences, 98*(20), 11818–11823.

Brach, T. (2003). *Radical acceptance: Embracing your life with the heart of a Buddha.* Bantam Books.

Bradford, D., & Robin, C. (2021). *Connect*. Penguin Random House.

Bruscia, K. (1987). *Improvisational models of music therapy*. Thomas.

Calhoun, L., & Tedeschi, R. (2006). Expert companions: Posttraumatic growth in clinical practice. In L. Calhoun & R. Tedeschi (Eds.), *Handbook of posttraumatic growth* (pp. 291–310). Psychology Press.

Carrera, P., Oceja, L., Caballero, A., Muñoz, D., López-Pérez, B., & Ambrona, T. (2013). I feel so sorry! Tapping the joint influence of empathy and

personal distress on helping behavior. *Motivation and Emotion, 37*(2), 335–345.
Chaitin, J., & Steinberg, S. (2008). "You should know better": Expressions of empathy and disregard among victims of massive social trauma. *Journal of Aggression, Maltreatment & Trauma, 17*(2), 197–226.
Christensen, J. (2018). *Sounds and the aesthetics of play: A musical ontology of constructed emotions.* Palgrave Macmillan.
Cialdini, R. B., Brown, S. L., Lewis, B. P., Luce, C., & Neuberg, S. L. (1997). Reinterpreting the empathy–altruism relationship: When one into one equals oneness. *Journal of Personality and Social Psychology, 73*(3), 481.
Clements-Cortés, A. (2010). The role of music therapy in facilitating relationship completion in end-of-life care. *Canadian Journal of Music Therapy, 16*(1).
Collins, R. (2004). *Interaction ritual chains.* Princeton University Press.
Coplan, A. (2011). Will the real empathy please stand up? A case for a narrow conceptualization. *The Southern Journal of Philosophy, 49*, 40–65.
Corrigall, K., & Schellenberg, E. (2013). Music: The language of emotion. In C. Mohiyeddini, E. Eysenck, & S. Bauer (Eds.), *Handbook of psychology of emotions* (pp. 299–325). Nova Science Publishers.
Cross, I., Laurence, F., & Rabinowitch, T. C. (2012). Empathy and creativity in group musical practices: Towards a concept of empathic creativity. In G. McPherson & G. Welch (Eds.), *The Oxford handbook of music education* (Vol. 2, pp. 337–353). Oxford University Press.
Cuff, B., Brown, S., Taylor, L., & Howat, D. (2016). Empathy: A review of the concept. *Emotion Review, 8*(2), 144–153.
Dalla Bella, S., Peretz, I., Rousseau, L., & Gosselin, N. (2001). A developmental study of the affective value of tempo and mode in music. *Cognition, 80*(3), B1–10.
de Vignemont, F., & Singer, T. (2006). The emphatic brain: How, when, and why? *Trends in the Cognitive Sciences, 10*, 435–441.
Decety, J., & Jackson, P. L. (2004). The functional architecture of human empathy. *Behavioral and Cognitive Neuroscience Review, 3*(2), 406–412.
Decety, J., & Lamm, C. (2011). Empathy vs. personal distress. In J. Decety, & W. Ickes (Eds.), *The social neuroscience of empathy* (pp. 199–214). MIT Press.
Dekeyser, M., Elliott, R., & Leijssen, M. (2009). Empathy in psychotherapy: Dialogue and embodied understanding. In J. Decety & W. Ickes (Eds.), *The social neuroscience of empathy* (pp. 113–124). The MIT Press.
Dileo, C. (2021). *Ethical thinking in music therapy* (2nd ed.). Jeffrey Books.

Dimaio, L. (2010). Music therapy entrainment: A humanistic music therapist's perspective of using music therapy entrainment with hospice clients experiencing pain. *Music Therapy Perspectives, 28*(2), 106–115.
Durkheim, E. (1912/1964). *The elementary forms of religious life.* Free Press.
Eisenberg, N. (2000). Emotion, regulation, and moral development. *Annual Review of Psychology, 51*, 665–697.
Elfenbein, H. (2014). The many faces of emotional contagion: An affective process theory of affective linkage. *Organizational Psychology Review, 4*(4), 326–362.
Ellis, T. E., Schwartz, J. A., & Rufino, K. A. (2018). Negative reactions of therapists working with suicidal patients: A CBT/Mindfulness perspective on "countertransference." *International Journal of Cognitive Therapy, 11*(1), 80–99.
Gutsell, J., & Inzlicht, M. (2010). Empathy constrained: Prejudice predicts reduced mental simulation of actions during observation of outgroups. *Journal of Experimental Social Psychology, 46*(5), 841–845.
Hart, T. (1997). Transcendental empathy in the therapeutic encounter. *The Humanistic Psychologist, 25*(3), 245–270.
Hatfield, E., Cacioppo, J., & Rapson, R. (1994). *Emotional contagion.* Cambridge University Press.
Haviland, J. (2014). *Exploring empathy in music therapy* [Masters Dissertation, Molloy College].
Hoffer, E. (1955). *The passionate state of mind.* Harper and Brothers.
Hoffman, M. (2000). *Empathy and moral development*. Cambridge University Press.
Hoffman, M. L. (1991). Is empathy altruistic? *Psychological Inquiry, 2*(2), 131–133.
Hollan, D. (2008). Being there: On the imaginative aspects of understanding others and being understood. *Ethos, 36*(4), 475–489.
Hornstein, H. A. (1991). Empathic distress and altruism: Still inseparable. *Psychological Inquiry, 2*(2), 133–135.
Hunt, P., Denieffe, S., & Gooney, M. (2017). Burnout and its relationship to empathy in nursing: A review of the literature. *Journal of Research in Nursing, 22*(1–2), 7–22.
Ickes, W. (2009). Empathic accuracy: Its links to clinical, cognitive, developmental, social, and physiological psychology. In J. Decety & W. Ickes (Eds.), *The social neuroscience of empathy* (pp. 57–70). The MIT Press.
Jackendoff, R., & Lerdahl, F. (2006). The capacity for music: What is it, and what's special about it? *Cognition, 100*(1), 33–72.

Jackson, N. A. (2010). Models of response to client anger in music therapy. *The Arts in Psychotherapy, 37*(1), 46–55.
Jacobowitz, R. M. (1992). Music therapy in the short-term pediatric setting: Practical guidelines for the limited time frame. *Music Therapy, 11*(1), 45–64.
Jones, S. (2018). *Effect of vocal style on perceived empathy, rapport, patient engagement, and competency of music therapists* [Masters Thesis, Florida State University].
Juslin, P. (2013). From everyday emotions to aesthetic emotions: Towards a unified theory of musical emotions. *Physics of Life Reviews, 10*(3), 235–266.
Juslin, P., & Västfjäll, D. (2008). Emotional responses to music: The need to consider underlying mechanisms. *Behavioral and Brain Sciences, 31*(6), 751.
Kenny, C. (1999). Beyond this point there be dragons: Developing general theory in music therapy. *Nordic Journal of Music Therapy, 8*(2), 127–136.
King, R. (2018). *Mindfulness of race: Transforming racism from the inside out.* Sounds True.
Koch, S., & Fuchs, T. (2011). Embodied arts therapies. *The Arts in Psychotherapy, 38*, 278–280.
Koelsch, S., & Skouras, S. (2014). Functional centrality of amygdala, striatum and hypothalamus in a "small-world" network underlying joy: An fMRI study with music. *Human Brain Mapping, 35*(7), 3485–3498.
Krol, S., & Bartz, J. (2021). The self and empathy: Lacking a clear and stable sense of self undermines empathy and helping behavior. *Emotion, Advance Online Publication.* https://doi.org/10.1037/emo0000943
Krueger, J. (2013). Empathy, enaction, and shared musical experience: Evidence from infant cognition. In T. Cochrane, B. Fantini, & K. Scherer (Eds.), *The emotional power of music* (pp. 177–196). Oxford University Press.
Kwan, M. (2010). Music therapists' experiences with adults in pain: Implications for clinical practice. *Qualitative Inquiries in Music Therapy,* 5.
Lamm, C., Decety, J., & Singer, T. (2011). Meta-analytic evidence for common and distinct neural networks associated with directly experienced pain and empathy for pain. *NeuroImage, 54*, 2492–2502.
Laurence, F. (2008). Music and empathy. In O. Urbain (Ed.), *Music and conflict: Harmonies and dissonances in geopolitics* (pp. 85–98). I. B. Tauris.
Layman, D. L., Hussey, D. L., & Reed, A. M. (2013). The beech brook group therapy assessment tool: A pilot study. *Journal of Music Therapy, 50*(3), 155–175.
Lee, C. (1996). *Music at the edge: The music therapy experience of a musician with AIDS.* Routledge.

Lehne, M., Rohrmeier, M., & Koelsch, S. (2014). Tension-related activity in the orbitofrontal cortex and amygdala: An fMRI study with music. *Social Cognitive and Affective Neuroscience, 9*(10), 1515–1523.
Leman, M. (2007). *Embodied music cognition and mediation technology*. The MIT Press.
Levine, P. (2010). *In an unspoken voice: How the body releases trauma and restores goodness.* North Atlantic Books.
Levinson, J. (2006). *Contemplating art: Essays in aesthetics.* Clarendon Press.
Lindvang, C., & Frederiksen, B. (1999). Suitability for music therapy: Evaluating music therapy as an indicated treatment in psychiatry. *Nordic Journal of Music Therapy, 8*(1), 47–57.
López-Pérez, B., Carrera, P., Ambrona, T., & Oceja, L. (2014). Testing the qualitative differences between empathy and personal distress: Measuring core affect and self-orientation. *The Social Science Journal, 51*, 676–680.
Maibom, H. (2017). Affective empathy. In H. Maibom (Ed.), *The Routledge handbook of philosophy of empathy* (pp. 22–32). Routledge.
Marangoni, G., Garcia, S., Ickes, W., & Teng, G. (1995). Empathic accuracy in a clinically relevant setting. *Journal of Personality and Social Psychology, 68*, 854–886.
McIntosh, D. N., Druckman, D., & Zajonc, R. B. (1994). Socially induced affect. In D. Druckman & R. A. Bjork (Eds.), *Learning, remembering, believing: Enhancing human performance* (pp. 251–276). National Academy Press.
Menon, V., & Levitin, D. J. (2005). The rewards of music listening: Response and physiological connectivity of the mesolimbic system. *NeuroImage, 28*, 175–184.
Molnar-Szakacs, I. (2017). Music: The language of empathy. In E. King & C. Waddington (Eds.), *Music and empathy* (pp. 97–123). Routledge.
Moors, A., & Kuppens, P. (2008). Distinguishing between two types of musical emotions and reconsidering the role of appraisal. *Behavioral and Brain Sciences, 31*, 588–589.
Paden, C. E. (2018). *Adolescents with behavioral health concerns: Increasing empathy through lyric discussions of peer-chosen music.* Illinois State University.
Pedersen, I. (1997). The Music Therapist's listening perspectives as source of information in improvised musical duets with grown-up: Psychiatric patients, suffering from Schizophrenia. *Nordic Journal of Music Therapy, 6*(2), 98–111.

Petitta, L., Jiang, L., & Härtel, C. (2017). Emotional contagion and burnout among nurses and doctors: Do joy and anger from different sources of stakeholders matter? *Stress and Health, 33*(4), 358–369.

Petitta, L., Probst, T., Ghezzi, V., & Barbaranelli, C. (2019). Cognitive failures in response to emotional contagion: Their effects on workplace accidents. *Accident Analysis & Prevention, 125*, 165–173.

Pizer, S. (2018). Core competency three: Deep listening/affective attunement. In R. Barsness (Ed.), *Core competencies of relational psychoanalysis* (pp. 104–120). Taylor & Francis.

Préfontaine, J. (2006). On becoming a music therapist. *Voices: A World Forum for Music Therapy, 6*(2).

Preston, S. D., & de Waal, F. B. M. (2002). Empathy: Its ultimate and proximate bases. *Behavioral and Brain Sciences, 25*, 1–72.

Priestley, M. (1994) *Essays on analytical music therapy.* Barcelona Publishers

Robarts, J. (2006). Music therapy with sexually abused children. *Clinical Child Psychology and Psychiatry, 11*(2), 249–269.

Rogers, C. (1957). The necessary and sufficient conditions of therapeutic personality change. *Journal of Consulting Psychology, 21*, 95–103.

Rosan, P. (2012). The poetics of intersubjective life: Empathy and the other. *The Humanistic Psychologist, 40*, 115–135.

Scherer, K., & Zentner, M. (2008). Music evoked emotions are different—More often aesthetic than utilitarian. *Behavioral and Brain Sciences, 31*, 595–596.

Sena Moore, K. (2017). Understanding the influence of music on emotions: A historical review. *Music Therapy Perspectives, 35*(2), 131–143.

Shlien, J. (1997). Empathy in psychotherapy: A vital mechanism? Yes. Therapist's conceit? All too often. By itself enough? No. In A. C. Bohart & L. S. Greenberg (Eds.), *Empathy reconsidered: New directions in psychotherapy* (pp. 63–80). American Psychological Association.

Siebert, D., Siebert, C., & Taylor-McLaughlin, A. (2007). Susceptibility to emotional contagion: Its measurement and importance to social work. *Journal of Social Service Research, 33*(3), 47–56.

Small, C. (1998). *Musicking: The meanings of performing and listening.* Wesleyan University Press.

Smith, J. (2017). What is empathy for? *Synthese, 194*(3), 709–722.

Spinrad, T., & Eisenberg, N. (2009). Empathy, prosocial behaviour, and positive development in schools. In R. Gilman, E. Huebner, & M. Furlong (Eds.), *Handbook of positive psychology in schools* (pp. 119–129). Routledge.

Stein, E. (1989). *On the problem of empathy* (W. Stein, Trans.). ICS Publishers.

Swanick, R. (2019). What are the factors of effective therapy? Encouraging a positive experience for families in music therapy. *Approaches: An Interdisciplinary Journal of Music Therapy.* https://approaches.gr/wp-content/uploads/2019/11/Approaches_FirstView_a20191109-swanick-1.pdf

Thompson, R. (1994). Emotion regulation: A theme in search of definition. *Monographs of the Society for Research in Child Development, 59*, 25–52.

Trentini, C., Tambelli, R., Maiorani, S., & Lauriola, M. (2021). Gender differences in empathy during adolescence: Does emotional self-awareness matter? *Psychological Reports*, 0033294120976631.

Valentino, R. (2006). Attitudes towards cross-cultural empathy in music therapy. *Music Therapy Perspectives, 24*, 108–114.

Van der Schyff, D., & Schiavio, A. (2017). The future of musical emotions. *Frontiers in Psychology, 8*, 988.

Viega, M. (2016). Exploring the discourse in hip hop and implications for music therapy practice. *Music Therapy Perspectives, 34*(2), 138–146.

Vuilleumier, P., & Trost, W. (2015). Music and emotions: From enchantment to entrainment. *Annals of the New York Academy of Sciences, 1337*(1), 212–222.

Wigram, T., & Elefant, C. (2009). Therapeutic dialogue in music: Nurturing musicality of communication in children with autistic spectrum disorder and Rett syndrome. In S. Malloch & C. Trevathen (Eds.), *Communicative musicality* (pp. 423–445). Oxford University Press.

Wilkerson, A., Dimaio, L., & Sato, Y. (2017). Countertransference in end-of-life music therapy. *Music Therapy Perspectives, 35*(1), 13–22.

Wilkinson, H., Whittington, R., Perry, L., & Eames, C. (2017). Examining the relationship between burnout and empathy in healthcare professionals: A systematic review. *Burnout Research, 6*, 18–29.

Winter, P. (2013). *Effects of experiential music therapy education on student's reported empathy and self-esteem: A mixed methods study* [Doctoral dissertation, Temple University].

Xu, X., Zuo, X., Wang, X., & Han, S. (2009). Do you feel my pain? Racial group membership modulates empathic neural responses. *Journal of Neuroscience, 29*(26), 8525–8529.

Zahavi, D. (2008). Simulation, projection and empathy. *Consciousness and Cognition, 17*, 514–522.

Zahavi, D. (2017). Phenomenology, empathy, and mindreading. In H. Maibom (Ed.), *The Routledge handbook of philosophy of empathy* (pp. 33–42). Routledge.

Zurek, P. P., & Scheithauer, H. (2017). Towards a more precise conceptualization of empathy: An integrative review of literature on definitions, associated functions, and developmental trajectories. *International Journal of Developmental Science, 11*(3–4), 57–68.

5

Understanding Emotions

Consciously sharing in another's emotions is most commonly referred to as affective empathy and understanding their emotions as cognitive empathy. These are separate constructs associated with distinct brain regions, but they seem highly intertwined (Cuff et al., 2016). De Waal (2008) argued that cognitive empathy builds on affective empathy.

Knowing Other Minds

In philosophy, whether we can know others' mental/emotional states is referred to as the problem of other minds (Avramides, 2001). Many suggestions have been offered to address this problem. Export theories, for example, claim that individuals naturally and primarily know their own inner experiences and then use this to derive what the experiences of others may be (Shoemaker, 1996; Zahavi, 2005). We use inference to make the leap from observing others' behaviours to assessing their mental states so frequently and routinely that we're often unaware that we're

A. dos Santos, *Empathy Pathways*,
https://doi.org/10.1007/978-3-031-08556-7_5

doing so (Epley & Waytz, 2009). This process rests on the "shared manifold hypothesis": interpersonal relationships depend on a shared space involving a pre-reflective assumption that others are like us (Gallese, 2003).

Theory of Mind

Within the field of psychology, theory of mind (ToM) refers to a person's capacity to attribute cognitive mental states (for example, beliefs, desires, intentions, and knowledge) to the self and others (Gallagher & Frith, 2003). Synonyms for ToM include "mind-reading" (Apperly, 2011) and "mentalising" (Saxe et al., 2004). When you're attempting to make sense of what's going on in your mind or someone else's mind, you are mentalising. Abrahams and Rohleder (2021) wrote that mentalising "is the ability to see yourself from the outside and others from the inside" (p. 150). First-person mentalising refers to attributing mental states to oneself, while a third-person mentalising entails attributing mental states to others (Goldman, 2012). Mental states are infused with emotions. Often the term "mentalisation" is used to refer to reflection on affective mental states, and "theory of mind" is employed to refer to reflection on beliefs and intentions (Wyl, 2014). We are mentalising when we reflect on how our client is playing and how it is to be part of the musicking with them. According to Fonagy and Allison (2014), experiencing being thought about in therapy allows us to feel safe enough to explore ourselves in relation to our world and learn new ways of being in the world.

We are mentalising all the time as we make sense of ourselves and the people around us. However, it's a dynamic function and can be negatively impacted by stress and distress. As therapists, the best route for encouraging clients to mentalise more freely is to assume a stance of curiosity and empathy. We wonder with them: can you help me understand this better? Mentalising through musical improvisation is similar to mentalising through verbal content, although the emphasis can lie more strongly on emotional experience than logical understanding. We monitor our thoughts as we reflect on what's going on. Within the improvisation, as we reflect on our feelings (of enmeshment, abandonment, being chased after, or suspended, for example), we may receive new ideas about the client's relationship patterns (Strehlow & Hannibal, 2019).

Theory of mind has received some criticism, and two main alternative explanations have been proposed, namely theory theory and simulation theory (although some authors argue that these are complementary) (Frith, 2007). Proponents of theory theory hold that we explain and predict other people's mental states and behaviours by drawing on sets of concepts and principles to form causal explanations (Apperly, 2008). Humans use generalisations to link what they observe about others' behaviours to specific mental states (we may think that a person who exercises without drinking fluids will feel thirsty); we link mental states to particular behaviours (we may hold the view that an angry person tends to frown); and we link certain mental states to other mental states (we believe that a person in pain tends to want to alleviate that pain) (Goldman, 2012). We are less likely to become angry with someone who pushes us if we see that they did so in error or to prevent us from being harmed (Ganeri, 2017).

Robert Gordon (1986) first developed simulation theory (ST), suggesting that we can predict another's behaviour by exploring the question, "What would I do if I were in that person's situation?" At least a portion of the task of thinking about another mind can be accomplished by using the model of one's own mind. Gordon explained how chess players visualise the board from their opponent's perspective, simulating what they would do and then projecting this decision onto their opponent. Similarly, researchers within the field of embodied music cognition (EMC) suggest that the way we infer the musical feelings and intentions of others could be driven by processes of mentally simulating or mirroring their bodily actions in relation to the sounds they produce. We then project such simulations back onto the music to make sense of it (Bamford & Davidson, 2017; Iacoboni, 2009). Simulation theory has received support from neuroscience research on mirror neurons (Goldman, 2012).

Many contemporary researchers consider pure theory theory and pure simulation theory to be inadequate, endorsing a hybrid explanation instead. At times we use theorising to understand another person's perspective (particularly when we differ from them). On other occasions, we mentally simulate (particularly when the other is relatively similar to ourselves). Theorising appears to be a superior strategy when we want to ensure that our perspective doesn't cloud our understanding of the other's perspective (Spaulding, 2015).

The Direct Social Perception Thesis

Those who argue for the direct social perception (DSP) thesis believe it is (at least sometimes) possible to perceive others' mental states directly (Reddy, 2008; Zahavi & Gallagher, 2008). A person's expressive behaviour can richly reveal their mental state. As Spaulding (2015) explained, when a baby cries, we immediately see that they're upset. As our friend reaches for a cake, we see that they want a piece. The argument from the DSP thesis is that we don't first observe the behaviour and then infer the mental state. We see the mental state in the behaviour.

When we observe another person acting in a certain way, not only do we observe physical movement, but we also recognise intention (Colle et al., 2008). However, empirical results on whether it's possible to know another's intentions (whether they are being cooperative or competitive, for example) have been mixed (Naish et al., 2013; Stapel et al., 2012). As discussed in Chapter 3, phenomenological positions counter the notion that we can directly understand others' emotions. Husserl used the term "givenness" to explain how something is given to one's conscious awareness: "When I feel the other's joy, I do not experience it primordially as my joy but as a joy experienced by the other" (Moran, 2017, p. 33). The contention is that while we may not have direct experience of another's inner world, we do have direct experience of them as experiencing their own inner world.

Psychodynamic Understandings of Other Minds

According to Freud (1922/1949), empathy "plays the largest part in our understanding of what is inherently foreign to our ego in other people" (p. 77). There are a wide variety of psychodynamic approaches to attempting to know another. While export theories contend that we use our own subjective experience to understand others, import theories hold that individuals first perceive subjectivity in others and then use that to build their own sense of having an inner world (Carruthers, 2009; Graziano, 2013; Habermas, 2008). Others can help one understand oneself in ways that would be impossible to accomplish alone (Jurist, 2010).

Fonagy (1991) explained how a mind comes into being by receiving appropriate input from a caregiver. Infants gain a sense of security from their caregivers through early attachment bonds, and their minds develop in a way that reflects the quality of that relationship (Fonagy, 2001). Secure attachment generates the ability to regulate affects, assisting a child transitioning from co-regulation to self-regulation. Through affect regulation, the child grows able to grasp the intentions of others, which precedes and fosters their capacity to read and understand their own mind.

In psychoanalytic psychotherapy, empathy applies not only to what the patient communicates consciously, but also to feelings and ideas outside their conscious awareness. The flow of empathy here takes place unidirectionally from the "knowing" analyst to the patient because the analyst is seen to bypass the patient's defences (Bennett, 2001). Olden (1953) wrote that the therapist may "trespass the…screen of defences" (p. 116) of the patient through empathy. The analyst moves between introspective and cognitive modes to attune to and understand the patient's affective state through careful attempts to confirm, disconfirm, or clarify interpretations elicited through immediate unconscious resonance.

The Nature of Cognitive Empathy

Cognitive empathy entails attempting to understand the emotional experience of another. This includes identifying the other person's emotion and comprehending why they may be feeling this way within this situation. We can think of this as constructing a working model of another's emotional state (Reniers et al., 2011). As Egan (2014) argued, empathy as a value entails a therapist's radical commitment to understanding their client as fully as possible, including the further commitment to understanding the context of their lives.

In music therapy, authors have referred to empathy as recognising and understanding another's internal reality and experience (Chan, 2014; Haviland, 2014; Jones, 2018), understanding what others are feeling (McGann, 2012; Paden, 2018), a mode of understanding that entails emotional resonance (Winter, 2013), and as "the capacity to understand, be aware of, or be sensitive to the thoughts and feelings of another person" (Gardstrom & Hiller, 2010, p. 149). Valentino (2006) described cognitive empathy as the capacity to sense and intellectually understand the client's meaning. This can include the importance of understanding clients' personal-cultural experiences to better understand their emotions (Brown, 2002; Ip-Winfield & Grocke, 2011; Seri & Gilboa, 2017).

Some empathy researchers use "perspective-taking" and "cognitive empathy" synonymously (for example, Davis [1994]) and a close relationship between these two terms is also present at times in music therapy literature. Winter (2013) described empathy as involving perspective-taking, and Spiro et al. (2013) articulated empathy as "seeing the world from another person's perspective" (p. 1). Priestley (1994) referred to "true empathy" (p. 123) as seeing life from the point of view of the client as far as possible (without losing one's own orientation). For Valentino (2006), empathy entails understanding the client's worldview, and Haviland (2014) discussed it in terms of viewing the client's life through the lens of knowing what it is like to be in their world. For Layman et al. (2013) and Trolldalen (1997), empathising involves seeing a situation through another's perspective and relating to their emotions from their perspective. According to Bruscia (2001), empathic listening to improvisation involves listening from the client's perspective. Nye

(2014) referred to understanding the client's world as they experience it. At least to a degree, the notion of placing oneself in another's shoes (terminology used by, for example, Hald et al. [2017], Ruud [2003], and Winter [2013]) relates to perspective-taking.

As discussed in Chapter 1, however, general perspective-taking (we could use ToM or mentalising synonymously here) is an expansive process, including emotions, thoughts, beliefs, intentions, and so on. When we engage in *empathic* perspective-taking, we specifically focus on the other person's emotional state (Zurek & Scheithauer, 2017). Some authors have considered ToM to be an aspect of cognitive empathy (for example, Zaki & Ochsner, 2012). However, according to Zurek and Scheithauer, ToM includes affective theory of mind (understanding another's emotions) and cognitive theory of mind (understanding another's thoughts). Cognitive empathy is then synonymous only with affective theory of mind.

Drawing on Self-Knowledge to Understand Another's Emotions

Self-observation and self-inquiry are central features of our reflexive process as music therapists (Bruscia, 2014a). These include how we draw on our own emotional experiences when attempting to understand a client's emotions (Abbott, 2018; Bruscia, 2014b; Jackson & Gardstrom, 2012; Lindvang, 2013; Murphy, 2007; Préfontaine, 2006), bearing in mind that self-knowledge is an ever-evolving process (Moonga, 2019). According to some authors, it's harder to understand another person's emotional experience if you haven't had a similar experience yourself. For example, in a study by Nordgren et al. (2011), participants who had not experienced social exclusion consistently underestimated the social pain experienced by those who had been excluded.

One's appreciation of another's experience can be enhanced when one "relives" one's own past experiences, as long as the experience remains other-directed (Ratcliffe, 2012). Tapping into one's own personal emotional experiences when attempting to understand another's can be most effective when one has an adequate and complex self-concept and when one can recognise similarities and differences between oneself and the other (Redmond, 2018).

In Chapter 3, we encountered the example of Short's (2017) music therapy process with Karly, a female offender who had a personality disorder at a secure unit in the United Kingdom. Karly felt anger towards the nursing staff, whom she viewed as persecuting her. Short suggested that these feelings may stem from the institutional dynamics of control, in addition to Karly's past attachment difficulties. Short then also reflected upon her own past experiences of perceived persecutory interactions that had not, in fact, been so. As she challenged Karly, she could do so empathically by demonstrating that she had shared such experiences and understood Karly's feelings, and Karly was able to receive this feedback.

Sorel (2010) offered an example of her work with a mother (Carly) and son (Elliot) with autism spectrum disorder at the Nordoff-Robbins Center for music therapy in New York. Sorel explained how she empathised with Carly as a mother of two sons. After becoming a parent, Sorel developed a deeper empathic understanding of the experience of the parents she encountered. Her parental role enabled her to take in the parents' stories and relate to them in ways she couldn't access before having her own child. Sorel felt less distance between herself and her clients' parents, seeing as she now had a greater understanding of the complexity of their lives.

Lee (2016) sought to explore how five American music therapists who had survived cancer and worked with patients in hospital or hospice settings described the potential influence of their cancer experiences on their clinical work. The participants expressed how they didn't need to use their imagination to empathise with patients but could now draw on "lived empathy" (p. 12) as they had experienced the needs of a person with cancer. They then used this experience of "really knowing what it's like to be there" (p. 12) to assist them in empathising with other situations that they had not personally experienced, because of expanded self-awareness and self-knowledge, more holistic awareness of the client, and shifts in their understanding of music, the environment, time, relationships, and resources.

While familiarity with what a client is going through may indeed enhance our capacity for cognitive empathy, it's also possible that familiarity with a situation could lead us to withhold empathy. Our past experience can decrease our curiosity; we may feel that the other's situation is "not that bad"; we may find ourselves judging them for having a different emotional response to our own when we were in a similar situation; and an aversion to previous events in our lives could make us resistant to relive them (Breithaupt, 2019). In a study on music therapists' empathic experiences of shared and different orientations to religion and spirituality in the client–therapist relationship, some participants articulated difficulties empathising with clients whose religious orientation mirrored how they used to identify (dos Santos & Brown, 2021). They felt they had shifted their views and now (often vehemently) disagreed with their previous perspectives. Pushing against these views inside themselves created some challenges when empathising with a client who currently held those views.

It appears that what is termed "surface-level similarities" (such as gender expression, race, age, and religion) increase our motivation and ability to empathise accurately with others. However, when we perceive another person as different from ourselves or having an experience that we have not had, we become more reliant on information gathering, imagination, listening and curiosity, which may actually bring us closer to empathic understanding. Not being able to see oneself in the other could be dangerous, though, when a sense of "otherness" then prevents empathy from emerging at all (Aucoin & Kreitzberg, 2018).

With no shared reference points or similar experiences, we may struggle to wrap our heads around what another person is going through. Food eaten with gusto in one culture may appear entirely unpalatable and nonsensical to eat for a person within a different culture. The sense of relief that self-harm brings some individuals may seem completely illogical and aversive to another person who has never had this urge. For some, the act of a child taking their own life through suicide may seem completely unfathomable, as if reality cannot hold any form of sense-making that could sufficiently explain it. Cognitive empathy requires a foundation of a shared reality within which the other's experience can be seen as meaningful to them in some way, even if we are still working to understand the details of the meaning of their experience for them. If we are struggling to enter a shared reality frame, we can still work towards this by gaining more knowledge about the other person's experience and what they are going through to expand our understanding of what they perceive as possible, reasonable, and sensible within their context.

Drawing on Knowledge of the Specific Other and Their Context to Understand Their Emotions

The greater one's knowledge of the other person (how did this client respond to this piece of music in the last session? How do they react emotionally to significant others in their lives?), the more likely one is to understand their emotional responses at the moment (Redmond, 2018). This process is enhanced through frequent interaction, careful observation (preferably in multiple settings, as can be the case in a community music therapy process for example [Wood, 2006]), when there has been self-disclosure from the other person (which, in music therapy, can occur cross-modally), and when the empathiser's responses motivate continued sharing (Redmond, 2018). Eyal et al. (2018) argued that if we want to feel more connected to another person, imagining ourselves "in their shoes" can be helpful. Still, if we really want to know what's going on in their minds, the best way to gain their perspective is to ask them. Instead of "taking" their perspective, we can "get" their perspective. As Frank (2015) explained, "to be empathic is to know someone else's story" (p. 152). Ma-Kellams and Lerner (2016) found that systematic (careful and effortful) thinking enhances empathic accuracy more than intuitive thinking.

As discussed in Chapter 3, perceiving another in context is crucial. Beyond shared lived experiences, practical knowledge of and engagement with another's cultural context (Brown, 2002), circumstances, and position in life (Hollan, 2008) can enhance empathy. As Solomon (1995) explained, when empathising with someone from a different culture, one needs to know the conditions, the rules, local myths, and popular expectations. To understand the emotion, one must understand the society. Segal (2018) distinguished between interpersonal empathy and social empathy, the second of which refers to the ability "to understand people and other social groups by perceiving and experiencing their life situations" (p. 10). A music therapist's knowledge of their client's history, musical experiences, diagnosis, developmental level, life issues, responses to musical cues, and transference-countertransference dynamics informs their ability to meet their musical-emotional needs (Austin, 2001).

Brown (2002) defined cultural empathy as the process of gaining an understanding of the client's personal cultural experiences with the goal of conveying this understanding. Therapists also need to pay attention to how empathy is expressed in a client's culture (Ip-Winfield & Grocke, 2011). Notably, we need to be careful not to assume that we don't have to work as hard to understand a client who seems similar to ourselves. I would argue that the approach described by those who contend for cultural empathy is in fact crucial in every empathic encounter.

In Steinhardt and Ghetti's (2020) approach to music therapy for procedural support, they described how they explicitly gathered holistic information about their clients and the context. As an example, they presented the case study of Lisa, a 12-year-old girl who was in hospital receiving treatment for a brain tumour. Lisa lived in a small town with her parents and two older siblings; she loved listening to music, reading, and playing soccer. Lisa found it difficult to adjust to being in hospital and the treatment demands. She missed her friends, independence, home, and her dog. Lisa frequently experienced overwhelming fear and struggled to regulate her feelings or articulate what she was specifically afraid of. Her eyes filled with panic when the music therapist first met her, and she was sobbing as her father stroked her hair to try to calm her. The music therapist started by acknowledging Lisa's fear and then led her through a music relaxation technique. Once her breathing was slower and her body tension had loosened, Lisa could explain her fears, interests, musical tastes, and love of stories. An adapted form of Guided Imagery and Music (GIM) became Lisa's preferred way to relax and stay focused. One procedure that Lisa needed to receive entailed the injection of chemotherapy into a reservoir that had been surgically inserted under the skin on her head while she remained completely still. Although not very painful, the procedure could be anxiety-provoking and uncomfortable. The music therapist gained information from Lisa about what aspects of the procedure she found most frightening (such as being held down and having people standing over her). She was also part of helping to prepare Lisa for what the procedure entailed. Through Lisa's previous relaxation experiences during adapted GIM techniques in sessions, she chose to use this technique during the procedure. Lisa decided she would like to create imagery of being at the beach. The music therapist asked Lisa what she saw at her beach, how the ocean was at that moment and played louder "swooshing" sounds on the ocean drum that Lisa had selected for her to play in response to Lisa's replies that the ocean was chaotic and the waves were enormous. The music therapist validated her feelings during the procedure and reassured her that she was doing an excellent job of remaining still and helping the process go smoothly.

While we can imagine that the music therapist may have experienced affective empathy during this event (and perhaps some emotional contagion as well), she also clearly attempted to understand the client's emotional flow moment by moment through drawing on the specific knowledge she had of Lisa (gained through intentional assessment and prior and current interaction) as well as an understanding of the client's contextual world, and the context related to the procedure itself. Through her knowledge of the procedure process, the music therapist could also anticipate what Lisa may feel as the event unfolded and could help prepare her for emotions that could emerge.

Drawing on Generalised Theories to Understand Another's Emotions

Humans often draw on concepts of the "generalised other" when attempting to understand another's emotional state ("teenagers are usually…"; "men tend to…."; "children in this area usually….''; "a person with this diagnosis will often…"). This process is often unconscious. We pull on both social-centred theories (such as the implicit personality theory that an extraverted person will also be friendly) and group-centred theories (like stereotyping). Drawing on generalised theories can be a helpful strategy only if we critique our theories and stereotypes, use effortful intentionality to examine our analyses, and are motivated to reach fair and thorough conclusions (Redmond, 2018). Notably, the generalised theories we use tend to have been obtained within our cultural frame of reference, and this needs to be critiqued as well, particularly when engaging in cross-cultural empathy (Brown, 2002; Chan, 2014; Grimmer & Schwantes, 2018; Haviland, 2014; Stensæth, 2018; Valentino, 2006).

Expanding Empathy Through Gathering Knowledge

We can grow in our ability to understand other people's emotions by learning about their lived experiences, contexts, struggles, and successes in many ways, such as through books and documentaries, as well as songs, stories, artworks, and films that offer fictional narratives. We can draw on these tools in our journeys as music therapists, and we may choose to integrate these types of experiences into sessions to enhance clients' empathy towards others.

Songs provide clinical material for therapists and group members to understand the lived experiences of songwriters (Baker, 2014). In counselling training, Ohrt et al. (2009) used music videos to enhance empathy by arousing emotional responses towards characters that could represent future clients and stimulate discussions about the characters' emotional experiences. The students first read the song's lyrics and thought about the situation being portrayed. They then watched the music video. Afterwards, they were given five to ten minutes for reflection, during which they could journal their thoughts and feelings. The instructor then facilitated a time of discussion where the students processed their responses, including what emotions emerged for them, the images and words that stood out to them, their physical reactions, and thoughts. They were also invited to reflect on any characters they identified with and why that may have been the case. The discussion then moved into how these experiences related to the therapeutic process.

Using fictional stories within therapy sessions can allow listeners to adopt characters' perspectives. We can be deeply engaged in the character's life we hear or read about without losing our separate sense of self. Literary narratives offer simulations or models of the social world through abstraction, compression, or simplification (Mar & Oatley, 2008). Bal and Veltkamp (2013) found that we can grow in understanding others through fiction mainly when we are emotionally transported into a story. Transportation refers to absorption into a narrative's affect, imagery, and attentional focus (Green & Brock, 2000). It appears that persons who generate higher levels of imagery when reading or listening to narrative fiction are significantly more transported into the story and feel higher empathy for the characters (Johnson et al., 2013).

There is a rich body of literature regarding the role and mechanisms of visual arts in eliciting empathy (Kesner & Horáček, 2017). Potash (2010) developed a process called "guided relational viewing" where viewers' experience relating to artworks can afford increased empathy. Potash et al. (2014) found that medical students who took part in art-making and poetry writing in response to witnessing a patient's suffering experienced heightened empathic understanding. Potash and Chen (2014) observed how medical students' empathy increased when they reflected on paintings and drawings created by other medical students in response to witnessing the suffering of their patients. Viewing art created by people living with mental illnesses can allow viewers to gain an increased empathic understanding of the artists, society, and themselves (Potash & Ho, 2011; Potash et al., 2013).

In films, multiple characters are presented with rich narrative layers, expressive close-ups that reveal interior emotional experiences, and voice-over narration in contexts that include past, present, and potential future (Stadler, 2017). It's rare to encounter such a complete picture of another person's life, and it's for this reason that film can help therapists and counsellors build their empathic skills (bell, 2018). As sound designer Ward (2015) explained, "cinema recruits our body's innate capacity for 'feeling into' another's affective state" (p. 185). In *The Empathic Screen*, Gallese and Guerra (2015) explained how films enable us to transcend the kinds of simulation scenarios available to us in our concrete encounters in the natural, tangible world. They offer us opportunities to simulate the experiences of real others in situations far from us in space and time and imaginary others in fictional places. This expands our empathic horizons.

Understanding Complex Emotional Expressions

In therapy, we encounter situations where it may be challenging to understand a client's emotions. We may feel confused, lost, or overwhelmed by the vast amount and variety of information that a client reveals, and it can be hard to identify what is most essential at that given moment (Dekeyser et al., 2009). It can also be difficult to understand another person's emotional experience and the cause of the experience when they have blunted affect. Understanding another person's emotions may be challenging when there is a discrepancy between what we perceive from their outer behaviour, how they verbally articulate their experience, and their inner emotional state. A range of psychopathologies and neurological disorders can cause incongruence between behavioural expressions of emotions and internal experiences, such as Parkinson's disease (Simons et al., 2004), schizophrenia, and obsessive–compulsive disorder (Bersani et al., 2012).

There are also many functional and adaptive reasons people dissociate their expressive emotional behaviour from their subjective, internally felt emotional experience. Our self-appraisal systems continuously attempt to create experiences that protect and strengthen rather than undermine the self. Dealing with interpersonal conflict requires regulation of subjective emotional experiences and communicative expression of emotion. Sometimes we may not align our emotional expression with our inner emotional experience to sustain a reasonably good relationship with another person, obtain validation or support from others, or as a strategy to influence others through impression management or persuasion. Children might do this to avoid getting into trouble (Saarni et al., 2008). We may purposefully hide our emotions (Gross & Cassidy, 2019) or displace the emotions we feel towards one person onto our interactions with another (McLeod, 2013). Emotional dysregulation may be due to past trauma, sparked by a trigger in the moment that is not the primary object of the emotion (Kerig, 2020).

There may be times when a therapist questions whether a client is expressing emotions authentically. If the music therapist and/or other members of a multidisciplinary team perceive a client's behaviour as manipulative (for example, when a client has borderline personality disorder (BPD) [Potter, 2006]), the task of empathising may appear complex. Potter argued that while some clients with BPD may be manipulative, it is often more appropriate to interpret their behaviour in a different light. Persons with BPD are suffering and need therapists to respond in ways that don't intensify their distress. The pervasive and pejorative attribution of manipulation does not enhance their healing process. Care and clarity in applying the term "manipulation" aid healthcare practitioners in intercepting their own negative perceptions, which is crucial to being empathetic.

Empathising with a client whose experience of reality may appear significantly different from one's own, during psychosis for example, may require significant cognitive effort (Egan, 2014). Disorganised behaviour and "inappropriate" affect may make it challenging for a therapist to accurately understand their client's internal emotional experience (Foley & Gentile, 2010). In 2014, the British Psychological Society (BPS, 2014) concluded no clear threshold or division between psychosis and other feelings, thoughts, and behaviours. Psychosis can be understood and addressed in the same way as other psychosocial struggles such as nervousness, fear, or sadness. Kamens (2019) asked what it would mean to de-other schizophrenia, conceptualising schizophrenia as an understandable and non-enigmatic human experience? She posited that social defeat theory enables us to conceptualise how the experiences and behaviours commonly classified as symptoms of schizophrenia tend to emerge in interpersonal or social contexts characterised by experiences of social marginalisation, where persons feel profoundly misunderstood or rejected. She argued that being treated as "ununderstandable" may precede psychosis rather than arising from it. Schizophrenia is then a social configuration in which there has been a failed struggle for empathy.

Individuals who encounter extraordinary experiences, such as hearing voices and holding delusional beliefs, perhaps with very little awareness of their own emotional experience or how this relates to their intrapsychic experience, require as much validation and acknowledgement as any other person (Lakeman, 2020; Lotter, 2017). Psychotic experiences may startle and even terrify the person experiencing them. An empathic therapist's task is to validate these emotions in the here and now. Developing rapport with a person experiencing psychosis is crucial. They may tend to jump to conclusions by being in a heightened state of arousal, blaming others, and attributing internal experiences to external sources. Empathic communication, therefore, needs to be unambiguous.

Perceptual experiences and delusions generate emotions, or an emotional state can precede an automatic thought or belief. A person experiencing hypomania may develop thoughts of grandeur, or a person who feels depressed may develop delusions of guilt, for example. These very real emotions can prime people to seek evidence to confirm their feelings, fuelling a cycle of emotional reasoning. While the beliefs and thoughts may not be rooted in consensual reality, the associated emotions need genuine acknowledgement. Lakeman (2020) offered an approach for verbally expressing empathy in such situations while acknowledging one's own understanding. One aims to understand the person's emotions accurately, which necessitates a tentative and flexible approach because unusual beliefs and thoughts can be associated with a wide range of emotions. Lakeman's steps involve: (1) acknowledging what the individual has said; (2) acknowledging your understanding; (3) trying to imagine how the other person may feel and checking your intuitions; and (4) exploring their feelings and methods of coping. Lakeman gave an example of a helper responding by saying: You often hear the banging noises on your roof, especially when you're home alone at night. You start to feel frightened and think that someone is breaking into your house. It would be easy to jump to that conclusion when you feel scared, but it doesn't mean someone is breaking into your home. How can we help you feel safer and less nervous at night?

Music therapists have more than verbal reflections and processing to draw on. Sometimes talking can be inappropriate, ineffectual, and enhance confusion (Pavlicevic, 1997). Lotter (2017) showed how music therapy welcomes the varied expressive potentials of people experiencing major depressive disorder and schizophrenia-spectrum psychotic disorders. Through a combined process of active music-making, music listening, working with imagery and symbols, visual art processes, and verbal reflection, a range of emotions were elicited from participants in her study. These ranged from emptiness and indifference to deeply unmanageable and painful emotions, as well as joy and hope. Lotter explained that she needed to exercise a very particular kind of listening in her work with clients who had these diagnoses. It can be challenging to listen to and understand the expressions of a client experiencing disorganisation, rigidity, emotional emptiness, and articulation challenges. If we only rely on verbal information, we may struggle to truly hear what the client wants to say.

When we listen to the stories of distressed and disturbed clients, not only do we grow in our understanding of their experiences, but we rediscover aspects of ourselves. Atwood (2012) describes madness as "the abyss"; it is an experience of utter annihilation and the frightening fall into nonbeing. This is a universal possibility of human existence. We are all capable of this fall. Something may happen in relation to which the centre cannot hold. She urged us to think of responses and human meaning rather than treatment and illness, as well as erasing the sharp distinction between madness and sanity, bringing severe psychological disturbance back into the circle of human intelligibility.

We may find that we are not experiencing "pure" empathy in a session. For example, when working with a client who has severely harmed others, we may experience empathy and discomfort. When in a session with a client experiencing psychosis, we may feel empathy and confusion. When working with a person we experience as being untruthful, we may experience empathy and suspicion. If a group member angrily insults another person in the group, we may feel empathy in light of knowing where their anger stems from and, in all honesty, perhaps a moment of dislike for them as we see the hurt that their behaviour has caused another group member (with whom we are empathising as well). Notably, the discomfort, confusion, suspicion, or even dislike do not negate the empathy that we are also experiencing. Being empathic is often not an "all or nothing" phenomenon. It is one thread within a complex interpersonal encounter. I suggest that as we lean into these other coexisting feelings (instead of denying them in the service of protecting our identity as "an empathic therapist"), we give our empathy (for others and ourselves) a chance to deepen even further.

References

Abbott, E. A. (2018). Subjective observation in music therapy: A study of student practicum logs. *Music Therapy Perspectives, 36*(1), 117–126.

Abrahams, D., & Rohleder, P. (2021). *A clinical guide to psychodynamic psychotherapy*. Routledge.

Apperly, I. (2008). Beyond simulation–theory and theory–theory: Why social cognitive neuroscience should use its own concepts to study "Theory of Mind." *Cognition, 107*(1), 266–283.

Apperly, I. (2011). *Mindreaders: The cognitive basis of "theory of mind."* Psychology Press.

Atwood, G. E. (2012). The abyss of madness and human understanding. *Pragmatic Case Studies in Psychotherapy, 8*(1), 49–59.

Aucoin, E., & Kreitzberg, E. (2018). Empathy leads to death: Why empathy is an adversary of capital defendants. *Santa Clara Law Review, 58*, 99–136.

Austin, D. (2001). In search of the self: The use of vocal holding techniques with adults traumatized as children. *Music Therapy Perspectives, 19*(1), 22–30.

Avramides, A. (2001). *Other minds*. Routledge.

Baker, F. (2014). *Therapeutic songwriting: Developments in theory, methods, and practice*. Palgrave Macmillan.

Bal, P. M., & Veltkamp, M. (2013). How does fiction reading influence empathy? An experimental investigation on the role of emotional transportation. *PLoS One, 8*(1), e55341.

Bamford, J. M. S., & Davidson, J. W. (2017). Trait empathy associated with agreeableness and rhythmic entrainment in a spontaneous movement to music task: Preliminary exploratory investigations. *Musicae Scientiae, 23*(1), 5–24.

bell, h. (2018). Creative interventions for teaching empathy in the counseling classroom. *Journal of Creativity in Mental Health, 13*(1), 106–120.

Bennett, M. (2001). *The empathic healer: An endangered species*. Academic Press.

Bersani, G., Bersani, F. S., Valeriani, G., Robiony, M., Anastasia, A., Colletti, C., Liberati, L., Capra, E., Quartini, A., & Polli, E. (2012). Comparison of facial expression in patients with obsessive-compulsive disorder and schizophrenia using the Facial Action Coding System: A preliminary study. *Neuropsychiatric Disease and Treatment, 8*, 537.

Breithaupt, F. (2019). *The dark sides of empathy*. Cornell University Press.

British Psychological Society (BPS). (2014). *Understanding psychosis and schizophrenia*. British Psychological Society. https://www1.bps.org.uk/system/files/Public%20files/rep03_understanding_psychosis.pdf

Brown, J. M. (2002). Towards a culturally centered music therapy practice. *Voices: A World Forum for Music Therapy, 2*(1).

Bruscia, K. E. (2001). A qualitative approach to analyzing client improvisations. *Music Therapy Perspectives, 19*(1), 7–21.

Bruscia, K. (2014a). *Defining music therapy* (3rd ed.). Barcelona.

Bruscia, K. (2014b). *Self-experiences in music therapy education training, and supervision*. Barcelona.

Carruthers, P. (2009). How we know our own minds: The relationship between mindreading and metacognition. *Behavioral and Brain Sciences, 32*(2), 121–182.

Chan, G. (2014). Cross-cultural music therapy in community aged-care: A case vignette of a CALD elderly woman. *Australian Journal of Music Therapy, 25*, 92–102.

Colle, L., Becchio, C., & Bara, B. G. (2008). The non-problem of the other minds: A neurodevelopmental perspective on shared intentionality. *Human Development, 51*(5–6), 336–348.

Cuff, B., Brown, S., Taylor, L., & Howat, D. (2016). Empathy: A review of the concept. *Emotion Review, 8*(2), 144–153.

Davis, M. (1994). *Empathy: A social psychological approach*. Westview Press.

de Waal, F. (2008). Putting the altruism back into altruism: The evolution of empathy. *Annual Review of Psychology, 59*, 279–300.

Dekeyser, M., Elliott, R., & Leijssen, M. (2009). Empathy in psychotherapy: Dialogue and embodied understanding. In J. Decety & W. Ickes (Eds.), *The social neuroscience of empathy* (pp. 113–124). The MIT Press.

dos Santos, A., & Brown, T. (2021). Music therapists' empathic experiences of shared and differing orientations to religion and spirituality in the client-therapist relationship. *The Arts in Psychotherapy, 74*, 101786.

Egan, G. (2014). *The skilled helper: A problem-management and opportunity-development approach to helping* (10th ed.). Cenage Learning.

Epley, N., & Waytz, A. (2009). Mind perception. In S. T. Fiske, D. T. Gilbert, & G. Lindzey (Eds.), *The handbook of social psychology* (5th ed., pp. 498–541). Wiley.
Eyal, T., Steffel, M., & Epley, N. (2018). Perspective mistaking: Accurately understanding the mind of another requires getting perspective, not taking perspective. *Journal of Personality and Social Psychology, 114*(4), 547.
Foley, G. N., & Gentile, J. P. (2010). Nonverbal communication in psychotherapy. *Psychiatry, 7*(6), 38.
Fonagy, P. (1991). Thinking about thinking: Some clinical and theoretical consideration in the treatment of a borderline patient. *International Journal of Psychoanalysis, 72*, 639–656.
Fonagy, P. (2001). *Attachment theory and psychoanalysis.* Other Press.
Fonagy, P., & Allison, E. (2014). The role of mentalizing and epistemic trust in the therapeutic relationship. *Psychotherapy, 51*, 372–380.
Frank, A. (2015). Practicing dialogical narrative analysis. In J. A. Holstein & J. F. Gubrium (Eds.), *Varieties of narrative analysis* (pp. 33–52). Sage.
Freud, S. (1922/1949). Group psychology and the analysis of the ego. In E. Jones (Eds), *International Psychoanalytic Library, No. 6.* (J. Strachey, Trans.). Hogarth Press.
Frith, C. D. (2007). The social brain? *Philosophical Transactions of the Royal Society, Series B, Biological Sciences, 362*, 671–678.
Gallagher, H., & Frith, C. (2003). Functioning imaging of 'theory of mind.' *Trends in Cognitive Sciences, 7*(2), 77–83.
Gallese, V. (2003). The roots of empathy: The shared manifold hypothesis and the neural basis of intersubjectivity. *Psychopathology, 36*(4), 171–180.
Gallese, V., & Guerra, M. (2015). *The empathic screen: Cinema and neuroscience.* Oxford University Press.
Ganeri, J. (2017). *Attention, not self*. Oxford University Press.
Gardstrom, S., & Hiller, J. (2010). Song discussion as music psychotherapy. *Music Therapy Perspectives, 28*(2), 147–156.
Goldman, A. (2012). Theory of mind. In E. Margolis, R. Samuels, & S. Stich (Eds.), *The Oxford handbook of philosophy and cognitive science* (pp. 1–25). Oxford University Press.
Gordon, R. M. (1986). Folk psychology as simulation. *Mind and Language, 1*, 158–171.
Graziano, M. (2013). *Consciousness and the social brain.* Oxford University Press.

Green, M., & Brock, T. (2000). The role of transportation in the persuasiveness of public narratives. *Journal of Personality and Social Psychology, 79*(5), 701–721.

Grimmer, M. S., & Schwantes, M. (2018). Cross-cultural music therapy: Reflections of American music therapists working internationally. *The Arts in Psychotherapy, 61*, 21–32.

Gross, J., & Cassidy, J. (2019). Expressive suppression of negative emotions in children and adolescents: Theory, data, and a guide for future research. *Developmental Psychology, 55*(9), 1938.

Habermas, J. (2008). *Between naturalism and religion.* Polity Press.

Hald, S. V., Baker, F. A., & Ridder, H. M. (2017). A preliminary evaluation of the interpersonal music-communication competence scales. *Nordic Journal of Music Therapy, 26*(1), 40–61.

Haviland, J. (2014). *Exploring empathy in music therapy* [Masters Dissertation, Molloy College].

Hollan, D. (2008). Being there: On the imaginative aspects of understanding others and being understood. *Ethos, 36*(4), 475–489.

Iacoboni, M. (2009). Imitation, empathy, and mirror neurons. *Annual Review of Psychology, 60*, 653–670.

Ip-Winfield, V., & Grocke, D. (2011). Group music therapy methods in cross-cultural aged care practice in Australia. *Australian Journal of Music Therapy, 22*, 59–80.

Jackson, N., & Gardstrom, S. (2012). Undergraduate music therapy students' experiences as clients in short-term group music therapy. *Music Therapy Perspectives, 30*(1), 65–82.

Johnson, D. R., Cushman, G. K., Borden, L. A., & McCune, M. S. (2013). Potentiating empathic growth: Generating imagery while reading fiction increases empathy and prosocial behavior. *Psychology of Aesthetics, Creativity, and the Arts, 7*(3), 306.

Jones, S. (2018). *Effect of vocal style on perceived empathy, rapport, patient engagement, and competency of music therapists* [Masters Thesis, Florida State University].

Jurist, E. L. (2010). Mentalizing minds. *Psychoanalytic Inquiry, 30*(4), 289–300.

Kamens, S. (2019). De-othering "schizophrenia." *Theory & Psychology, 29*(2), 200–218.

Kerig, P. (2020). Emotion dysregulation and childhood trauma. In T. Beauchaine & S. Crowell (Eds.), *The Oxford handbook of emotion dysregulation* (pp. 265–282). Oxford University Press.

Kesner, L., & Horáček, J. (2017). Empathy-related responses to depicted people in art works. *Frontiers in Psychology, 8*, 228.

Lakeman, R. (2020). Advanced empathy: A key to supporting people experiencing psychosis or other extreme states. *Psychotherapy and Counselling Journal of Australia, 8*(1).

Layman, D. L., Hussey, D. L., & Reed, A. M. (2013). The beech brook group therapy assessment tool: A pilot study. *Journal of Music Therapy, 50*(3), 155–175.

Lee, J. H. (2016). A qualitative inquiry of the lived experiences of music therapists who have survived cancer who are working with medical and hospice patients. *Frontiers in Psychology, 7*, 1840.

Lindvang, C. (2013). Resonant learning: A qualitative inquiry into music therapy students' self-experiential learning processes. *Qualitative Inquiries in Music Therapy, 8*, 1–31.

Lotter, C. (2017). *The qualitative affordances of active and receptive music therapy techniques in major depressive disorder and schizophrenia-spectrum psychotic disorders* [Doctoral Thesis, University of Pretoria].

Ma-Kellams, C., & Lerner, J. (2016). Trust your gut or think carefully? Examining whether an intuitive, versus a systematic, mode of thought produces greater empathic accuracy. *Journal of Personality and Social Psychology, 111*(5), 674.

Mar, R. A., & Oatley, K. (2008). The function of fiction is the abstraction and simulation of social experience. *Perspectives on Psychological Science, 3*, 173–192.

McGann, H. S. (2012). *Finding a place for music therapy practice in a hospital child development service* [Masters Thesis, Victoria University of Wellington].

McLeod, J. (2013). *An introduction to counselling*. Open University Press.

Moonga, N. U. (2019). *Exploring music therapy in the life of the batonga of Mazabuka Southern Zambia* [Master's dissertation, University of Pretoria].

Moran, D. (2017). *Mindfulness and the music therapist: An approach to self-care* [Doctoral dissertation, Concordia University].

Murphy, K. (2007). Experiential learning in music therapy: Faculty and student perspectives. *Qualitative Inquiries in Music Therapy, 3*, 31–61.

Naish, K., Reader, A., Houston-Price, C., Bremner, A., & Holmes, N. (2013). To eat or not to eat? Kinematics and muscle activity of reach-to-grasp movements are influenced by the action goal, but observers do not detect these differences. *Experimental Brain Research, 225*, 261–275.

Nordgren, L., Banas, K., & MacDonald, G. (2011). Empathy gaps for social pain: Why people underestimate the pain of social suffering. *Journal of Personality and Social Psychology, 100*(1), 120.

Nye, I. (2014). A response to Christine Atkinson's essay "'Dare we speak of love?' An exploration of love within the therapeutic relationship." *British Journal of Music Therapy, 28*(1), 36–39.

Ohrt, J. H., Foster, J. M., Hutchinson, T. S., & Ieva, K. P. (2009). Using music videos to enhance empathy in counselors-in-training. *Journal of Creativity in Mental Health, 4*(4), 320–333.

Olden, C. (1953). On adult empathy with children. *The Psychoanalytic Study of the Child, 8*(1), 111–126.

Paden, C. E. (2018). *Adolescents with behavioral health concerns: Increasing empathy through lyric discussions of peer-chosen music.* Illinois State University.

Pavlicevic, M. (1997). *Music therapy in context.* Jessica Kingsley.

Potash, J. (2010). *Guided relational viewing: Art therapy for empathy and social change to increase understanding of people living with mental illness* [Doctoral dissertation, University of Hong Kong, Pokfulam, Hong Kong].

Potash, J., & Chen, J. (2014). Art-mediated peer-to-peer learning of empathy. *The Clinical Teacher, 11*, 327–331.

Potash, J., Chen, J., Lam, C., & Chau, V. (2014). Art-making in a family medicine clerkship: How does it affect medical student empathy? *BMC Medical Education, 14*, 247–256.

Potash, J., & Ho, R. (2011). Drawing involves caring: Fostering relationship building through art therapy for social change. *Art Therapy, 28*(2), 74–81.

Potash, J., Ho, R., Chick, J., & Au Yeung, F. (2013). Viewing and engaging in an art therapy exhibit by people living with mental illness: Implications for empathy and social change. *Public Health, 127*, 735–744.

Potter, N. (2006). What is manipulative behavior, anyway? *Journal of Personality Disorders, 20*(2), 139–156.

Préfontaine, J. (2006). On becoming a music therapist. *Voices: A World Forum for Music Therapy, 6*(2).

Priestley, M. (1994). *Essays on analytical music therapy.* Barcelona Publishers.

Ratcliffe, M. (2012). Phenomenology as a form of empathy. *Inquiry, 55*(5), 473–495.

Reddy, V. (2008). *How infants know minds.* Harvard University Press.

Redmond, M. (2018). *Social decentering: A theory of other-orientation encompassing empathy and perspective taking.* De Bruyter.

Reniers, R. L., Corcoran, R., Drake, R., Shryane, N. M., & Völlm, B. A. (2011). The QCAE: A questionnaire of cognitive and affective empathy. *Journal of Personality Assessment, 93*(1), 84–95.
Ruud, E. (2003). "Burning scripts" self psychology, affect consciousness, script theory and the BMGIM. *Nordic Journal of Music Therapy, 12*(2), 115–123.
Saarni, C., Campos, J., Camras, L., & Witherington, D. (2008). Principles of emotion and emotional competence. In W. Damon & R. Lerner (Eds.), *Child and adolescent development: An advanced course* (pp. 361–405). Wiley.
Saxe, R., Carey, S., & Kanwisher, N. (2004). Understanding other minds: Linking developmental psychology and functional neuroimaging. *Annual Review of Psychology, 55*, 87–124.
Segal, E. (2018). *Social empathy: The art of understanding others.* Columbia University Press.
Seri, N., & Gilboa, A. (2017). When music therapists adopt an ethnographic approach: Discovering the music of ultra-religious boys in Israel. *Approaches: An Interdisciplinary Journal of Music Therapy*, First View, 1–15.
Shoemaker, S. (1996). *The first-person perspective and other essays.* Cambridge University Press.
Short, H. (2017). It feels like Armageddon: Identification with a female personality-disordered offender at a time of cultural, political and personal attack. *Nordic Journal of Music Therapy, 26*(3), 272–285.
Simons, G., Pasqualini, M. C. S., Reddy, V., & Wood, J. (2004). Emotional and nonemotional facial expressions in people with Parkinson's disease. *Journal of the International Neuropsychological Society, 10*(4), 521–535.
Solomon, R. C. (1995). The cross-cultural comparison of emotions. In J. Marks & R. T. Ames (Eds.), *Emotions in Asian thought: A dialogue in comparative philosophy* (pp. 253–308). SUNY Press.
Sorel, S. (2010). Presenting Carly and Elliot: Exploring roles and relationships in a mother-son dyad in Nordoff-Robbins music therapy. *Qualitative Inquiries in Music Therapy, 5.*
Spaulding, S. (2015). On direct social perception. *Consciousness and Cognition, 36*, 472–482.
Spiro, N., Schofield, M., & Himberg, T. (2013). *Empathy in musical interaction.* The 3rd International Conference on Music & Emotion, Jyväskylä, Finland, June 11–15. Department of Music, University of Jyväskylä.
Stadler, J. (2017). Empathy in film. In H. Maibom (Ed.), *Routledge handbook of philosophy of empathy* (pp. 317–326). Routledge.
Stapel, J. C., Hunnius, S., & Bekkering, H. (2012). Online prediction of others' actions: The contribution of the target object, action context and

movement kinematics. *Psychological Research Psychologische Forschung, 76*(4), 434–445.
Steinhardt, T., & Ghetti, C. (2020). Resonance between theory and practice: development of a theory-supported documentation tool for music therapy as procedural support within a biopsychosocial frame. In L. Ole bonde & K. Johansson (Eds.), *Music in paediatric hospitals: Nordic perspectives* (pp. 109–140). Norwegian Academy of Music.
Stensæth, K. (2018). Music therapy and interactive musical media in the future: Reflections on the subject-object interaction. *Nordic Journal of Music Therapy, 27*(4), 312–327.
Strehlow, G., & Hannibal, N. (2019). Mentalizing in improvisational music therapy. *Nordic Journal of Music Therapy, 28*(4), 333–346.
Trolldalen, G. (1997). Music therapy and interplay: A music therapy project with mothers and children elucidated through the concept of "appreciative recognition." *Nordic Journal of Music Therapy, 6*(1), 14–27.
Valentino, R. (2006). Attitudes towards cross-cultural empathy in music therapy. *Music Therapy Perspectives, 24*, 108–114.
Ward, M. (2015). Art in noise: An embodied simulation account of cinematic sound design. In M. Coëgnarts & P. Kravanja (Eds.), *Embodied cognition and cinema* (pp. 155–186). Leuven University Press.
Winter, P. (2013). *Effects of experiential music therapy education on student's reported empathy and self-esteem: A mixed methods study* [Doctoral dissertation, Temple University].
Wood, S. (2006, November). "The Matrix": A model of community music therapy processes. *Voices: A World Forum for Music Therapy, 6*(3).
Wyl, A. (2014). Mentalization and theory of mind. *Praxis Der Kinderpsychologie und Kinderpsychiatrie, 63*(9), 730–737.
Zahavi, D. (2005). *Subjectivity and selfhood: Investigating the first-person perspective*. MIT Press.
Zahavi, D., & Gallagher, S. (2008). The (in) visibility of others: A reply to Herschbach. *Philosophical Explorations, 11*(3), 237–244.
Zaki, J., & Ochsner, K. N. (2012). The neuroscience of empathy: Progress, pitfalls and promise. *Nature Neuroscience, 15*(5), 675–680.
Zurek, P. P., & Scheithauer, H. (2017). Towards a more precise conceptualization of empathy: An integrative review of literature on definitions, associated functions, and developmental trajectories. *International Journal of Developmental Science, 11*(3–4), 57–68.

6

Responding with Action

It is insufficient for a therapist to have an inner experience of perceiving a client's emotions, sharing in them, and understanding them without expressing this carefully to the client in a non-judgemental and congruent manner that makes a difference to them (Lakeman, 2020). Communicating empathy may be transformational in itself or it may be the spark that leads to additional compassionate action.

Empathy Needs to Be Communicated

Receiving a client's emotional expression and responding to it (often musically) for the purpose of communicating resonance with their emotional and musical universe is a core empathic task of a music therapist (Préfontaine, 2006). In Barrett-Lennard's (1981) influential formulation of the empathy cycle in therapy, the client and therapist search together for an accurate expression and understanding of the client's experience by cycling through four phases: (1) the client expresses their experience; (2) the therapist empathically resonates; (3) the therapist expresses their empathy; and (4) the client receives the therapist's

A. dos Santos, *Empathy Pathways*,
https://doi.org/10.1007/978-3-031-08556-7_6

empathy. The communication of empathy from the therapist encourages the client to share further, and every new response from the client enables the therapist to understand them better and become more empathically accurate.

In Rogers' (1957) description of empathy as sensing the client's private world, entering into it and moving about in it freely, he included the idea of endeavouring to communicate this understanding to the client. We do this with caution as the process is exploratory and cooperative. As discussed in Chapter 3, empathic communication frequently occurs cross-modally in music therapy (Dimitriadis & Smeijsters, 2011). Short (2017b), for example, described how the stability of the music conveyed empathy to her client, Marcus (who had recently experienced a psychotic episode). During music-making centred on Hip Hop, they collaborated to create a one-bar percussive beat that looped continuously and predictably throughout the song. Short then suggested simple one-bar motifs as hooks that could be repeated with slight variations. This offered Marcus a sense of being held and anchored. She explained how the reliability of this musical intervention communicated her empathy. Music therapists in Kwan's (2010) study who worked with adults in pain described how they conveyed empathy through attention, listening, posture, affect, and responsive tailoring of the music so that their clients experienced their supportive presence.

Bruscia's (1987) empathy techniques within improvisational music therapy are explicitly communicative. He included the following: imitating (when the therapist reproduces or echoes the client's response after they have offered it); synchronising (simultaneously copying or matching what the client is doing, or aspects of what they are doing, unimodally or cross-modally); incorporating (using a motif offered by the client within their improvisation); pacing (matching the client's energy level expressed through speed and intensity); reflecting (conveying the emotion(s) that the client is communicating, unimodally or cross-modally, at the same time or straight afterwards), and exaggerating (enhancing a distinct aspect of the client's expression).

When Communicated Empathy Is Beneficial in Itself

When we communicate that we are perceiving, sharing in, and understanding our client's emotions, and when group members are offering this expression to one another, we know that this can bring about therapeutic benefits that enhance well-being. Empathy itself can be a therapeutic intervention (Haviland, 2014). For example, when a music therapist accurately attunes to a client's experience and reflects this back to them within a song or improvisation, the client may sense, "Yes! That's how I feel. You know me," which can be transformational.

Wigram (2004) explained empathic improvisation as a technique first applied by Juliette Alvin. At the beginning of a session, Alvin would typically improvise on her cello to complement the client's way of being, taking their facial expression, posture, and attitude into account, as well as her knowledge of their characteristics and personality as expressed in previous sessions. Alvin used this technique to empathise, not to shift the client's emotions or behaviour. For example, if a client arrived distressed, she would incorporate this emotion into the empathic improvisation without trying to ease or reduce its intensity. Through her music, she expressed empathic confirmation and support.

Through experiencing the therapist's empathy, clients may grow in their ability to identify (Preston-Roberts, 2011) and differentiate their emotions (Austin, 2001), accept and take ownership of their emotions (Ruud, 2003), and express and shape their emotions more clearly (Pedersen, 1997). Clients can become more attentive to their own experiences through experiencing the therapist's empathy. This is important because, as Préfontaine (2006) highlighted, awareness of what "is" precedes transformation. Through the empathic support of the music therapist, a client can engage in exploratory work, grow self-awareness, and experience a process of transformation that can be carried into everyday life outside of sessions (Leite, 2003). When a client feels heard, understood, and wondered about on their own terms, they may experience satisfaction, tension relief, and a sense of security, as opposed to the experiences of anxiety, stress and fear that accompany threat (Price & Caouette, 2018).

North (2014) described her music therapy work with Julianne, a two-year-old child with multiple and profound learning disabilities and visual impairment. Julianne would become intensely distressed when separated from her mother, crying loudly, exhausting herself, and becoming unable to concentrate on anything else. Initially, North tried to calm her with soothing music, but this had little impact. She then shifted her focus to using music to meet Julianne in her distress. She tried to feel the ebb and flow of tension and energy and aligned herself to the rhythm of Julianne's crying, following the volume, pitch, and shape of her sounds. As she did this, the quality of Julianne's vocalisations changed. Her pitching became more defined, the tension in her voice softened, and they could sing two longer phrases together. While Julianne was referred to music therapy because she could not listen to or focus on anything other than her anxiety and distress when she was separated from her mother, North could penetrate this by the empathy she conveyed through music. Julianne was then able to experience a sudden and vivid relationship with another person. As Nordoff and Robbins (2007) wrote, when one plays or sings to express the quality or intensity of a child's screaming or crying, the child hears something that resonates with what they are feeling. Their experience of themselves at that moment becomes related to their knowledge of the music. The music meets and accepts their state while also accompanying and enhancing their expression. Their screaming or crying becomes less isolated, and the child moves towards an experience of intercommunication.

As mentioned in Chapter 1, communication of empathy contributes to developing rapport and building a therapeutic alliance (Corey, 2005; Maltsberger, 2011). Rapport developed through the therapist's empathy has been described as leading to positive therapeutic outcomes (Rogers, 1957; Swanick, 2019) and improved quality of life (Kim, 2013) through experiences of psychological safety (Short, 2017a), security (Ruud, 2003), support (Kwan, 2010), trust (Lindvang & Frederiksen, 1999), increased vulnerability (Dindoyal, 2018), willingness to take risks (Behrens, 2012), and to explore more deeply (So, 2017). The therapist's empathy can provide dyadic affective regulation, through which the client may experience enhanced resilience, especially in the aftermath of traumatic experiences (Kim, 2014). Through experiencing empathy, clients may find validation, self-acceptance, and belonging (Bennett, 2001; Potvin et al., 2015; Valentino, 2006) as their music-making is heard and valued (Gardstrom & Hiller, 2016) and their worldview (Brown, 2002), as well as their experiences of various styles of music, is understood (Valentino, 2006). Through being empathised with, a client may be more able to weave fragmented memories into narratives (Radoje, 2014). Being empathised with can decrease stress (So, 2017) and penetrate resistance in therapy (Gardstrom & Hiller, 2016). Experiencing an empathic therapist can enhance an individual's sense of hope for the future (Swanick, 2019). While this list of benefits is extensive, it's important to bear in mind that not all clients will necessarily respond positively to a therapist's empathy. It may be met with rejection, suspicion, or ambivalence if a client is not yet ready to receive it or if empathy was communicated in an insensitive, irrelevant, or inappropriate way (Dileo, 2021).

Music therapy groups can facilitate the development of empathy between their members (Bullard, 2011). When group members express empathy towards one another, a sense of shared experience can be developed, as Rosado (2019) observed in her work with adolescents in an inpatient crisis stabilisation unit. Through empathy, members can respectfully recognise one another's vulnerabilities (McCaffrey et al., 2021). The validation that members receive can encourage them to share their difficult emotions further. In their work with women who were refugees, Ahonen and Mongillo Desideri (2014) explained three experiential elements of trauma: devastating physical and/or emotional pain, perceiving oneself as entirely helpless, and a lack of empathy from others. While the event is profoundly impactful, a lack of empathy shown by important people in the individual's life during the event, and the concrete loneliness experienced, can cause even more significant psychological injury. While trauma isolates, stigmatises, humiliates, and shames, the role of therapy groups is to create belonging, validation, affirmation, and witnessing.

When Empathy Is a Starting Point

Let's step outside of the therapy room for a moment and consider an example from daily life. Suppose I (in a motivated position to be empathic) encountered a young child sitting on the side of the road, crying and hungry with no caregiver in sight. In that case, I could perceive her distress, share in and understand her emotions, and communicate my empathy to her. However, it would be a difficult case to make that this is a sufficient response. Additional action would be required.

There is indeed broad consensus that empathy leads to compassionate action, thereby fostering greater socialwell-being within societies (Da Silva et al., 2004). Hoffman (2000), for example, wrote that "empathy's congruence with caring is obvious" (p. 14). A large body of research has affirmed the value of empathy in relation to prosocial behaviour, defined as "voluntary behaviour intended to benefit another" (Spinrad & Gal, 2018, p. 40). For example, individuals who show empathy for others tend to display less aggression and interact more constructively with their peers (Decety & Lamm, 2006; DeWall et al., 2012). Outward prosocial behaviours such as sharing resources, cooperating, and helping are generated by internal prosocial preferences for outcomes that will benefit others and affirm prosocial norms (Bolton & Ockenfels, 2000).

However, the notion that empathy always leads to prosocial behaviour has been critiqued. Empathy may be neither inherently good nor bad. According to Nussbaum (2003), empathy is morally neutral. It can be used to help others or dominate them (Bayram & Holmes, 2020), to respond with kindness or manipulation (Hart et al., 1995). For example, imagining another's situation and taking pleasure in their pain ("schadenfreude") could be considered a form of empathy.

Within the process of insightful empathy, if I share in and understand another's emotions, *and* I value this person's well-being and the enhancement thereof, then my empathy can lead to a response of empathic concern. Batson (2017) defined empathic concern as "other-oriented emotion elicited by and congruent with the perceived welfare of a person" (p. 57). It entails a sense of being moved and feeling tender and soft-hearted towards the other person (Batson, 2011). This is different from affective empathy because the emotional state need not match what the other person is feeling. Empathic concern is closely related to (and the term is often used synonymously with) sympathy (Feldman-Hall et al., 2015). Sympathy is defined by Eisenberg et al. (1994) as "an emotional reaction that is based on the apprehension of another's emotional state or condition, and that involves feelings of concern and sorrow for the other person" (p. 776). (For the sake of clarity, the term "pity" entails feeling concern for someone who is considered inferior to oneself [Goetz et al., 2010].)

If we return to the example of the crying, hungry child, if my empathy has extended into empathic concern and I'm now feeling tender, soft-hearted, and "moved," is this now sufficient? The question is whether this feeling of being "moved" turns into actual *movement* that will alleviate her distress. As a result of my empathic concern, I may feel more motivated to engage in compassionate action.

The empathy-altruism hypothesis was developed by Batson (2011). Building on over 40 years of research, Batson claimed that empathic concern leads to altruistic motivation. Other researchers have also concluded that empathic concern is the mediator between empathy and compassionate action (for example, de Vignemont and Singer [2006], Eisenberg et al. [1994], and Lishner et al. [2011]). After reanalysing several previous studies, Cialdini et al. (1997) found, however, that empathic concern only facilitates helping when substantial costs are not demanded of the helper. In other words, empathic concern can help motivate us towards superficial helping. Substantial, meaningful helping appears to be influenced by aspects such as the willingness to use the term "we" when describing the relationship with the person in need.

The word "compassion" originates from the Latin "com-pati," meaning "to suffer with" (Tirch et al., 2014). Compassion has been defined by Gilbert (2009) as a "basic kindness" with a deep awareness of and sensitivity to the presence and causes of suffering combined with the desire and effort to alleviate it. It crucially involves an authentic desire to help (Trzeciak et al., 2019) and a behavioural response to the other's suffering or pain (Price & Caouette, 2018).

Within the field of music therapy, while empathy is often described as beneficial in itself (as we saw above), several authors have explained how empathy can lead the therapist to engage in additional, helpful actions as a compassionate responder. For example, in Kwan's (2010) work with adults in pain, she showed how empathy allowed the therapist to extend herself into the client's world to then work with it. Empathy can enable the therapist to assist the client in meaning-making (Valentino, 2006). Rickson (2003) described how she communicated her empathy to her clients through music and then, once contact had been made, she used a variety of musical techniques to invite her clients in a non-threatening way to participate and to support their contributions, which then led to their skill development and increased expressive freedom.

Through empathy, therapists both communicate their presence to their clients and gain a deeper understanding of how to shape their interventions so the experience can be as meaningful as possible. Empathy involves gathering "psychological data" that can guide clinical decision-making (So, 2017), including music selection (Gonzalez, 2011) and the use of silence (Nolan, 2005). Nordoff-Robbins music therapists in Cooper's (2010) research on clinical improvisation explained how their empathy provided clarity regarding their treatment approach, a meaningful musical experience for both client and therapist, and an essential source of information upon which to base musical choices. Empathy can also guide ethical decision-making (Dileo, 2021).

Pavlicevic and Fouché (2014) wrote about community music therapy work at a tuberculosis ward in a hospital where the patients were far from their homes. Everything in the hospital environment was unfamiliar to them. In their home contexts, they would have frequently been singing (at church services, burials, births), yet here in the hospital, no singing could be heard. Pavlicevic and Fouché considered the importance of providing opportunities for these individuals to sing as a way to reconnect with their sense of identity and with their homes. However, they reflected on the complexity of this idea and used empathy to do so. The hospital context is one of strangeness, illness, and Western medicine. Pavlicevic and Fouché asked: "Were we to ask them to uproot their songs too, and reinsert them into this difficult place of illness?" (p. 62). This recontextualising could help the patients reframe the present space as a more hopeful one, but it could also taint their home space by poaching, removing, and culling songs from those places. Based on this empathic "tuning in," Pavlicevic and Fouché decided, rather than "doing music therapy," which may impose inappropriate values and norms and reinforce a foreign healthcare frame, to invite the participants to choose and perform the songs that they wanted to in a manner that felt useful for them in reclaiming their identities. They did this along with community musicians who supported and witnessed the participants as people with resources, knowledge, and rootedness, who had tuberculosis and who also had health and resilience. Pavlicevic and Fouché explained how they were by no means dismissing the value and role of conventional music therapy practice. Still, they were mindful of how conventional approaches can be based on certain frameworks of problems and care, which were not necessarily relevant here. They described their stance as one of "musical caring."

Inaction or Distancing

The work of therapy is complex, the feelings evoked in the therapist are vast, and the therapist's responsive options are plentiful. Therapists who experience empathy towards their clients may not necessarily proceed to compassionate action (Vega & Ward, 2016; Winczewski et al., 2016). They may, in fact, retreat from their client's emotions.

Helping behaviour is the result of a cost/benefit analysis. When cost is deemed greater than benefit, humans tend to direct their attention away from the other's distress and subvert their empathic processes, lowering the desire to help (Preston & De Waal, 2002). In a study by Cassidy et al. (2018), when participants first recalled hurt feelings, they produced defensive barriers to helping a stranger who was suffering. When the stranger was viewed as unfamiliar, off-putting, dirty, and unlike themself, the protective barrier involved a tendency to be avoidant and dismissive. When the stranger was viewed as similar to themself and was suffering psychological pain (caused by a relationship partner), the barrier entailed caregiving anxiety (a concern that one does not have the adequate ability to help). However, when the participants were first primed to feel a sense of attachment security, these barriers were weakened. While music therapists have received specific training to reflect on the impact that their life histories may have on their interactions with clients, I would argue that it's naïve to assume we are immune to the kinds of processes identified in a study such as this. We are multifaceted human beings who also encounter personal hurt in our own lives. Therefore, considering and enhancing our sense of attachment security may assist us when engaging with clients experiencing emotional hurt so that we can respond consistently with compassionate action.

When Empathy Leads to Harm

Sometimes empathic concern can lead us to offer what we consider to be compassionate action, when in fact, the action is not actually helpful for the other person and may even be harmful. In her study of children receiving music therapy as procedural support, Yinger (2016) found that empathic statements made to paediatric patients well in advance of their procedure in an attempt to offer reassurance may heighten a child's distress, whereas this may not be distress-promoting at other times, such as during the procedure itself. We may also find ourselves entering into another's experience motivated by the desire to "take over" (to help, guide, save, or rescue them). We then delegitimise their experience and their ownership of that experience.

Oakley et al. (2012) defined pathological altruism as behaviours or personal tendencies that entail a stated aim or implied motivation of promoting the welfare of another; however, instead of producing beneficial outcomes, this "altruism" results in substantial negative consequences for the other or the self. An example is a co-dependency where, for example, a parent continuously forgives the physical and emotional abuse that is directed towards themselves, thereby supporting the dysfunctional behaviour (McGrath & Oakley, 2012). Brin (2012) discussed self-righteous indignation as a form of pathological altruism. We may gain pleasure in believing that an opponent is wrong and we are right or that our method of helping is correct and driven by pure motivation, no matter what contradictory evidence may be presented to us. Empathy-based guilt can also be the driving force behind pathological altruism (O'Connor et al., 2012). People may falsely think that they are to blame for another's misfortune or have the means to alleviate the other's suffering. A woman may believe that her partner's violence towards her is her fault, or a young adult with depression may commit suicide believing that he will be alleviating the burden on his family.

There are times when altruism can be a threat to the common good. In a classic experiment, participants were told about Sheri Summers, a 10-year-old girl with a fatal disease who was low on the waiting list for pain-relieving treatment. Participants were given the option of moving Sheri up to the top of the list, even though this would result in a potentially more deserving child not getting the treatment. The majority of participants declined. However, if they were first given an empathy prompt (asked to feel what Sheri felt), a majority chose to move Sheri up the list. Empathy then clashed with fairness (Batson et al., 1995). In another study on the identifiable victim effect, when people were told that eight children needed a life-saving drug, they gave less money than when they were shown the name and picture of one child in need (Kogut & Ritov, 2005). It has been argued, therefore, that optimal moral decision-making requires impartiality above empathy (Bloom, 2017; Greene & Haidt, 2002).

Breithaupt (2019) proposed a three-person model of empathy (victim-perpetrator-observer). When observing a conflict, humans tend to pick a side and then see the situation from that person's perspective. This then creates a positive feedback loop. As a victim's perspective is selected, they are considered increasingly worthy of more empathy. Simultaneously, the person perceived to be the perpetrator tends to be seen as less appealing, wrong, or even hostile, and empathy towards them decreases. The observer justifies their selection of sides through rational arguments and believes their own justifications, even if they are wrong.

Strong empathy for a victim corresponds with a desire for the perpetrator(s) to be punished (Bloom, 2017). Buffone and Poulin (2014) showed that people were even willing to inflict pain on innocent others when this would help the individuals in distress they were empathising with. This applies to groups as well as individuals. Greater empathic concern for members of one's in-group can motivate one to be more aggressive towards out-group members (Cikara et al., 2011). Through this process, relatively moderate tensions can escalate into intense conflicts. Governments commonly use empathic rage to justify war (Aucoin & Kreitzberg, 2018) as moral outrage predicts third-party punishment (Pfattheicher et al., 2019; Smith, 2010). Even terrorists can act out of empathy for those they kill for. As Breithaupt (2019) wrote, human beings can commit atrocities not out of a failure of empathy but as a direct result of successful empathy. Torture requires empathy, too: one cannot deliberately cause such harm to another without realising how the other will experience the harm (Aucoin & Kreitzberg, 2018).

Breithaupt (2019) discussed a case in Northern Ireland where a curriculum was developed to assist children in grades six to eight understand historical events in their country through both Catholic and Protestant perspectives. While the children became successful in their abilities to do this, they grew even more strongly polarised than learners who had not been exposed to this curriculum. While a variety of hypotheses were proposed for why this was the case, Breithaupt argued that the process of reinforcing the two perspectives resulted in empathy being interlocked with side-taking. It was emphasised to the students that all events in their history could be viewed from both a Protestant and a Catholic side, and each student knew precisely which side was theirs. While they had learned to assume the other's perspective cognitively, the curriculum reinforced that the other perspective was just that: the perspective of the other side. In South Africa, the process engaged after Apartheid through the Truth and Reconciliation Commission (TRC) offered a different approach in that forgiveness was sought as a judicial verdict, independent of side-taking or empathy. The goal was to develop a new national identity.

What could the risk of perspective-taking leading to increased polarisation mean for music therapists? When we listen to a client share about another person in their life whom they perceive has harmed them, do we continue to attempt to hold space for this other person's possible experiences too, or do we move straight into the three-person model where we experience increased empathy for our client, and decreased empathy for the "other" whom we hear about? What may the implications of this be for the client's exploration process? When working in contexts where our goal is community-level enhancement of well-being, how do we intentionally resist being drawn into one-sided perspective-taking at the expense of another "side"? How do we create spaces where the view of one doesn't define the belonging of another (Leonard, 2020) and where solidarity can hold radical difference, interdependency, and mutual obligation (Gaztambide-Fernandez, 2012), but without reinforcing the kind of perspective-taking that serves ongoing polarisation?

Let's consider a final dimension in which empathy may bring about harm. In advocative exploitative empathy, the advocate takes the side of and empathises with a person suffering so that they can find satisfaction and pleasure in this advocacy role. The other's suffering is necessary for the existence of their role. As therapists, we may feel some discomfort reflecting on this, as the very nature of our profession depends on suffering. Breithaupt (2019) argued that what may occur when we think we are empathically identifying with a suffering person is that we are in fact identifying with a benevolent helper figure. This self-identification may boost our ego at the expense of the person who is or whom we perceive to be in need. He referred to this as "filtered empathy." In the ultimate fantasy, the observer-helper receives gratitude from the victim for their intervention. There could also be a third imagined witness who praises the helper for their goodness. The problem is that once the structure of this scene (involving observer-helper-victim-appreciator) starts to unravel, compassion for the victim disappears. Filtered empathy promotes the well-being (and superiority) of the helper more than the well-being of the victim (and can lock the other into their role as a victim through facilitation of dependency, for example). If the victim fails to deliver to the helper what they require (such as gratitude, recognition, praise, or approval), the helper may start to experience resentment. This kind of identification-based empathy can be viewed as a form of narcissism.

Pavlicevic (1997) wrote about how our role as therapists is to assist our clients on their journey towards healing, yet our own needs are consistently present. The less attention we pay to those needs, the less willing we may be to embrace and ethically manage the "shadow" part of our work. In Chapter 2, we discussed the importance of self-awareness. We see again how essential this is throughout the process of empathy. Empathy needs to be employed through careful, critical reflection (Morris, 2019).

Self-Empathy and Self-Compassion

Much of our focus thus far has been on directing empathy towards others. Empathy is also a gift we can give ourselves. Empathy can be extended towards one's "past self," a facet of self (such as one's "inner child," the superego, an archetype, a layer of one's identity, a part of one's body that may be ill, and so on). Self-empathy involves taking up a perspective towards oneself that allows for a degree of narrative distance (Goldie, 2011). When empathising with ourselves, we may respond in a range of ways—just as we might when engaging with others—by helping, distancing, or harming. Humans often respond to themselves in harsher and more unkind ways than how they would treat another person in the same situation. This severity towards the self can stem from a fear of egotism, self-indulgence, or a harsh inner critic (Neff, 2003). Considerations of self-empathy and potential responses towards it are essential both for how we treat ourselves as therapists and how we understand our clients' journeys of self-empathy.

Self-compassion has been defined as "compassion turned inward" (Neff & McGehee, 2010, p. 226). It's the capacity to hold one's feelings of suffering from warmth, concern, and connection. Just as compassion involves being moved to action by encountering another's suffering, self-compassion entails being open to and touched by one's own suffering in a non-judgemental manner. We recognise how our suffering is part of the broader human condition; we don't disconnect from it; and we tap into our desire to alleviate our suffering with kindness (Neff, 2003). This is relevant both for situations where the suffering came about through no fault of our own and when we brought about the suffering through our actions, inadequacies, or poor decisions. Through self-compassion, "failure" can be experienced as an opportunity for growth, not an indicator of one's worth (Neff & Vonk, 2009).

This process is distinct from self-pity, where an individual can become immersed in their problems and ignore their interconnectedness with others. Self-compassion allows one to see how one's experiences are related to the experiences of others. In self-pity, individuals can also become over-identified with their emotional reactions so that alternative perspectives are inaccessible. When faced with suffering, self-compassion entails self-kindness, sensing common humanity, and mindfulness, where one can hold one's pain in balanced awareness rather than over-identification (Neff, 2003).

Sherman (2014) viewed self-empathy in a manner that is synonymous with the current discussion of self-compassion. She argued that self-empathy doesn't only entail brushing over our faults and weaknesses to make ourselves feel good because this isn't useful for growth. Empathy towards the self includes dwelling in or preserving access to difficult emotions, but in a compassionate manner that can lead to integration. Without self-empathy, we can hold unrealistic expectations of what we are capable of.

The goals of self-empathy are protection and transformation, assisting oneself to safely remember and revisit feelings and events, reconstructing what took place with fairer and more flexible self-judgement, deconstructing self-destructive patterns, and engaging in self-critique with honesty, goodwill, compassion, and the desire to improve (Gilbert, 2014; Gilbert & Woodyatt, 2017; Goldie, 2011). Acknowledging failure as a therapist is critical. Exceptional therapists can honestly admit their limitations, welcome constructive feedback, work through their errors and learn from them (Kottler & Carlson, 2014).

References

Ahonen, H., & Mongillo Desideri, A. (2014). Heroines' journey-emerging story by refugee women during group analytic music therapy. *Voices: A World Forum for Music Therapy, 14*(1). https://doi.org/10.15845/voices.v14i1.686

Aucoin, E., & Kreitzberg, E. (2018). Empathy leads to death: Why empathy is an adversary of capital defendants. *Santa Clara Law Review, 58*, 99–136.

Austin, D. (2001). In search of the self: The use of vocal holding techniques with adults traumatized as children. *Music Therapy Perspectives, 19*(1), 22–30.

Barrett-Lennard, G. T. (1981). The empathy cycle: Refinement of a nuclear concept. *Journal of Counseling Psychology, 28*, 91–100.

Batson, C. (2011). *Altruism in humans*. Oxford University Press.

Batson, C. (2017). The empathy-altruism hypothesis: What and so what? In E. Seppala, E. Simon-Thomas, S. Brown, M. Worline, C. Cameron, & J. Doty (Eds.), *The Oxford handbook of compassion science* (pp. 56–73). Oxford University Press.

Batson, C., Klein, T. R., Highberger, L., & Shaw, L. L. (1995). Immorality from empathy-induced altruism: When compassion and justice conflict. *Journal of Personality and Social Psychology, 68*(6), 1042.

Bayram, A. B., & Holmes, M. (2020). Feeling their pain: Affective empathy and public preferences for foreign development aid. *European Journal of International Relations, 26*(3), 820–850.

Behrens, G. A. (2012). Use of traditional and nontraditional instruments with traumatized children in Bethlehem, West Bank. *Music Therapy Perspectives, 30*(2), 196–202.

Bennett, M. (2001). *The empathic healer: An endangered species*. Academic Press.

Bloom, P. (2017). *Against empathy: The case for rational compassion*. Random House.

Bolton, G., & Ockenfels, A. (2000). ERC: A theory of equity, reciprocity, and competition. *American Economic Review, 90*, 166–193.

Breithaupt, F. (2019). *The dark sides of empathy*. Cornell University Press.

Brin, D. (2012). Self-addiction and self-righteousness. In B. Oakley, A. Knafo, G. Madhavan, & D. Sloan Wilson (Eds.), *Pathological altruism* (pp. 77–84). Oxford University Press.

Brown, J. M. (2002). Towards a culturally centered music therapy practice. *Voices: A World Forum for Music Therapy, 2*(1).

Bruscia, K. (1987). *Improvisational models of music therapy*. Thomas.

Buffone, A. E., & Poulin, M. J. (2014). Empathy, target distress, and neurohormone genes interact to predict aggression for others–even without provocation. *Personality and Social Psychology Bulletin, 40*(11), 1406–1422.

Bullard, E. (2011). Music therapy as an intervention for inpatient treatment of suicidal ideation. *Qualitative Inquiries in Music Therapy, 6*, 75–121.

Cassidy, J., Stern, J. A., Mikulincer, M., Martin, D. R., & Shaver, P. R. (2018). Influences on care for others: Attachment security, personal suffering, and similarity between helper and care recipient. *Personality and Social Psychology Bulletin, 44*(4), 574–588.

Cialdini, R. B., Brown, S. L., Luce, C., Sagarin, B. J., & Lewis, B. P. (1997). Does empathy lead to anything more than superficial helping? Comment on Batson et al. (1997). *Journal of Personality and Social Psychology, 73*(3), 510–516.

Cikara, M., Bruneau, E. G., & Saxe, R. R. (2011). Us and them: Intergroup failures of empathy. *Current Directions in Psychological Science, 20*(3), 149–153.

Cooper, M. (2010). Clinical-musical responses of Nordoff-Robbins music therapists: The process of clinical improvisation. *Qualitative Inquiries in Music Therapy, 5*, 86–115.

Corey, G. (2005). *Theory and practice of counselling and psychotherapy* (7th ed.). Brooks/Cole—Thomson Learning.

Da Silva, L., Sanson, A., Smart, D., & Toumbourou, J. (2004). Civic responsibility among Australian adolescents: Testing two competing models. *Journal of Community Psychology, 32*(3), 229–255.

de Vignemont, F., & Singer, T. (2006). The emphatic brain: How, when, and why? *Trends in the Cognitive Sciences, 10*, 435–441.

Decety, J., & Lamm, C. (2006). Human empathy through the lens of social neuroscience. *The Scientific World Journal, 6*, 1146–1163.

DeWall, C., Lambert, N., Pond, R., Kashdan, T., & Fincham, F. (2012). A grateful heart is a nonviolent heart: Cross-sectional, experience sampling, longitudinal, and experimental evidence. *Social Psychological and Personality Science, 3*(2), 232–240.

Dileo, C. (2021). *Ethical thinking in music therapy* (2nd ed.). Jeffrey Books.

Dimitriadis, T., & Smeijsters, H. (2011). Autistic spectrum disorder and music therapy: Theory underpinning practice. *Nordic Journal of Music Therapy, 20*(2), 108–122.

Dindoyal, L. (2018). 'In the therapist's head and heart': An investigation into the profound impact that motherhood has on the work of a music therapist. *British Journal of Music Therapy, 32*(2), 105–110.

Eisenberg, N., Fabes, R. A., Murphy, B., Karbon, M., Maszk, P., Smith, M., O'Boyle, C., & Suh, K. (1994). The relations of emotionality and regulation to dispositional and situational empathy-related responding. *Journal of Personality and Social Psychology, 66*(4), 776.

FeldmanHall, O., Dalgleish, T., Evans, D., & Mobbs, D. (2015). Empathic concern drives costly altruism. *NeuroImage, 105*, 347–356.

Gardstrom, S., & Hiller, J. (2016). Resistances in group music therapy with women and men with substance use disorders. *Voices: A World Forum for Music Therapy, 16*(3).

Gaztambide-Fernandez, R. (2012). Decolonization and the pedagogy of solidarity. *Decolonization: Indigeneity, Education and Society, 1*(1), 41–67.

Gilbert, P. (2009). *The compassionate mind: A new approach to life's challenges.* Constable and Robinson.
Gilbert, P. (2014). The origins and nature of compassion focused therapy. *British Journal of Clinical Psychology, 53*(1), 6–41.
Gilbert, P., & Woodyatt, L. (2017). An evolutionary approach to shame-based self-criticism, self-forgiveness, and compassion. In L. Woodyatt, E. Worthington, M. Wenzel, & B. Griffin (Eds.), *Handbook of the psychology of self-forgiveness* (pp. 29–41). Springer.
Goetz, J., Keltner, D., & Simon-Thomas, E. (2010). Compassion: An evolutionary analysis and empirical review. *Psychological Bulletin, 136*(3), 351–374.
Goldie, P. (2011). Self-forgiveness and the narrative sense of self. In C. Fricke (Ed.), *The ethics of forgiveness: A collection of essays* (pp. 81–94). Routledge.
Gonzalez, P. J. (2011). The impact of music therapists' music cultures on the development of their professional frameworks. *Qualitative Inquiries in Music Therapy, 6*, 1–13.
Greene, J., & Haidt, J. (2002). How (and where) does moral judgment work? *Trends in Cognitive Sciences, 6*, 517–523.
Hart, S., Cox, D., & Hare, R. (1995). *The Hare psychopathy checklist: Screening version.* Multi-Health Systems.
Haviland, J. (2014). *Exploring empathy in music therapy* [Masters Dissertation, Molloy College].
Hoffman, M. (2000). *Empathy and moral development*. Cambridge University Press.
Kim, J. (2014). The trauma of parting: Endings of music therapy with children with autism spectrum disorders. *Nordic Journal of Music Therapy, 23*(3), 263–281.
Kim, S. (2013). *Multimodal quantification of interpersonal physiological synchrony between non-verbal individuals with severe disabilities and their caregivers during music therapy* [Doctoral dissertation, University of Toronto].
Kogut, T., & Ritov, I. (2005). The "identified victim" effect: An identified group, or just a single individual? *Journal of Behavioral Decision Making, 18*(3), 157–167.
Kottler, J., & Carlson, J. (2014). *On being a master therapist*. Wiley.
Kwan, M. (2010). Music therapists' experiences with adults in pain: Implications for clinical practice. *Qualitative Inquiries in Music Therapy, 5*, 43.

Lakeman, R. (2020). Advanced empathy: A key to supporting people experiencing psychosis or other extreme states. *Psychotherapy and Counselling Journal of Australia, 8*(1).
Leite, T. (2003). Music, metaphor and "being with the other". *Voices: A World Forum for Music Therapy, 3*(2).
Leonard, H. (2020). The arts are for freedom: Centering Black embodied music to make music free. *Journal of Performing Art Leadership in Higher Education, 11*, 4–25.
Lindvang, C., & Frederiksen, B. (1999). Suitability for music therapy: Evaluating music therapy as an indicated treatment in psychiatry. *Nordic Journal of Music Therapy, 8*(1), 47–57.
Lishner, D. A., Batson, C. D., & Huss, E. (2011). Tenderness and sympathy: Distinct empathic emotions elicited by different forms of need. *Personality and Social Psychology Bulletin, 37*, 614–625.
Maltsberger, J. (2011). Empathy and the historical context, or how we learned to listen to patients. In K. Michel & D. Jobes (Eds.), *Building a therapeutic alliance with the suicidal patient* (pp. 29–48). American Psychological Association.
McCaffrey, T., Higgins, P., Monahan, C., Moloney, S., Nelligan, S., Clancy, A., & Cheung, P. S. (2021). Exploring the role and impact of group songwriting with multiple stakeholders in recovery-oriented mental health services. *Nordic Journal of Music Therapy, 30*(1), 41–60.
McGrath, M., & Oakley, B. (2012). Codependency and pathological altruism. In B. Oakley, A. Knafo, G. Madhavan, & D. Sloan Wilson (Eds.), *Pathological altruism* (pp. 49–74). Oxford University Press.
Morris, S. (2019). Empathy on trial: A response to its critics. *Philosophical Psychology, 32*(4), 508–531.
Neff, K. D. (2003). Self-compassion: An alternative conceptualization of a healthy attitude toward oneself. *Self and Identity, 2*(2), 85–102.
Neff, K., & McGehee, P. (2010). Self-compassion and psychological resilience among adolescents and young adults. *Self and Identity, 9*(3), 225–240.
Neff, K., & Vonk, R. (2009). Self-compassion versus global self-esteem: Two different ways of relating to oneself. *Journal of Personality, 77*, 23–50.
Nolan, P. (2005). Verbal processing within the music therapy relationship. *Music Therapy Perspectives, 23*(1), 18–28.
Nordoff, P., & Robbins, C. (2007). *Creative music therapy: A guide to fostering clinical musicianship* (2nd ed.). Barcelona Publishers.

North, F. (2014). Music, communication, relationship: A dual practitioner perspective from music therapy/speech and language therapy. *Psychology of Music, 42*(6), 776–790.

Nussbaum, M. (2003). *Upheavals of thought: The intelligence of emotions.* Cambridge University Press.

Oakley, B., Knafo, A., & McGrath, M. (2012). Pathological altruism—An introduction. In B. Oakley, A. Knafo, G. Madhavan, & D. Sloan Wilson (Eds.), *Pathological altruism* (pp. 3–9). Oxford University Press.

O'Connor, L., Berry, J., Lewis, T., & Stiver, D. (2012). Empathy-based pathogenic guilt, pathological altruism, and psychopathology. In B. Oakley, A. Knafo, G. Madhavan, & D. Sloan Wilson (Eds.), *Pathological altruism* (pp. 10–30). Oxford University Press.

Pavlicevic, M. (1997). *Music therapy in context.* Jessica Kingsley.

Pavlicevic, M., & Fouché, S. (2014). Reflections from the market place–community music therapy in context. *International Journal of Community Music, 7*(1), 57–74.

Pedersen, I. (1997). The music therapist's listening perspectives as source of information in improvised musical duets with grown-up: Psychiatric patients, suffering from Schizophrenia. *Nordic Journal of Music Therapy, 6*(2), 98–111.

Pfattheicher, S., Sassenrath, C., & Keller, J. (2019). Compassion magnifies third-party punishment. *Journal of Personality and Social Psychology, 117*(1), 124.

Potvin, N., Bradt, J., & Kesslick, A. (2015). Expanding perspective on music therapy for symptom management in cancer care. *Journal of Music Therapy, 52*(1), 135–167.

Préfontaine, J. (2006). On becoming a music therapist. *Voices: A World Forum for Music Therapy, 6*(2).

Preston, S. D., & de Waal, F. B. M. (2002). Empathy: Its ultimate and proximate bases. *Behavioral and Brain Sciences, 25*, 1–72.

Preston-Roberts, P. (2011). An interview with Dr Diane Austin. *Voices: A World Forum for Music Therapy, 11*(1).

Price, C., & Caouette, J. (2018). Introduction. In J. Caouette & C. Price (Eds.), *The moral psychology of compassion* (pp. ix–xviii). Rowman & Littlefield.

Radoje, M. (2014). Where were you born? A music therapy case study. *British Journal of Music Therapy, 28*(2), 25–35.

Rickson, D. (2003). The boy with the glass flute. *Voices: A World Forum for Music Therapy, 3*(2).

Rogers, C. (1957). The necessary and sufficient conditions of therapeutic personality change. *Journal of Consulting Psychology, 21*, 95–103.
Rosado, A. (2019). Adolescents' experiences of music therapy in an inpatient crisis stabilization unit. *Music Therapy Perspectives, 37*(2), 133–140.
Ruud, E. (2003). "Burning scripts" self psychology, affect consciousness, script theory and the BMGIM. *Nordic Journal of Music Therapy, 12*(2), 115–123.
Sherman, N. (2014). Recovering lost goodness: Shame, guilt, and self-empathy. *Psychoanalytic Psychology, 31*(2), 217–235.
Short, H. (2017a). It feels like Armageddon: Identification with a female personality-disordered offender at a time of cultural, political and personal attack. *Nordic Journal of Music Therapy, 26*(3), 272–285.
Short, H. (2017b). "Big up West London Crew": One man's journey within a community rap/music therapy group. *Music Therapy Perspectives, 35*(2), 151–159.
Smith, A. (2010). *The theory of moral sentiments*. Penguin.
So, H. (2017). US-trained music therapists from East Asian countries found personal therapy during training helpful but when cultural disconnects occur these can be problematic: A qualitative phenomenological study. *The Arts in Psychotherapy, 55*, 54–63.
Spinrad, T., & Gal, D. (2018). Fostering prosocial behavior and empathy in young children. *Current Opinion in Psychology, 20*, 40–44.
Swanick, R. (2019). What are the factors of effective therapy? Encouraging a positive experience for families in music therapy. *Approaches: An Interdisciplinary Journal of Music Therapy*. https://approaches.gr/wp-content/uploads/2019/11/Approaches_FirstView_a20191109-swanick-1.pdf
Tirch, D., Schoendorff, B., & Silberstein, L. (2014). *The ACT practitioner's guide to the science of compassion: Tools for fostering psychological flexibility*. New Harbinger Publications.
Trzeciak, S., Mazzarelli, A., & Booker, C. (2019). *Compassionomics: The revolutionary scientific evidence that caring makes a difference* (pp. 287–319). Studer Group.
Valentino, R. (2006). Attitudes towards cross-cultural empathy in music therapy. *Music Therapy Perspectives, 24*, 108–114.
Vega, M., & Ward, J. (2016). The social neuroscience of power and its links with empathy, cooperation and cognition. In P. Garrard & G. Robinson (Eds.), *The intoxication of power: Interdisciplinary insights* (pp. 155–174). Palgrave Macmillan.
Wigram, T. (2004). *Improvisation: Methods and techniques for music therapy clinicians, educators and students*. Jessica Kingsley.

Winczewski, L. A., Bowen, J. D., & Collins, N. L. (2016). Is empathic accuracy enough to facilitate responsive behavior in dyadic interaction? Distinguishing ability from motivation. *Psychological Science, 27*(3), 394–404.

Yinger, O. S. (2016). Music therapy as procedural support for young children undergoing immunizations: A randomized controlled study. *Journal of Music Therapy, 53*(4), 336–363.

Part II

Translational Empathy

A quality of presence in which a sense of withness is generated through a situated and productive process of emotion translation moves of expression and response.

7

A Stance of Honouring Opacity

Is it always possible or desirable to share and understand another's emotions accurately? Perhaps that would be presumptuous? Human beings frequently struggle to identify what is "really happening" within others and to judge what other people need to flourish, especially across cultural differences (Boler, 1997). In this chapter, we'll examine why that may be the case and begin to explore an alternative.

Our Understanding of Others Is Always Incomplete

Perspective-taking is a highly cognitively costly mechanism for everyday communication. Humans often employ heuristics instead as a shortcut (Epley et al., 2004; Galati & Brennan, 2010; Shintel & Keysar, 2009). Through the anchoring-and-adjustment heuristic, for example, we consider our own experiences when we judge other peoples' behaviour. This is also called self-referencing (Akgün et al., 2015). When we start from our own point of view, we need to make a set of adjustments to

A. dos Santos, *Empathy Pathways*,
https://doi.org/10.1007/978-3-031-08556-7_7

consider the likely differences between ourselves and the other person. This takes time, motivation, and resources. When these are in short supply, the adjustment process falls short, and our assessment of another's perspective merely remains aligned with our own (Epley et al., 2004).

Humans engage in many other errors and biases when attempting to understand others. For example, fundamental attribution error (Ross, 1977) describes the process that comes into play when I do something wrong and interpret it as a mistake but observe another person doing something wrong and interpret this as resulting from a flaw in their character or personality. When falling into the trap of the just-world bias, we assume that if something terrible has happened to someone, they must have done something wrong to deserve it (Lerner, 1980).

When we've never personally experienced what another is going through or the context they live in, we might be more ready to acknowledge that we don't fully know what they're feeling. However, when we have had similar experiences, we often assume that we have greater access to and understanding of their experience. Hodges (2005), a psychologist at the University of Oregon, explored whether having similar life experiences enhanced empathy. She reported on three studies of empathy and shared experiences that she and her colleagues were involved in. One concerned new motherhood; another focussed on alcoholism; and the third was about parental divorce. Across the studies, there was a striking similarity. When perceivers had experienced the same life events as the targets described, their empathic accuracy was no greater than those who had not had these life experiences. This research showed, in Hodges' words, "an utter lack of evidence that shared experience improved empathic accuracy" (p. 301).

Humans even struggle to have empathy for themselves. Loewenstein (2005) described what he termed the hot–cold empathy gap. When hot, we struggle to imagine what it's like to feel cold and vice versa. In addition, when we are in a hot emotional state (such as anger), we struggle to imagine how we feel when "cool" (calm). This applies both retrospectively (we struggle to connect back to the emotions we felt in a particular situation in the past) and prospectively (when our stomachs are full, for example, we imagine that we'll happily want to eat salads for the whole of the following week). We make affective forecasting errors when we imagine how we'll feel in the future.

Our understanding of clients may be incomplete for a wide range of reasons. A client's feelings may seem incomprehensible if they are too terrible to imagine (for example, if they relate to an experience of surviving torture). When a client is experiencing psychosis, their trajectory of behaviour, narratives, or emotional expressions may be difficult to follow. Therapists may struggle to relate to unusual experiences such as bizarre hallucinations, as they may be unable to find a correlation in their own everyday life (Kirmayer, 2008). A person with BPD may experience intense and abrupt affective instability in response to idiosyncratic cues, which may feel unpredictable to someone attempting to empathise. It may also be challenging to understand their experience when it is characterised by feelings of absence and emptiness. Individuals with BPD may exhibit exquisite interpersonal sensitivity, which can heighten their awareness of the limitations of those around them to understand what they are experiencing (Starcevic & Piontek, 1997). Many first-person accounts of depression highlight how aspects of the experience can be indescribable (Ratcliff, 2012). An inability to share one's experiences with others can exacerbate the sense of estrangement that is so central to depression.

Ratcliffe (2012) argued for a kind of radical empathy that involves engaging with another's experiences by suspending the common assumption that we both share the same experiential space. In our everyday encounters, we recognise that our own experiences differ in many ways from others, but we still tend to take much for granted as being part of a shared world. When I see someone running towards a departing bus and waving, I may perceive something of their frustration and sense of urgency. I experience this against a shared backdrop: *we* co-inhabit a world of objects, functions, social roles, and norms, a world that includes buses, bus drivers, bus stops, and departure times. The assumptions we tend to make about what comprises "our world" include not only what is "there" but how we encounter what is there. Researchers in phenomenological psychopathology (for example, Blankenburg [2001], Stanghellini [2004], Sass [2007], and Sass and Parnas [2007]) have shown how psychiatric illness can involve alterations in the form of experience rather than just the content of experience.

Suppose, for example, that one lost one's sense that life unfolds relatively predictably. Every object and event would appear as unexpected. One's all-encompassing experience of the world would be surprise and bewilderment. The space of experiential possibilities would have been transformed. Ratcliffe (2012) highlighted the account of a man with depression who described a change in his general structure of experience. The world had lost its welcoming quality and its usual range of practical possibilities, and action seemed impossible. These kinds of changes to how we encounter the world are not only present in extreme psychiatric illnesses. From time to time, we all experience shifts in how we experience the world (think of profound grief, jet lag, or a hangover). The way clients experience time may vary. De Kock (2003) discussed how, in Western societies, time is perceived as infinitely spanning a range from past infinity to future infinity. This range is then subdivided into regular "subsections" and measured as such. Many non-Western societies perceive time in a more cyclical way and as existing through reference to natural phenomena such as the rising and setting sun and the passing of the seasons. Continuously repetitive patterns may be interpreted within Western frameworks such as Nordoff-Robbins music therapy as indicating a degree of rigidity and even pathology. The meaning of musical repetition is very different in other contexts, however. Repetition could entail generating stability, belonging, spiritual transcendence, comfort, and affirmation of cyclical processes. When encountering another person, I can work towards neutralising a natural attitude that leads me to assume that the other's experience took place in a world like my own (Stanghellini & Rosfort, 2013). I could approach their world as I would do when exploring an unknown, foreign country.

According to Starcevic and Piontek (1997), while a therapist may have some available strategies to understand their client, in many circumstances, understanding is ultimately determined by the client. In *The anthropology of empathy*, Hollan and Throop (2011) wrote that empathy is an accomplishment "that depends very much on what others are willing or able to let us understand about them" (p. 8). A client may not want to be understood and may not allow others to understand them. Many are afraid of being understood, particularly at the start of the therapeutic process. Nordoff and Robbins (2007) used the term "resistiveness" to indicate how a child may need time to become comfortable with the intimacy experienced through clinical improvisation. In music therapy work with traumatised refugees, Orth (2005) highlighted how clients who speak a different language to the therapist might find this gap helpful for gaining a sense of extra psychological and emotional safety when expressing themselves in their mother tongue in musical improvisations.

In these situations, the therapist may still hope that the client will grow increasingly comfortable revealing more of their inner world. Perhaps, however, a client may remain opaque. Anthropologists Robbins and Rumsey (2008) wrote about the notion of "the opacity of other minds" (p. 407) that is widespread in societies of the Pacific. The other's mental space belongs to them alone and trying to simulate another person's mind within one's own is considered highly invasive and unethical. While the opacity doctrine is unusually well-developed in the Pacific, it's a view not limited to this area. In most societies, one can find people who contend that it's difficult to see into others' hearts and minds. Another person is an "other."

In Bradford and Robin's (2021) book *Connect*, they acknowledged that it's difficult to resist the assumption that you can know another's inner world. However, when one person acts towards another, three different realities (or areas of understanding) are present: person A's intention (only person A knows this); person A's behaviour (person A and person B know this); and the impact of person A's behaviour on person B (only B knows this). Sticking with one's own reality is more complicated than one may imagine. To convert an example that these authors give into a music therapy scenario, let's imagine that Mmusi and Ryan are teenagers within a music therapy session. They are in the process of collaboratively writing a song about being bullied. Ryan is offering many ideas for the song, forcefully and persistently, hardly giving Mmusi a chance to get a word in edgeways. Near the end of the session, Mmusi becomes frustrated and shouts out, "Hey, Ryan, give it a break! You really like listening to your own voice! Can I actually have a chance to give my idea?" Suppose we view this scenario in terms of Ryan's contribution. In that case, the first reality we can consider is Ryan's intent (he has had personal experience with bullying and excitedly believes that the knowledgeable lyrics he is offering will produce an excellent song). Only Ryan knows his own personal intention (unless he chooses to directly and transparently share it). The second reality is Ryan's behaviour, which both he and Mmusi are aware of. The third area is the impact of Ryan's behaviour on Mmusi, who now feels frustrated and angry. Only Mmusi is fully aware of his own feelings. Note how each party only knows two of the three realities. If Mmusi assumes that he knows Ryan's intentions, he risks coming across as accusatory, and chances are that Ryan will feel misunderstood and even attacked. If, however, Mmusi remains rooted in the two aspects of reality that he knows (he can point to the behaviour and share his reaction), then he can offer a comment that is non-accusatory and assist them both in gaining an understanding of what is unfolding. Emotional meaning can be negotiated as they both remain aware of which realities they do and do not have knowledge of. Bradford and Robin used the metaphor of a tennis net between the realities of intent and behaviour. In tennis, you can't play on the other person's side of the court.

Damaging Forms of Othering

At the outset of a conversation on otherness, we must exercise a weighty pause. Specific processes of Othering have been and still are dangerous and devastating. For example, as Drakulic (1993) wrote, "I understand now that nothing but 'otherness' killed Jews, and it began with naming them, by reducing them to the other. Then everything became possible…even the worst atrocities like concentration camps" (p. 145). Let's first examine what we do *not* mean by otherness in translational empathy before exploring how we will use this term in our current empathy pathway.

"Othering" was coined as a theoretical term by literary theorist and feminist critic Spivak (1985). She specifically explained the process that produces positionally superior, colonising subjects and marginalised, colonised Others. By referring to individuals or groups as "Other," projections of apparent negative difference are enforced and magnified (Johnson et al., 2004) as dominance is exercised (Said, 1978).

Emanuel (2016) referred to individuals with disabilities (especially with intellectual disabilities) as "a group of most othered Others" (p. 5). The notion of "normalcy" was created by statisticians in nineteenth-century Europe who developed the bell curve, categorising people along with a spread from "normal" to "deviant" (Davis, 2010). Before this, no culture had the concept of normal that is used now. These statisticians were not only interested in identifying deviance but eradicating it. Music therapist Metell (2019) explained how both disabled and queer people experience forms of exclusion that label "them" as different from white, non-disabled, heterosexual, cisgender men and, therefore, as "other." Children are also often constructed as "other" in terms of knowledge production (Klyve, 2019). From an adult's perspective, the inner world of a young child may seem distant and mysterious, and the younger they are, the more "other" a child may seem.

Patriarchal systems of exclusion and oppression are deeply entrenched in music discourse, particularly through naturalised practices that are adhered to without question (Scrine, 2016). Music therapists and community musicians can perpetuate marginalisation and exclusion through the very medium they seek to use for purposes of liberation. Dominant narratives still abound in music therapy, resulting in subjugation and oppression (Hadley, 2013). As Thomas and Norris (2021) wrote, "Those who have historically occupied dominant space and those who come from the 'right' kind of outside maintain a presumably 'benign' sense of control over the way the profession functions" (p. 6). The historical practice of othering non-European perspectives in art is evident within music therapy aesthetics (Norris, 2019). Improvisation practices rooted in a Eurocentric model of Western music are still emphasised, and this requires critique and urgent change (Sajnani et al., 2017). If we want to understand how music is communicative in music therapy, we need to look beyond the Western model, as it is only one among many notions of music and is remarkably recent in comparison (Cross, 2014).

Benevolent othering has been described as another hostile form of othering, involving self-serving and overly simplistic representations of others that gloss over the diversity and complexity of their lives (Grey, 2016). This process constructs a self-affirming image of oneself as a benevolent subject who is masterful and superior and a needy other on whom benevolence can be bestowed. While presenting a positive representation of mental health consumers (through, for example, anti-stigma discourses of recovery, co-production, and consumer participation), benevolent othering re-inscribes structures of subordination.

Respect for Radical Otherness

"Otherness" does not only operate oppressively. There are alternative ways of employing the idea of "otherness," distinct from these highly problematic forms that rely on negative difference, dominance, dehumanisation, and benevolence. The importance of valuing difference as otherness is vital. Zondi (2021) called for fundamental differences to be welcomed and encouraged as a necessary part of being human. Difference doesn't need to equate to inequality, and diversity doesn't imply inferiority and superiority. Diversity is a strength of human society. It increases choices and opens possibilities for new ways of being human. Multilogue (dialogue that comes from all directions) is fundamental for the revival of indigenous ways of being and doing human that entail mutual recognition. Carolyn Kenny (1999) wrote that empathic sensibilities can and must take "difference" into account, not merely surface-level differences, but differences at deeper levels of values and beliefs. We must come to terms with radical differences, and this needs to be reflected in music therapy theory. When we engage in translational empathy, we begin with a stance of honouring otherness.

French philosopher Emmanuel Levinas (1948) proposed a notion of "the other" as an individual who is (largely) unknowable. The Other who stands before me is radically "Other," and I must let them remain in their infinite otherness (Lim, 2007). Jaspers (1997), a psychopathologist and philosopher, articulated a (highly criticised) "theorem of incomprehensibility," insisting that understanding fails when we encounter someone radically and insurmountably unfamiliar. The specific lifeworld of another person is always beyond our full comprehension. This doesn't need to be viewed negatively or as an opposition (I am different from you). It can be seen as an affirmation (I am unique, and so are you) (Warren, 2008). Levinas's Other is not an enemy or an inferior presence; they are a neighbour (Brons, 2015). The relation with the Other is not determined by power (Levinas, 1948). I can only be aware of another person as that which is not me. I am then free to enjoy the other person without dissolving them into my sense of self (Lim, 2007).

We can choose to stay in an (often unpleasant) state of confusion and uncertainty rather than trying to pin down the Other's "true nature" or "precise experience." When faced with the complexity of another human being, we are not called to choose one acceptable version of them while dismissing all the other facets of their character to ease our own discomfort and uncertainty and force them into a neat and coherent story. An ethical response welcomes the Other as she is, with all her undetermined possibilities (Amiel-Houser & Mendelson-Maoz, 2015). As Biehl and Locke (2010) argued, the time has come to attribute to the people we encounter the complexities we acknowledge in ourselves.

This kind of empathy recognises the ontological distance between myself and the person I'm attempting to empathise with because their experience is not my own (LaCapra, 2001). Cummings (2016) wrote of the goal of empathy as not always closing gaps but also acknowledging them. On the pathway of translational empathy, we are only empathising with another when we acknowledge how deeply mysterious they are. We can only enter a relationship with another if we allow them to be truly Other. A genuine relationship requires difference. If we merely identify with the other, then no real relationship is possible, as there is no "Other" to have a relationship with. If I care for another simply because they are like me, this is no form of care at all (Slife, 2004).

As Sajnani (2012) wrote, "To improvise is to risk stepping into the unknown" (p. 81). All therapeutic encounters are improvisational (Kindler & Gray, 2010). The jazz alto saxophonist Benny Golson poetically explained how a creative individual "walks two steps into the darkness. Everybody can see what's in the light. They can imitate it, they can underscore it, they can modify it, they can reshape it. But the real heroes delve in the darkness of the unknown" (as cited in Green, 2003, p. 53).

In music therapy, we might assume that the broad range of expressive modes we are using enables us to meet and understand a client who may not be as comprehensible through verbal means and body language alone (Pavlicevic, 1997). While this may be the case, we can't assume that we can (or should necessarily attempt to) bypass all incomprehensibility. No therapist can experience a client's full range of emotions and states of mind at any given moment (Starcevic & Piontek, 1997). In Karasu's (1992) words, "the therapist who 'completely understands' the patient has stopped listening" (p. 71) and "there will always be more to know if [the therapist] keeps looking" (p. 80).

McKearney (2021) wrote about care workers' experiences at an organisation for people with intellectual disabilities in Britain. The new care workers struggled to understand people who communicated atypically during their first encounters. When they asked more experienced carers questions about how the residents may be feeling or what their behavioural expressions mean, the staff declined to answer them or addressed the residents directly, refusing to dispel opacity with transparency. The carers learned how to interact with the residents by waiting for them to communicate in unexpected ways rather than probing for "objective knowledge" about what may be going on inside their minds. They learned how to relate to opacity through assuming a stance that allowed others to be "mysterious," waiting with fine attention to see what a resident chose to reveal. This nourished an interactive space where people with disabilities emerged as agents and communicators in their own right with the ability to influence others.

The approach described by McKearney (2021) resonates with the perspective on music therapy in non-medical mental health provision articulated by Procter (2002). He critiqued medical mental health care that is structured such that diagnosis is the gateway to services, which potentially offers a framework to make sense of experiences, but is also laden with immense authority, potentially eclipsing the individual's experiences with the psychiatric system's view. Music therapists who work in multidisciplinary teams in medical settings can often read clients' files before meeting them and add their own documentation. Patients in psychiatric settings lose much of their privacy, putting them at a disadvantage to the experts who wield authority.

There are times in music therapy when our work affords a participant the space to have an experience that there is an Other with them. Verney (in Verney & Ansdell, 2010) discussed the case study of Edward, initially presented in *Creative music therapy* by Nordoff and Robbins (1977/2007). In Verney's view, as Edward screamed and the music therapist heard the intensity, rhythm, and tonal qualities of his scream and matched this musically, Edward gained an experience of being met by someone who was almost him, but not quite him. This afforded a gradual dawning awareness of otherness, which he then responded to. In her reflection on music therapy with children with ASD, Subiantoro (2018) commented on how therapists usually start by attuning to elements such as melody, rhythm, dynamics, contour, texture, and colour that the child initiates, to raise the child's awareness of the presence of another in the room with them. When this awareness dawns ("how I am is being reflected back to me by an Other"), interaction can begin to unfold.

Curious Humility

Many of us proceed through our lives assuming that we are essentially correct (about our assessment of situations, impressions of others, intellectual ideas, political convictions, moral and religious beliefs, memories, and grasp of facts). Our feeling of being right may move into the foreground (in an argument, for example, or when we make a prediction), but it often remains as psychological backdrop (Schulz, 2011). In contrast, translational empathy rests on humility. We may be wrong when assessing how another person feels. Elliott et al. (2011) wrote that therapists should neither assume that they are mind-readers nor that their clients should feel understood by their experience and expressions of understanding.

It requires humility to recognise that one's understanding of emotion is far from universal. Before the start of the nineteenth century, the word "emotion" was barely used. Emotions weren't discovered; they were created as a psychological category (Dixon, 2003). The Western psychology of emotions offers an authoritative scientific translation of emotions that rests on reification and embodies specific values and forms of social selfhood associated with capitalism, colonialism, and constructions of gender. This influences how emotions are described and understood and how they are experienced and performed.

Cultural psychology has emphasised that emotions are culturally constructed. Markus and Kitayama (1994) defined emotions as socially shared scripts, including various subjective, physiological, and behavioural processes that develop as people actively adjust to their immediate socio-cultural and semiotic environment. Different emotions are encountered, expressed, understood, and valued differently across cultures. Languages conceptualise emotions differently, and some emotions have no translated equivalent in other languages. Solomon (1995) contended that we would need to translate an entire culture, not merely a word, to understand how an emotion (and the term used for that emotion) belongs within a language, worldview, and way of life.

In a range of studies, participants from different cultures have inconsistently identified anger in music (Argstatter, 2016; Kwuon, 2009). Laukka et al. (2013) proposed that specific culture-specific cues serve as a kind of dialect communication. Individuals within the same culture who can understand this dialect have a greater chance of appreciating and communicating with the emotional meaning of music compared to listeners who have no experience with that culture. Susino and Schubert (2017) suggested that, when emotions are communicated in music, listeners perceive psychophysical cues through a stereotyping filter that primes a particular mental representation of that culture expressed through the music.

Cultural competence has been described as developing and maintaining culture-specific skills needed to function well in a new cultural context and interact effectively with individuals from different cultural backgrounds (Wilson et al., 2013). In therapy, cultural competence refers to a therapist's ability to recognise cultural and social contexts and the impact of colonialism, slavery, racism, and discrimination on communication between themselves and their clients (Kirmayer, 2008). To grow in cultural competence entails increasing one's knowledge of different cultures and worldviews and developing skills to implement music therapy processes that involve culturally sensitive techniques used in culturally sensitive ways (Sue & Sue, 2016). Showing respect and engaging skilfully when working with people from cultures different from one's own demands more than studying a range of instruments, learning songs in other languages, or reading up on various music cultures. Therapists also need to learn how each of their clients understands music and how their clients formed and are forming their musical identities (Hadley & Norris, 2016).

Critics of the notion of cultural competence have argued that it's impossible to learn about the entirety of the cultures of all the clients and groups one encounters in a diverse music therapy practice. In discourses of cultural competence, "culture" is often reduced to race and ethnicity, which are seen as residing in the "other," while dominant cultures are left unproblematised. Practitioners are positioned as members of dominant groups, and minority professionals can be made invisible. Power relations are considered to lie in the hands of an individual to alter. Proponents of cultural competence are criticised for understanding culture as uniform, unchanging, and overdetermining (in the lives of Others, not in the lives of the professionals attempting to be culturally competent) (Beagan, 2018). The political imperative for understanding "others" from the perspective of their "own" culture can also serve to reify "cultures" as fixed, explicable, and bound (Pedwell, 2014).

Comte (2016) identified how persons classified as refugees are usually referred to as a homogenous collective in music therapy literature, unified by a shared trauma narrative that is considered the most salient aspect of their experience. Therefore, it is assumed that all refugees will have similar needs, regardless of their religion or nationality. Understanding culture in terms of group membership risks gross over-generalisations, as it can ignore enormous heterogeneity within groups (Clark, 2000). Pedersen et al. (2008) understood culture more broadly as including ethnographic (nationality and ethnicity), demographic (age, residence, gender, lifestyle), status (social, economic, educational), and affiliation (formal and informal) categories. These aren't static, and an individual might change their cultural referent group within the course of a single interaction, from emphasising socioeconomic status to gender, to age, to ethnic affiliation. Each potentially salient and changing identity requires a different interpretation of that person's behaviour. Also, the same culturally learned behaviour could have varied meanings across people and even for the same person within different situations and times.

Competence is frequently measured by the confidence and comfort of the practitioner, which may not correlate with their actual ability to work inter-culturally. The notion of cultural competence places the practitioner in the powerful position of "expert knower" who needs to master information about other cultures (Kirmayer, 2008). A language of competence implies control rather than vulnerability and uncertainty within the music therapy encounter. "Competency" also suggests that we seek some kind of arrival. In a world of rapidly shifting dynamics, however, and in the ongoing and critical self-reflection required of us as therapists, such objective arrival points rarely apply (Schimpf, 2021).

Instead of attaining cultural competence, the idea of working towards cultural humility is being used more frequently (King, 2021). This refers to a person's ability and commitment to treating clients as experts on their own experience, recognising the limits of one's cultural knowledge, sustaining a lack of superiority, developing an awareness of cultural biases and values, and building an attitude of willingness to learn rather than presuming expertise (Wright, 2019). As an example of this stance, we can refer to Short's (2017) work with Marcus, whom she described as "a 35-year-old black British male with African heritage" (p. 153). Marcus showed a strong interest in rap; he lived in social housing and was unemployed, and he had experienced a psychotic episode. Short explained that "as a white, middle-class female" (p. 157) who had recently developed an interest in the genre of rap, she was aware of her need to remain aware of how her relationship to rap may differ from Marcus's. She worked on gaining a greater understanding of the complex social, political, racial, and emotional issues intertwined within this art form but stayed cognisant that her lived experience of the genre was different to that of Marcus, who had a deeply embedded and internalised relationship with the values and cultural roots of rap, as reflected in his lyrics.

Cultural humility permits therapists to consider and negotiate the value of alternative points of view, including those in opposition to their own perspectives (Wright, 2019). Spelman (1997) argued for an understanding of empathy rooted in the notion of apprenticeship. As an apprentice, we need to be prepared to receive new information constantly, change our actions accordingly, and develop in response to what the other person is doing, whether or not we like what they're doing.

Conventional forms of empathy can fix some people in place as "empathisers" through their privilege and ability to help while fixing others as those who need and receive help. Pedwell (2012) contended, instead, for the importance of mutual vulnerability that opens possibilities for both parties to influence and be influenced by each other. In the realm of translational empathy, it is not by practising "empathic accuracy" as a form of mastery that we enhance our sense of affective understanding, but by losing a certain amount of control and diminishing our authority (Pedwell, 2014). We give up the pursuit of cultural mastery and give in to being affected by what is other in the other person so that we can form a relationship of solidarity that ultimately transforms us both (LaCapra, 2001).

It is a form of imperialism to try to enter another's world imaginatively while holding our own as the norm (Eagleton, 2000). Alternatively, learning to engage in "loving perception" (Lugones, 1987) involves a challenging and disorienting process of leaning into the unfamiliar, different, and strange, embracing pluralities and not knowing. While music therapists claim that working through the medium of musicking can allow for intimate knowledge of another (Pavlicevic, 1999), there is also an acknowledgement that "in the realm of the non-verbal there is always an element of not knowing" (Warnock, 2012, p. 91). As music is ineffable, De Backer and Sutton (2014) urged music therapists to remain open to not knowing the complex, delicate, and intricate musical phenomena we are working with. We should never forget to respect what we cannot and do not know. In Salmon's (2008) words, music therapy is a "not-knowing-explorative-improvisational-play-space" (p. 8). As music therapists engaging in translational empathy, we need to be driven by continuous curiosity.

References

Akgün, A. E., Keskin, H., Ayar, H., & Erdoğan, E. (2015). The influence of storytelling approach in travel writings on readers' empathy and travel intentions. *Procedia-Social and Behavioral Sciences, 207*, 577–586.

Amiel-Houser, T., & Mendelson-Maoz, A. (2015). Against empathy: Levinas and ethical criticism in the 21st century. *Journal of Literary Theory, 8*(1), 199–218.

Argstatter, H. (2016). Perception of basic emotions in music: Culture-specific or multicultural? *Psychology of Music, 44*(4), 674–690.

Beagan, B. L. (2018). A critique of cultural competence: Assumptions, limitations, and alternatives. In C. Frisby & W. O'Donohue (Eds.), *Cultural competence in applied psychology* (pp. 123–138). Springer.

Biehl, J., & Locke, P. (2010). Deleuze and the anthropology of becoming. *Current Anthropology, 51*(3), 317–351.

Blankenburg, W. (2001). First steps toward a psychopathology of 'common sense' (A. L. Mishara, Trans.). *Philosophy, Psychiatry & Psychology, 8*, 303–315.

Boler, M. (1997). The risks of empathy: Interrogating multiculturalism's gaze. *Cultural Studies, 11*(2), 253–273.

Bradford, D., & Robin, C. (2021). *Connect*. Penguin Random House.

Brons, L. L. (2015). Othering, an analysis. *Transcience, A Journal of Global Studies, 6*(1), 69–90.

Clark, J. (2000). *Beyond empathy: An ethnographic approach to cross-cultural social work practice.* Unpublished manuscript, Faculty of Social Work, University of Toronto.

Comte, R. (2016). Neo-colonialism in music therapy: A critical interpretive synthesis of the literature concerning music therapy practice with refugees. *Voices: A World Forum for Music Therapy, 16*(3).

Cross, I. (2014). Music and communication in music psychology. *Psychology of Music, 42*(6), 809–819.

Cummings, L. (2016). *Empathy as dialogue in theatre and performance*. Palgrave Macmillan.

Davis, L. (2010). Constructing normalcy. In L. Davis (Ed.), *The disability studies reader* (3rd ed., pp. 3–19). Routledge.

De Backer, J., & Sutton, J. (2014). Therapeutic interventions in psychodynamic music therapy. In J. De Backer & J. Sutton (Eds.), *The music in*

music therapy: Psychodynamic music therapy in Europe: Clinical, theoretical and research approaches (pp. 338–350). Jessica Kingsley.
De Kock, K. (2003). *Experiencing time and repetition: Finding common ground between traditional and modern music therapy practices* [Master's dissertation, University of Pretoria].
Dixon, T. (2003). *From passions to emotions.* Cambridge University Press.
Drakulic, C. (1993). *The Balkan Express: Fragments from the other side of the war.* Norton.
Eagleton, T. (2000). *The idea of culture.* Blackwell.
Elliott, R., Bohart, A., Watson, J., & Greenberg, L. (2011). Empathy. In J. Norcross (Ed.), *Psychotherapy relationships that work: Evidence-based responsiveness* (2nd ed., pp. 132–152). Oxford University Press.
Emanuel, C. (2016). The disabled: The most othered others. In D. Goodman & E. Severson (Eds.), *The ethical turn: Otherness and subjectivity in contemporary psychoanalysis* (pp. 270–285). Routledge.
Epley, N., Keysar, B., Van Boven, L., & Gilovich, T. (2004). Perspective taking as egocentric anchoring and adjustment. *Journal of Personality and Social Psychology, 87*(3), 327.
Galati, A., & Brennan, S. E. (2010). Attenuating information in spoken communication: For the speaker, or for the addressee? *Journal of Memory and Language, 62*, 35–51.
Green, B. (2003). *The mastery of music: Ten pathways to true artistry.* Broadway Books.
Grey, F. (2016). Benevolent othering: Speaking positively about mental health service users. *Philosophy, Psychiatry, & Psychology, 23*(3), 241–251.
Hadley, S. (2013). Dominant narratives: Complicity and the need for vigilance in the creative arts therapies. *The Arts in Psychotherapy, 4*, 373–381.
Hadley, S., & Norris, M. S. (2016). Musical multicultural competency in music therapy: The first step. *Music Therapy Perspectives, 34*(2), 129–137.
Hodges, S. D. (2005). Is how much you understand me in your head or mine? In B. F. Malle & S. D. Hodges (Eds.), *Other minds* (pp. 298–309). The Guilford Press.
Hollan, D., & Throop, J. (2011). The anthropology of empathy: Introduction. In D. Hollan & J. Throop (Eds.), *The anthropology of empathy: Experiencing the lives of others in Pacific societies* (pp. 1–24). Berghahn Books.
Jaspers, K. (1997). *General psychopathology* (J. Hoenig & M. Hamilton, Trans.). Johns Hopkins University Press.

Johnson, J. L., Bottorff, J. L., Browne, A. J., Grewal, S., Hilton, B. A., & Clarke, H. (2004). Othering and being othered in the context of health care services. *Health Communication, 16*(2), 255–271.
Karasu, T. B. (1992). *Wisdom in the practice of psychotherapy*. Basic Books.
Kenny, C. (1999). Beyond this point there be dragons: Developing general theory in music therapy. *Nordic Journal of Music Therapy, 8*(2), 127–136.
Kindler, R. C., & Gray, A. A. (2010). Theater and therapy: How improvisation informs the analytic hour. *Psychoanalytic Inquiry, 30*, 254–266.
King, K. (2021). Musical and cultural considerations for building rapport in music therapy practice. In M. Belgrave & S. Kim (Eds.), *Music therapy in a multicultural context: A handbook for music therapy students and professionals* (pp. 43–74). Barcelona.
Kirmayer, L. J. (2008). Empathy and alterity in cultural psychiatry. *Ethos, 36*(4), 457–474.
Klyve, G. P. (2019). Whose knowledge? Epistemic injustice and challenges in hearing childrens' voices. *Voices: A World Forum for Music Therapy, 19*(3).
Kwuon, S. (2009). An examination of cue-redundancy theory in cross-cultural decoding of emotions in music. *Journal of Music Therapy, 46*(3), 217–237.
LaCapra, D. (2001). *Writing history, writing trauma.* Johns Hopkins University Press.
Laukka, P., Eerola, T., Thingujam, N. S., Yamasaki, T., & Beller, G. (2013). Universal and culture-specific factors in the recognition and performance of musical affect expressions. *Emotion, 13*(3), 434–449.
Lerner M. (1980). *The belief in a just world: A fundamental delusion.* Plenum.
Levinas, E. (1948). *Le Temps et l'Autre*. Presses Universitaires de France.
Lim, M. (2007). The ethics of alterity and the teaching of otherness. *Business Ethics: A European Review, 16*(3), 251–263.
Loewenstein, G. (2005). Hot-cold empathy gaps and medical decision making. *Health Psychology, 24*, S49.
Lugones, M. (1987). Playfulness, "world"—Travelling, and loving perception. *Hypatia, 2*(2), 3–19.
Markus, H. R., & Kitayama, S. (1994). The cultural shaping of emotion: A conceptual framework. In S. Kitayama & H. R. Markus (Eds.), *Emotion and culture: Empirical studies of mutual influence* (pp. 339–351). American Psychological Association.
McKearney, P. (2021). The limits of knowing other minds: Intellectual disability and the challenge of opacity. *Social Analysis, 65*(1), 1–22.

Metell, M. (2019). How we talk when we talk about disabled children and their families: An invitation to queer the discourse. *Voices: A World Forum for Music Therapy, 19*(3).

Nordoff, P., & Robbins, C. (2007). *Creative music therapy: A guide to fostering clinical musicianship* (2nd ed.). Barcelona Publishers.

Norris, M. (2019). *Between lines: A critical multimodal discourse analysis of Black aesthetics in a vocal music therapy group for chronic pain* [Doctoral dissertation, Drexel University].

Orth, J. (2005). Music therapy with traumatized refugees in a clinical setting. *Voices: A World Forum for Music Therapy, 5*(2).

Pavlicevic, M. (1997). *Music therapy in context.* Jessica Kingsley.

Pavlicevic, M. (1999). Thoughts, words and deeds: Harmonies and counter-points in music therapy theory: A response to Elaine Streeter's 'finding a balance between psychological thinking and musical awareness in music therapy theory: A psychoanalytic perspective'. *British Journal of Music Therapy, 13*(2), 59–62.

Pedersen, P., Crethar, H., & Carlson, J. (2008). *Inclusive cultural empathy: Making relationships central in counseling and psychotherapy.* American Psychological Association.

Pedwell, C. (2012). Affective (self-) transformations: Empathy, neoliberalism and international development. *Feminist Theory, 13*(2), 163–179.

Pedwell, C. (2014). *Affective relations: The transnational politics of empathy.* Palgrave Macmillan.

Procter, S. (2002). Empowering and enabling—Music therapy in non-medical mental health provision. In C. Kenny & B. Stige (Eds.), *Contemporary voices in music therapy: Communication, culture and community* (pp. 95–107). Unipub.

Ratcliffe, M. (2012). Phenomenology as a form of empathy. *Inquiry, 55*(5), 473–495.

Robbins, J., & Rumsey, A. (2008). Introduction: Cultural and linguistic anthropology and the opacity of other minds. *Anthropological Quarterly, 81*(2), 407–420.

Ross, L. D. (1977). The intuitive psychologist and his shortcomings: Distortions in the attribution process. In L. Berkowitz (Ed.), *Advances in experimental social psychology* (Vol. 10, pp. 174–221). Academic Press.

Said, E. (1978). *Orientalism: Western concepts of the Orient.* Pantheon.

Sajnani, N. (2012). Improvisation and art-based research. *Journal of Applied Arts & Health, 3*(1).

Sajnani, N., Marxen, E., & Zarate, R. (2017). Critical perspectives in the arts therapies: Response/ability across a continuum of practice. *The Arts in Psychotherapy, 54*, 28–37.
Salmon, D. (2008). Bridging music and psychoanalytic therapy. *Voices: A World Forum for Music Therapy, 8*(1).
Sass, L. (2007). Contradictions of emotion in schizophrenia. *Cognition & Emotion, 21*(2), 351–390.
Sass, L., & Parnas, J. (2007). Explaining schizophrenia: The relevance of phenomenology. In M. Chung, K. Fulford, & G. Graham (Eds.), *Reconceiving schizophrenia* (pp. 63–95). Oxford University Press.
Schimpf, M. (2021). Cultural humility in clinical music therapy supervision. In M. Belgrave & S. Kim (Eds.), *Music therapy in a multicultural context: A handbook for music therapy students and professionals* (pp. 157–184). Jessica Kingsley Publishers.
Schulz, K. (2011). *Being wrong: Adventures in the margin of error.* Granta Books.
Scrine, E. (2016). Enhancing social connectedness or stabilising oppression: Is participation in music free from gendered subjectivity? *Voices: A World Forum for Music Therapy, 16*(2).
Shintel, H., & Keysar, B. (2009). Less is more: A minimalist account of joint action in communication. *Topics in Cognitive Science, 1*(2), 260–273.
Short, H. (2017). "Big up West London Crew": One man's journey within a community rap/music therapy group. *Music Therapy Perspectives, 35*(2), 151–159.
Slife, B. D. (2004). Taking practice seriously: Toward a relational ontology. *Journal of Theoretical and Philosophical Psychology, 24*(2), 157.
Solomon, R. C. (1995). The cross-cultural comparison of emotions. In J. Marks & R. T. Ames (Eds.), *Emotions in Asian thought: A dialogue in comparative philosophy* (pp. 253–308). SUNY Press.
Spelman, E. (1997). *Fruits of sorrow: Framing our attention to suffering.* Beacon Press.
Spivak, G. C. (1985). The Rani of Sirmur: An essay in reading the archives. *History and Theory, 24*(3), 247–272.
Stanghellini, G. (2004). *Disembodied spirits and deanimated bodies: The psychopathology of common sense.* Oxford University Press.
Stanghellini, G., & Rosfort, R. (2013). Empathy as a sense of autonomy. *Psychopathology, 46*(5), 337–344.
Starcevic, V., & Piontek, C. M. (1997). Empathic understanding revisited: Conceptualization, controversies, and limitations. *American Journal of Psychotherapy, 51*(3), 317–328.

Subiantoro, M. (2018). The role of music therapy in promoting communication and social skills in children with autism spectrum disorder: A Pilot Study. *Advances in Social Science, Education and Humanities Research, 133*, 252–257.
Sue, D. W. & Sue, D. (2016). *Counseling the culturally diverse: Theory and practice* (7th ed.). Wiley.
Susino, M., & Schubert, E. (2017). Cross-cultural anger communication in music: Towards a stereotype theory of emotion in music. *Musicae Scientiae, 21*(1), 60–74.
Thomas, N., & Norris, M. (2021). "Who you mean 'we?'" Confronting professional notions of "belonging" in music therapy. *Journal of Music Therapy, 58*(1), 5–11.
Verney, R., & Ansdell, G. (2010). *Conversations on Nordoff-Robbins music therapy* (Vol. 5). Barcelona Publishers.
Warnock, T. (2012). Vocal connections: How voicework in music therapy helped a young girl with severe learning disabilities and autism to engage in her learning. *Approaches: Music Therapy & Special Music Education, 4*(2), 85–92.
Warren, J. T. (2008). Performing difference: Repetition in context. *Journal of International and Intercultural Communication, 1*(4), 290–308.
Wilson, J., Ward, C., & Fischer, R. (2013). Beyond culture learning theory: What can personality tell us about cultural competence? *Journal of Cross-Cultural Psychology, 44*(6), 900–927.
Wright, S. (2019). *Therapist cultural humility and the working alliance: Exploring empathy, congruence, and positive regard as mediators.* SUNY.
Zondi, S. (2021). A fragmented humanity and monologues: Towards a diversal humanism. In M. Steyn & W. Mpofu (Eds.), *Decolonising the human: Reflections from Africa on difference and oppression* (pp. 224–242). Wits University Press.

8

Awareness of Translational Processes

Along the pathway of translational empathy, while honouring the Other's opacity, we become aware of and tune into the translational process. Translation is always multidirectional, complex, and partial. Rather than reading the emotional truth that lies within our clients' expressions, we hear (through certain "filters") their emotional articulations as they communicate these in specific ways (consciously and/or unconsciously), and we are actively interpreting these as particular truths, situated within the given moment and a specific context. The client also actively receives and interprets our emotional communicative expressions as part of an ongoing translational and improvisational encounter.

Communicating Emotions

Emotional expressions can be richly communicative. Scarantino (2017) described how emotional expression accomplishes four main things. First, when we express emotion, we are providing information to others. Even if a person is clenching their fists and scowling while pretending to be angry, this nonverbal behaviour can still count as an expression of

A. dos Santos, *Empathy Pathways*,
https://doi.org/10.1007/978-3-031-08556-7_8

anger in the sense that it's a communicative move. Clenched fists and scowling make anger more probable as they hold natural information about anger, whether or not anger is actually instantiated. Natural information refers to a kind of meaning that refers to a state of affairs in the world. For example, smoke carries natural information about fire due to the correlation between smoke and fire upon which the recipient can infer the meaning (Scarantino, 2017). As a musical example, Wilce and Wilce (2009) reflected on Tolbert's (1990) ethnomusicological insights regarding the laments that were communally produced in Karelia (an area on the current Finnish-Russian border). Lament singers (or "cry-women") used micro-tonal and micro-rhythmic variations in their lamenting voices to manifest emotional intensity, which indexed spiritual power and the trance-like state of the lamenter's journey to the next world. The cry-women reported how, as lamenters, they projected their own experiences of grief through their voices as a way of carrying messages back and forth between the worlds of the living and the dead and orchestrating a collective experience of sorrow. Wilce and Wilce asked whether these cry-women were sincerely expressing a feeling. Was it their own? The laments that the cry-women performed didn't necessarily refer to feelings but were suffused with feelings. The women sobbed as they sang, with the "cry-break" in their voices functioning as an icon of real crying. Similarly, we may express emotions musically in music therapy as part of role-playing, which can still be considered emotional expression.

Second, humans express emotion to motivate another person to do something. When we express anger, for example, we demand that the recipient cease what they're doing and take us more seriously. To express happiness is to call on the recipient to celebrate our success with us. To express fear is to demand that the recipient protect and help us. Third, we use the emotional expression to declare how things are in the world. A "contempt face" may be a display of superiority, while a "sad face" could be a display of hurt. The emotional expression in both cases is a representation of what the world is like. Fourthly, we communicate emotions to commit to a subsequent course of action. Nonverbal bodily changes can display various "states of readiness" to act. The sender and receiver read and interpret each other's expressions within the exchange. If we are working from the premise that our knowledge of one another's inner worlds is always incomplete, then we can consider this a translational, not only interpretive, process.

The Vital and Responsive Contributions of All Participants

The transactional model of communication offered by Wood (2016) highlighted how interpersonal communication is always dynamic and collaborative. As relationships unfold, communication changes. Rather than thinking of discrete senders and receivers of messages, interpersonal communication involves people who play multiple simultaneous roles as both senders and receivers. No one is simply a recipient of another's message. We can send a powerful message even by silently listening to a client sing. Clients and therapists are always mutually influential communicators. Communicators are situated in fields of experience (which may have some overlaps but may also be highly diverse). Noise (literal noise, but also noise in the form of preconceived ideas, misinterpretations, stereotypes, and so on) plays a role in how messages—including emotional expressions—are conveyed and interpreted. Communicative moves take place in these complex contexts. Reddy (2001) suggested that when one expresses emotion in the presence of another, one is aware of one's own expression, one sees the other's reception, and one feels one's response to their reception. This process may result in confirmation or enhancement of the original emotional expression, or the encounter may produce the opposite effect, or no effect at all.

Some music therapy training programmes include immersion experiences. Pedwell (2014) critiqued the project of "immersions" in development training as an example of where mutuality can be neglected. A practitioner spends time (albeit three to four days) immersing themself in another community, hoping they will see (and feel) a particular "truth" or "reality" and will have a "moving," "challenging," "touching," or "disturbing" experience that will evoke empathy within them, leading to new ways of learning, self-transformation, the recognition of responsibility, and motivation for action that contributes to social change. There is an attempt to remove imagination from the process entirely in that through the experience of living and working with a low-income family in a developing context, one may experience their "reality" directly, thereby seeing more clearly how to proceed concerning practice and policy. In descriptions of immersions, there is often relatively little exploration of the hosts' feelings. The voices of immersion hosts are primarily framed as expressing gratitude to their guests for listening to them and paying attention to their lives. Hosts are not represented as cultivating empathy themselves, nor are they heard as articulating resentment, frustration, anger, or envy. Hosts tend to be attributed an affective register that is more limited than the emotions associated with their guests. Insufficient attention is paid to how hosts prefer to share their life stories and their "truths" to outsiders.

In translational empathy, we acknowledge all parties playing a mutually important role. While Carl Rogers' (1975) notion of empathy entailed "becoming thoroughly at home" (p. 3) in the private perceptual world of another person—which translational empathy does not assume is necessarily possible—his views also included an understanding of empathy as dialogue. Empathy involves communicative and cautious give and take, perceiving, reflecting one's understanding back for the other person to consider, and then engaging again, with sensitivity to changes in the other person. This entails continuous, multidirectional flow between two or more participants as they try to experience and understand their own and each other's emotions.

Cummings (2016) explored empathy in relation to theatre, explaining how the practice of theatre offers an understanding of empathy as a sense of activity, exchange, and play. She approached empathy as a collaboratively developed process, not as an emotional state to be achieved. The labour of empathy requires self-reflection and can be challenging, uncomfortable, and even scary. In this process, we don't necessarily experience the same emotion as the person we empathise with. Still, we consider their emotional experience while allowing their emotions to influence us. We draw information from that experience, reflecting on what we are feeling and why we may be feeling it. Our emotions, bodily sensations, and critical thinking intertwine and are mutually informing as we focus on the other and our responses to them. This framing of empathy is active and imaginative (not instantaneous and innate). When we think of empathy as dialogue, the question shifts from, "Did I empathise?" (in which empathy itself is the goal) to something like "What has this process of empathising provoked me to feel, think, wonder, or question?".

Aigen (2005) reminded music therapists not to forget that we're communicating, too. One can't base a session entirely on what the client is doing. While the client should undoubtedly inspire the music therapist's communications in the moment, we can only help clients develop communicativeness and relatedness if we are creatively offering them ourselves with whom they can communicate and relate. Clients hear our expressive and creative response to them (Verney & Ansdell, 2010). Swaney (2020) valued music therapy as a space where participants can be curious towards each other, offer each other invitations ("Shall we?" "Can we?" "Why not? "How about this?"), and celebrate and recognise one another. A communicative repertoire emerges with its own unique set of themes and variations.

If translational empathy is a collaborative process, what happens when the other party doesn't want to participate? In Chapter 11, we'll explore further the importance of being invited into an encounter of translational empathy (initially and in an ongoing manner). If another person does not wish to engage in translational empathy, then the attempt has failed, and so be it. Our task is not to impose an empathic encounter on another person.

Diversity in the Expression and Interpretation of Emotions

Human beings express and interpret emotions in diverse ways. In Chapter 3, we explored how mirror neurons are understood to play a role in the way that humans perceive and share in each other's emotions. Wahman (2008) argued that mirror neurons are not directly involved in communicating meaning about what the other experiences, feels, or intends. Instead, they enact a stimulus–response mechanism that is a kind of signalling. Meanings are not only determined by what we observe (the sign) but also crucially by our interpretations. We generate meanings through symbolic exchange and effortful deliberation as we translate our felt response towards a signal into something meaningful. If our mirror neurons fire at the sight of another's frown, for example, we need to add a layer of symbolic interpretation to that experience to gain some understanding of what the other is feeling (A frown could mean any number of things). We can't necessarily rely on our physiological responses to others' actions as a reliable basis for understanding their intentions. In fact, when we do so, we frequently misattribute others' intentions. The automatic firing of sensorimotor neurons serves as a condition for the possibility of understanding, but by itself, this will fall short of or stand in the way of shared meaning. Understanding is a joint, interpersonal achievement.

Music can mirror the structure and dynamics of emotional experiences. Langer's (1953) famous statement that music sounds the way emotions feel captures this view well. We can facilitate and structure emotional experiences through music to help an individual experience, identify, express, process, and modulate different emotions and receive the emotional communications of others (Wheeler & Thaut, 2011). Importantly, however, music therapists work with a wide range of clients who may engage in these processes differently due to diverse social, physical, and neurological features. As music therapists, we are also situated in social contexts that imbue understandings of how emotions "should" and "should not" be expressed and by whom.

Alexithymia is characterised by reduced emotional experience and expression, including an inability to talk about feelings (Cole, 2009). Almost ten per cent of people with no clinical diagnosis exhibit some alexithymic features, and alexithymic traits can be more pronounced in people suffering from a range of clinical conditions such as schizophrenia, ASD, neurodegenerative diseases, and eating disorders (Poquérusse et al., 2018). Individuals with end-stage dementia may still experience a wide range of emotions but have difficulties expressing them (Adams & Oliver, 2011). People with Parkinson's disease develop hypomimia (impairment of facial expressivity) and have difficulty recognising and interpreting others' emotional facial expressions, especially aversive ones (Ricciardi et al., 2017). Depression affects the recognition, expression, and regulation of emotions (Renneberg et al., 2005). A higher level of depressive symptoms is associated with more frequent use of expressive suppression (Campbell-Sills et al., 2006; Gross & John, 2003). Kan et al. (2004) found that people with depression interpret neutral prosodic emotive stimuli more negatively. There are also some indications that this negative bias may occur when people with depression evaluate music (Al'tman et al., 2000; Bodner et al., 2007).

There are debates in the literature regarding the experience and expression of emotions by individuals with intellectual disabilities. People with severe or profound intellectual disabilities may display more subtle facial expressions, which others may be poor at interpreting if they don't have experience doing so (Adams & Oliver, 2011). Children who are deaf and born to hearing parents may experience delays in emotional understanding, attributing emotions from situational cues, and deducing their causes (Jones et al., 2021). It's possible that the use of specific linguistic facial expressions as part of learning sign language may influence deaf children's expressive behaviour (Hosie et al., 1998). Some have argued that the appropriation of certain facial expressions for linguistic purposes may reduce their relevance for emotional expression (Most & Aviner, 2009), while others have contended that the use of certain facial expressions as linguistic markers enhances and refines deaf children's expressive knowledge and control (Gray et al., 2001).

Möbius syndrome is a rare congenital disorder leading to facial paralysis. It's sometimes associated with mild intellectual disability and comorbid ASD. Lack of facial movement then reduces embodied feedback. Also, without social reinforcement of emotional experience by others, an individual may struggle to recognise and internalise their own emotions (Adams & Oliver, 2011). Cole and Spalding (2008) offered an account of the experiences of Eleanor, who had Möbius syndrome. Eleanor described how she was a competent pianist by the age of 13 and found that playing the instrument unleashed emotions within her that could then find expression through music. As she played with the emotions she was expressing, she grew in her ability to experience them, too. Across all humans in our neurodiversity, we don't simply use music to express internal emotional states but to constitute those states reflexively. Music offers resources for knowing how we are feeling. As DeNora (2004) wrote, music provides material to build subjectivity. When listening to music, one may say to one's self, "Oh, yes, *this* is how I feel."

Socio-political factors play a role in relation to experiences, expressions, and interpretations of emotions, for example, anger (Zembylas, 2007). Holmes (2004) argued that there are three main anger discourses in Western societies: (a) Expressing anger is good, but this needs to be managed carefully; (b) it is dangerous to repress anger; and (c) it is good to repress anger because it shows that one is civilised. All three discourses share the assumption that emotion and reason are dichotomous, with anger denoting a lack of rational control and therefore needing to be managed. Anger management programmes tend not to make any clear distinction between various kinds of anger, particularly moral anger. In contrast, as Lyman (2004) wrote, anger can be defined as the essential political emotion, because it can motivate people to speak out against injustice and can be a tool for inspiring social transformation.

Status is a crucial factor for anger expression. Those with higher status are afforded more opportunities to express their anger (Sloan, 2004). Feminists have highlighted how anger frequently concerns men, not women. Men's anger is legitimated, but women's anger is criticised and controlled to maintain their subordination (Zembylas, 2007). Historically, women have been told not to express their anger overtly. Race and gender also intersect in shaping the performances of emotions like anger. As Jackson and Harvey Wingfield (2013) articulated, the stereotype of the "angry black man" is a common cultural representation that is frequently drawn on in media and popular culture. These authors explained how Black men in primarily white environments often manage this stereotyping to minimise its resonance with white peers.

Inequality is "written on the body" (Hawkesworth, 2016, p. 3). We can't explore how bodies express and interpret emotions within interpersonal exchanges without acknowledging that this process is intertwined in socio-political dynamics. Musicology has a history of failing to see how music is embedded in cultural contexts. This has led to essentialising music and separating it from the body, culture, dance, language, and so on (Rolvsjord, 2010). Music is not just situated in contexts but plays a role in constituting social class and social and cultural capital. All musicking is a political act (Small, 1998).

In Norris' (2019) study on Black aesthetics in a music therapy group for chronic pain, she highlighted how the conceptualisation of chronic pain has been widened to reflect how socialisation affects the pain experience. Black men, for example, are reported to experience lower pain intensity than other sociocultural groups. This is generally explained in relation to the socialisation of men to project strength, autonomy, and individuality and avoid emotional expression and vulnerability that may be interpreted as a weakness (Baker et al., 2017). Norris highlighted how Black masculinity has been historically situated in the United States within the socio-historical context of slavery, racial stereotypes, characterisations of Black men within the civil rights and black power movements, and current media representations of Black manhood. The male body racialised as Black aestheticises the stigmas of strength and bounded emotion. Healthcare disparities perpetuate low levels of access to pain management for Black populations. Norris worked with Black members of a music therapy group for persons with chronic pain, exploring with participants what it meant to them to be in this group as this body, marked as Black and physically disabled, and within systems of healthcare, therapy, and research.

Translation is a Mediated and Creative Process

Translation across languages is not simply about converting information represented in one language to precisely equivalent meaning in another. Equivalence would depend on the assumption that there is a shared, universal reality that stands outside of all languages and that language organises or fits this objective reality (Capan et al., 2021). Diverse cultures, however, often have very different notions of what reality, identity, childhood, ageing, health and illness, social isolation, or music-making mean (Kavaliova-Moussi, 2017; Ridder et al., 2017).

Instead, translation can be thought of as a canvas on which the translator paints as a creative process (Pound, 1963). When we engage in translation, we are constituting, not just conveying, information (Nergaard, 2020). It is always a mediated activity infused with issues of agency, power, ideology, identity, and representation. Through translation, different versions of fact, reality, and truth are enacted, reflected, mediated, (re)framed, (re)constructed, (re)narrated, contested, and even manipulated (Almanna & Gu, 2021).

Individuals position themselves in fluid ways during communicative exchanges (Gillespie et al., 2012). Instead of focussing on the idea of "the intercultural" (with an emphasis on cultural difference and meeting a cultural other), Dervin (2011) proposed an epistemological shift to "interculturality," which centres the *process* of these encounters. Interaction can't be determined in advance because it's an outcome of the process itself, which is always open-ended (Ferri, 2018). I don't know how we will be changed through our encounter. "Interculturality" unfolds through "the mangle of practice" (Pickering, 2010, p. 10). Multiple perspectives are continuously engaged in overt and covert ways. Intercultural music practices can activate diverse creativities, offering and supporting mutual reciprocity (Burnard et al., 2015). Songwriting, for example, can be a highly intercultural and cooperative creative process. Songs can hold the past through the power of identity and story; they can provoke the imagination and invite us to explore who we are and want to become; and they enable us to share these reflections with others through the translation of meaning (Bennett, 2015).

Empathy as translation can involve multiple and ongoing processes of musical, linguistic, temporal, cultural, and affective, differentiation, negotiation, and attunement, which is a far cry from striving to reach an accurate knowledge of the other, of direct equivalence or "dead repetition" (Deleuze, 1994). Through translational empathy, we are engaging in an imaginative production of new affective rhythms, languages, and relations (Pedwell, 2014). As we express and imaginatively interpret our own and each other's emotions, we produce ways of being together.

Translation in Music Therapy

Music therapists are familiar with the notion of translation. Bruscia (2001) described four relational layers within music therapy improvisations: intramusical (relationships between elements within the music itself); intrapersonal (relationships between a person's music and other aspects, such as mood, posture, movement, facial expressions); intermusical (relationships between the elements of one person's music and another person's music); and interpersonal (relationships between two or more persons taking place within and outside of the music). In addition, if one were to record one's musical improvisation and listen back afterwards, one creates a relationship between oneself and the recorded musical expression. When we incorporate other art forms, such as image-making or dance, into a session, relational layers expand further. We have also explored how such multi-layered interactions occur within social, political, and cultural contexts.

Within each of these numerous relational layers, we see translational possibilities (a rhythm could be translated into a melody; mood could be translated into a guitar improvisation; one musician plays a piano melody and another translates this onto the marimba; musicians translate each other's emotions through their music; a client moves their body expressively as a translation of their recorded improvisation). Turry and Marcus (2003) described how the beginnings of an improvisation can be a first translation as the music therapist converts what is felt, heard, and seen through aesthetic shaping. It is an act of intersemiotic translation (Pârlog, 2019) when a client draws aspects of their experience after a GIM journey, and the guide facilitates verbal processing of this image and experience (from music to pictures to words; and between two meaning makers). Priestley (1983/1995) wrote that translation between the language of music and the language of words is difficult and that "interpreters are few and far between" (p. 28). Music therapists, she suggested, have a role in building bridges between these two languages.

Sometimes we work with a client who speaks a language that is different from our first language as the therapist or from other group members. In their discussion of music in the cultures of Native Americans/first peoples, West and Kenny (2017) explained how meanings are difficult to understand fully when one has had little exposure to another's culture and no understanding of their language. Native music and language are intimately interwoven within the larger fabric of the culture. Shapiro (2005), an English-speaking music therapist in New York, described his work with Guy Chin, a 74-year-old woman in a nursing home who spoke Chinese. Shapiro attempted to imitate her singing phonetically. He explained that his prior experience listening to Chinese music helped him somewhat. Noticing her affinity for specific instruments and her enjoyment of dancing, Shapiro engaged with Guy Chin through his openness to appreciating, learning, and interacting with whatever she communicated in music.

In a discussion by Whitehead-Pleaux et al. (2017) on culturally sensitive music therapy assessments, one of the authors (Brink) presented her work with a 14-month-old girl from the Middle East who had suffered a severe burn from a roadside bomb. When Brink arrived, the girl was crying inconsolably in her mother's arms. After receiving consent from the girl's mother, Brink improvised with guitar and voice. She had no Arabic vocabulary, so she went back and forth between humming and singing common Western language syllables, such as "la." She played and sang while the child's bandages were changed. The girl appeared to respond positively and grew calmer. Later, Brink discovered that the syllable "la" directly translated to "no" in Arabic. Although this was a coincidence, she suggested that this may have resonated with the child's feelings in the moment.

There are times when a music therapist may work with a silent client. Ansdell and Pavlicevic (2005) described a music therapy process with Jay, who was detained in a locked psychiatric ward. Jay was unresponsive to attempts to communicate with her and was presumed to be electively mute. She had not spoken for weeks, being withdrawn from all social contact, her life reduced to stillness and silence. People who came into contact with Jay described her as unreachable, suggesting that the problem was not only that Jay wouldn't talk but that others couldn't find a way to talk with her. Attempts to communicate with Jay were verbal, took place within perceived professional relationships that were hierarchical, and centred on challenging questions. Ansdell began by listening to how she was and to the "music" of her stillness. Starting from where she was, he then gently invited her into a musical relationship. When Jay was musically invited instead of verbally questioned, the communication channel changed, and they found a communicative bridge. Jay received an invitation to participate rather than a demand to explain herself. The therapist offered to play with her, and Jay had the space and freedom to play around. The therapist's playing with Jay did not just entail attunement but also mis-attunement, allowing more explorative musical conversation possibilities.

References

Adams, D., & Oliver, C. (2011). The expression and assessment of emotions and internal states in individuals with severe or profound intellectual disabilities. *Clinical Psychology Review, 31*(3), 293–306.

Aigen, K. (2005). *Being in music: Foundations of Nordoff-Robbins music therapy.* Barcelona Publishers.

Almanna, A., & Gu, C. (2021). *Translation as a set of frames.* Routledge.

Al'tman, Y. A., Alyanchikova, Y. O., Guzikov, B. M., & Zakharova, L. E. (2000). Estimation of short musical fragments in normal subjects and patients with chronic depression. *Human Physiology, 26*(5), 553–557.

Ansdell, G., & Pavlicevic, M. (2005). Musical companionship, musical community: Music therapy and the process and value of musical communication. In D. Miell, R. MacDonald, & D. Hargreaves (Eds.), *Musical communication* (pp. 193–214). Oxford University Press.

Baker, T. A., Clay, O. J., Johnson-Lawrence, V., Minahan, J. A., Mingo, C. A., Thorpe, R. J., Ovalle, F., & Crowe, M. (2017). Association of multiple chronic conditions and pain among older black and white adults with diabetes mellitus. *BMC Geriatrics, 17*(1), 255–259.

Bennett, J. (2015). Creativities in popular songwriting curricula: Teaching or learning? In P. Burnard & L. Haddon (Eds.), *Activating diverse musical creativities: Teaching and learning in higher music education* (pp. 186–199). Bloomington.

Bodner, E., Iancu, I., Gilboa, A., Sarel, A., Mazor, A., & Amir, D. (2007). Finding words for emotions: The reactions of patients with major depressive disorder towards various musical excerpts. *The Arts in Psychotherapy, 34*(2), 142–150.

Bruscia, K. E. (2001). A qualitative approach to analyzing client improvisations. *Music Therapy Perspectives, 19*(1), 7–21.

Burnard, P., Hasslen, L., Jong, O., & Murphy, L. (2015). The imperative of diverse and distinctive musical creativities as practices of social justice. In C. Benedict, P. Schmidt, G. Spruce & P. Woodford (Eds.), *The Oxford handbook of social justice in music education* (pp. 357–371). Oxford University Press.

Campbell-Sills, L., Barlow, D., Brown, T., & Hofmann, S. (2006). Acceptability and suppression of negative emotion in anxiety and mood disorders. *Emotion, 6*, 587–595.

Capan, Z. G., dos Reis, F., & Grasten, M. (2021). The politics of translation in international relations. In Z. G. Capan, F. dos Reis, & M. Grasten (Eds.), *The politics of translation in international relations* (pp. 1–19). Palgrave Macmillan.

Cole, J. (2009). Impaired embodiment and intersubjectivity. *Phenomenology and the Cognitive Sciences, 8*(3), 343–360.

Cole, J., & Spalding, H. (2008). *The invisible smile*. Oxford University Press.

Cummings, L. (2016). *Empathy as dialogue in theatre and performance*. Palgrave Macmillan.

Deleuze, G. (1994). *Difference and repetition.* The Athlone Press.

DeNora, T. (2004). *Music in everyday life.* Cambridge University Press.

Dervin, F. (2011). A plea for change in research on intercultural discourses: A 'liquid' approach to the study of the acculturation of Chinese students. *Journal of Multicultural Discourses, 6*(1), 37–52.

Ferri, G. (2018). *Intercultural communication: Critical approaches and future challenges.* Palgrave Macmillan.

Gillespie, A., Howarth, C. S., & Cornish, F. (2012). Four problems for researchers using social categories. Culture Psychology, 18(3), 391–402

Gray, C. D., Hosie, J. A., Russell, P. A., & Ormel, E. A. (2001). Emotional development in deaf children: Facial expressions, display rules, and theory of mind. In D. Clark, M. Marschark, & M. Karchmer (Eds.), *Context, cognition, and deafness* (pp. 135–160). Gallaudet University Press.

Gross, J., & John, O. (2003). Individual differences in two emotion regulation processes: Implications for affect, relationships, and well-being. *Journal of Personality and Social Psychology, 85*(2), 348–362.

Hawkesworth, M. (2016). *Embodied power: Demystifying disembodied politics.* Routledge.

Holmes, M. (2004). Introduction. The importance of being angry: Anger in political life. *European Journal of Social Theory, 7*, 123–132.

Hosie, J. A., Gray, C. D., Russell, P. A., Scott, C., & Hunter, N. (1998). The matching of facial expressions by deaf and hearing children and their production and comprehension of emotion labels. *Motivation and Emotion, 22*(4), 293–313.
Jackson, B. A., & Harvey Wingfield, A. (2013). Getting angry to get ahead: Black college men, emotional performance, and encouraging respectable masculinity. *Symbolic Interaction, 36*(3), 275–292.
Jones, A., Gutierrez, R., & Ludlow, A. (2021). Emotion production of facial expressions: A comparison of deaf and hearing children. *Journal of Communication Disorders,* 106113.
Kan, Y., Mimura, M., Kamijima, K., & Kawamura, M. (2004). Recognition of emotion from moving facial and prosodic stimuli in depressed patients. *Journal of Neurology, Neurosurgery & Psychiatry, 75*(12), 1667–1671.
Kavaliova-Moussi, A. (2017). Discovering Arab/Middle Eastern culture. In A. Whitehead-Pleaux & X. Tan (Eds.), *Cultural intersections in music therapy: Music, health, and the person* (pp. 91–104). Barcelona Publishers.
Langer, S. K. (1953). *Feeling and form: A theory of art.* Routledge.
Lyman, P. (2004). The domestication of anger: The use and abuse of anger in politics. *European Journal of Social Theory, 7*(2), 133–147.
Most, T., & Aviner, C. (2009). Auditory, visual, and auditory–visual perception of emotions by individuals with cochlear implants, hearing aids, and normal hearing. *Journal of Deaf Studies and Deaf Education, 14*(4), 449–464.
Nergaard, S. (2020). Living in translation. In F. Fernández & J. Evans (Eds.), *The Routledge handbook of translation and globalization* (pp. 147–160). Routledge.
Norris, M. (2019). *Between lines: A critical multimodal discourse analysis of Black aesthetics in a vocal music therapy group for chronic pain* [Doctoral dissertation, Drexel University].
Pârlog, A. C. (2019). *Intersemiotic translation.* Springer.
Pedwell, C. (2014). *Affective relations: The transnational politics of empathy.* Palgrave Macmilllan.
Pickering, A. (2010). *The mangle of practice: Time, agency and science.* University of Chicago Press.
Poquérusse, J., Pastore, L., Dellantonio, S., & Esposito, G. (2018). Alexithymia and autism spectrum disorder: A complex relationship. *Frontiers in Psychology, 9*, 1196.
Pound, E. (1963). *Translations.* New Directions.
Priestley, M. (1983/1995). The meaning of music. *Nordic Journal of Music Therapy, 4*(1), 28–32.

Reddy, W. (2001). *The navigation of feeling: A framework for the history of emotions.* University Press.
Renneberg, B., Heyn, K., Gebhard, R., & Bachmann, S. (2005). Facial expression of emotions in borderline personality disorder and depression. *Journal of Behavior Therapy and Experimental Psychiatry, 36*, 183–196.
Ricciardi, L., Visco-Comandini, F., Erro, R., Morgante, F., Bologna, M., Fasano, A., Ricciardi D., Edwards, M. J., & Kilner, J. (2017). Facial emotion recognition and expression in Parkinson's disease: An emotional mirror mechanism? *PloS One, 12*(1), e0169110.
Ridder, H. M., McDermott, O., & Orrell, M. (2017). Translation and adaptation procedures for music therapy outcome instruments. *Nordic Journal of Music Therapy, 26*(1), 62–78.
Rogers, C. (1975). Empathic: An unappreciated way of being. *The Counseling Psychologist, 5*(2), 2–10.
Rolvsjord, R. (2010). *Resource-oriented music therapy in mental health care.* Barcelona Publishers.
Scarantino, A. (2017). How to do things with emotional expressions: The theory of affective pragmatics. *Psychological Inquiry, 28*(2–3), 165–185.
Shapiro, N. (2005). Sounds in the world: Multicultural influences in music therapy in clinical practice and training. *Music Therapy Perspectives, 23*(1), 29–35.
Sloan, M. M. (2004). The effects of occupational characteristics on the experience and expression of anger in the workplace. *Work and Occupations, 31*(1), 38–72.
Small, C. (1998). *Musicking: The meanings of performing and listening.* Wesleyan University Press.
Swaney, M. (2020). Four relational experiences in music therapy with adults with severe and profound intellectual disability. *Music Therapy Perspectives, 38*(1), 69–79.
Tolbert, E. (1990). Magico-religious power and gender in the Karelian lament. In M. Herndon & S. Zigler (Eds.), *Music, gender, and culture* (pp. 41–56). Institute for Comparative Music Studies.
Turry, A., & Marcus, D. (2003). Using the Nordoff-Robbins approach to music therapy with adults diagnosed with autism. In D. Wiener & L. Oxford (Eds.), *Action therapy with families and groups* (pp. 197–228). American Psychological Association.
Verney, R., & Ansdell, G. (2010). *Conversations on Nordoff-Robbins music therapy* (Vol. 5). Barcelona Publishers.

Wahman, J. (2008). Sharing meanings about embodied meaning. *The Journal of Speculative Philosophy, 22*(3), 170–179.

West, T., & Kenny, C. (2017). The cultures of native Americans/first peoples: The voices of two indigenous woman scholars. In A. Whitehead-Pleaux & X. Tan (Eds.), *Cultural intersections in music therapy: Music, health, and the person* (pp. 125–136). Barcelona.

Wheeler, B., & Thaut, M. (2011). Music therapy. In P. Juslin & J. Sloboda (Eds.), *Handbook of music and emotion* (pp. 819–848). Oxford University Press.

Whitehead-Pleaux, A., Brink, S., & Tan, X. (2017). Culturally competent music therapy assessments. In A. Whitehead-Pleaux & X. Tan (Eds.), *Cultural intersections in music therapy: Music, health, and the person* (pp. 271–283). Barcelona Publishers.

Wilce, J. M., & Wilce, J. M. (2009). *Language and emotion* (No. 25). Cambridge University Press.

Wood, S. (2016). *A matrix for community music therapy practice.* Barcelona Publishers.

Zembylas, M. (2007). Mobilizing anger for social justice: The politicization of the emotions in education. *Teaching Education, 18*(1), 15–28.

9

Responsively Being with Each Other's Emotional Expressions

One of the most fundamental needs in music therapy is for the therapist to be with the person who is coming to therapy (Bunt & Hoskyns, 2002). This sounds simple, but we know it isn't always easy. It can be challenging to sit alongside and stay with whatever feeling the person is expressing. If they are confused, in pain, or distressed it can be hard to resist the urge to shift their emotional state (perhaps to escape our own discomfort) or to use the music to fill the gaps with sound. This can be in direct contrast to what the client actually needs. They may want us to wait and to attend to their emotions just as they are, even if this means remaining silent and "just" being with them. As Lipari (2010) wrote, I don't need to "feel" what you feel, or "know" what it feels like to be you, but what I need to do is stand in proximity to your pain. I need to be right next to you, to stand with you, and to be fully present to the continuous expression of you. This requires me to let go of my ideas about who I am, who you are, and what "should" be, while remaining present, attentive, and aware.

Musical experiences create opportunities for practising negotiating the dynamics of how to be with another in a reciprocal exchange by alleviating some anxieties that could accompany the experience of expressing

A. dos Santos, *Empathy Pathways*,
https://doi.org/10.1007/978-3-031-08556-7_9

parts of self to another (Swaney, 2020). The therapeutic relationship can function as a collaborative partnership (Rolvsjord, 2016). A good relationship in therapy is something that evolves between the client and therapist, both of whom have strengths and weaknesses; it is not an intervention that the therapist carries out. Musicking together draws participants into the present moment where we don't know what will emerge. As we surrender to this process in an engaged, emotionally present, critical, and exploratory way, we invite new, creative possibilities (Cummings, 2016). According to Ansdell (2014), to improvise is essentially to engage in hospitality through musicking, inviting, and accepting all offers. Musical improvisation entails welcoming the unexpected as normal and following where music and people may lead. Musical hosts can become guests, and guests can become hosts. We are equal throughout as we support and respect each other.

Being with Another as a Witness

One of our key roles as music therapists is to hold the explorative "space" open in a safe and supportive way for any emotional responses that may arise. We hold space that welcomes experience, expression, and processing of uncertainty, vulnerability, pain, accounts of oppression and trauma, of loss and grief, decline and renewal. We hold space for hope, resilience, care, confidence, celebration, and transformation. As we hold this space, we invite but do not demand contributions (Scrine & McFerran, 2018) and we respectfully witness the contributions that are offered. As Nelligan et al. (2020) described, it is the experience of being accepted and witnessed in the creative process that transforms it into one that is conducive to healing. In a Canadian study by Black et al. (2020) on experiences of music therapy within medically assisted dying, for example, all the caregivers who participated explained how music therapy supported them in being able to witness their loved ones' emotional expressions and narratives.

As Gottesman (2017) explained, we can establish empathy through deeply witnessing the stories and emotions of another. Witnessing is not merely seeing with our eyes or hearing with our ears. It entails being provoked to full presence as we confront and mourn our previous understanding of the other's world. It's in these transformative moments that we can develop empathy, challenging and disrupting oppressive languages, categories, and frameworks so that together we can generate new structures and more equal social relationships through learning and co-creating.

Kopf (2010) argued that the creative artist is the most fully realised practitioner of the witnessing imagination. The arts compel us to expand our capacity for witnessing and the arts can heal through restoring meaning where this has been destroyed, integrating painful and suppressed experiences into a collective memory and giving victims of violence agency, voice, and dignity.

Witnessing offers a sense of confirmation and validation. As a therapeutic function, witnessing is founded on the "struggle to try to know the other while still recognising the other's radical alterity and unknowability" (Benjamin, 1988, p. 101). A client can experience safety through being witnessed, as well as separation without a promised fantasy of union (Poland, 2000). Witnessing is therefore a specific way of listening through which otherness is not only recognised but is required for the expression of the story (Berman, 2014).

While empathic acknowledgement is a core tenet of most therapeutic frameworks, it's specifically emphasised in trauma therapy models (Kaminer, 2006). The presence of an empathic witness is key to recovery from trauma because the essential experience of trauma entails a disintegration of the relationship between self and empathic other. Herman (1997) described the dialectics of trauma as the conflict between the wish to tell and the will to deny. Narration of trauma is also a highly complex process due to the presence of both the difficulties of telling and the difficulties of listening (Kopf, 2010). The importance of witnessing trauma is that it affords the creation of the yet untold story, and it's the readiness of the other who does not yet know (who will take the risk of finding their own safe assumptions troubled) that makes the testimony possible. As Laub (2013) wrote, it's only in the process of narration or bearing witness that the story of the trauma comes into being.

It's incredibly difficult to face trauma without an "addressable other" (Herman, 1997). Trauma narratives offer a process for healing through a trustworthy community of listeners willing to bear witness to the story (Aigen, 2012; Kaminer, 2006; Storsve et al., 2010; Zharinova-Sanderson, 2004). Survivors need to be able to trust that listeners will have the strength not to be overwhelmed by the story, will sustain a respectful attitude towards them, and will show compassion for their painful feelings (Shay, 1994). The main role of the therapist, then, is to offer a relational space where the trauma survivor's experience can be heard. Sometimes they may assume the role of a relatively passive container for the client's trauma story, and in other cases, they play a more active role as a collaborator in reworking the trauma narrative. An empathic witness helps a survivor to rebuild trust in the benevolence of others (Kaminer, 2006).

Within a humanistic psychotherapeutic approach, music therapists in Bensimon's (2020) study found seven relational needs to be most significant in therapy for clients who had experienced trauma: recognition, emotional witnessing, acceptance, emotional responsiveness, trust, safety, and someone to reach out. Through musicking, we can validate the client as they are. Clients' music can reflect inner representations of self, while also becoming an outer auditory object that invites dialogue. Clients are therefore both participants and observers in a process of expression and reflection.

Therapists are not neutral witnesses of a client's experience of trauma. They take sides as part of their stance of "withness," as they actively affirm that the events the client has experienced should not happen (Ullman, 2006). The therapist is the representative of a moral community that acknowledges the injury and affirms the hope for things to be different. The therapist needs to remain an Other who enables the testimony, instead of a self-object who becomes merged with the survivor (Orange, 2003). To witness is to see and to be engaged with another person's experience of traumatisation, in all of its enormity and complexity. One witnesses not only the traumatic experience, but also a life that was shattered and a life that could be rebuilt (Weine, 1996).

Withness in Musical Ambiguity

How can I be emotionally moved by that which I may not be able to know or feel? (as Ahmed [2014] asked). When we neither fully apprehend nor comprehend another's emotions, we can still empathise. Verbal communication entails a shared reference system, where particular sounds consistently represent particular concepts. Music, in contrast, does not require such specificity and consistency in order to be shared meaningfully (North, 2014). When people engage in listening to or performing music, they tend to experience it as meaningful, even though they may not agree on what the music means. Cross (2014) referred to this attribute as "floating intentionality." In musicking, interacting participants can each interpret musical meanings idiosyncratically without this evoking conflict between them. Music is an optimal medium for navigating situations of social uncertainty because a sense of joint affiliation can grow between participants as they experience an event as shared and yet deeply personal.

While clinical improvisation has been argued to offer a way of intimately knowing a client (Pavlicevic, 1997), perhaps we could think about this differently. Clinical improvisation can offer a way to be in the appropriately ambiguous space between intimate knowing and the radical Otherness of another whom we cannot fully know. When we engage in affect attunement within musical improvisation, we are translating cross-modally. A child may run around the therapy room with rapid, bright, energetic footsteps. We hear the musical elements in this action and we reflect the quality of their energy and movement back to them through playing an energetic melody in a major key on the piano, in the same tempo as their running. Pavlicevic (1997) drew on Stern's notion of vitality affects to explain how we accomplish this translation. According to Stern (2010), when we experience a dynamic event (one that unfolds over time), we perceive a form of vitality within the event. Where there's movement through force and intention, there is a creation of flow through time and space. Stern focussed on the qualities of these aspects as forms of vitality, giving examples such as surging, exploding, slipping away, floating, fleeting, and attacking. They are multi-modal: we might have a rush of ideas, a rush of movement, a rush of emotion. Music therapists match the forms of vitality that the client presents (through their movement, music, silence, speech, facial expressions, and so on) with musical responses (such as on a piano, or guitar, or with their voice) that express the same form of vitality in a cross-modal manner.

When I match and meet a client's forms of vitality, I both know and do not know. I may encounter the client's forms of vitality and match this musically, meeting them in their experience, but I may also not know where this quality "comes from," what it is "about," the meaning that the client makes of it, or the context in which this affective quality has arisen for this person. It could be argued that similar processes take place through imagery in art therapy, through movement in dance and drama therapy, and through metaphor in verbal forms of therapy.

Clients develop a relationship with the music and they develop a relationship through the music with the therapist. Access to music in music therapy is, thus, always access to the musical/personal. Verney and Ansdell (2010) explained that music in music therapy is not a "thing" in itself, and neither is it a transparent medium that merely opens the window onto the personal client–therapist relationship. Music acts as a third person in the room with its own character, offering its own help to the situation. In their text, Verney recounted her music therapy process with a young woman who was experiencing a mental illness and had little motivation or ability to relate to others. When they started their improvisation, the quality of their musical relatedness was separate and distant, fragmented and stagnant. Suddenly, the music and the quality of their relatedness became fluid and more dynamic. Verney argued that their personal relationship had not automatically been converted into one that was "full of flow." Working in the music offered this client the freedom to practise being in and out of the relationship. She could explore experiences in the music that were not directly associated with specific personal matters in her life. In musicking, familiarity is there and not there. Verney and Ansdell reflected on how this is qualitatively different from many interpersonal relationships that emphasise familiarity, shared experience, and known and dependable patterns. Verney's aim was to offer a state of meetings and provide a musical/personal process that could ebb and flow. Through this, the client could receive a deep experience of being accompanied.

North (2014) facilitated a music therapy process with a group of five young women, Jane, Clare, Mary, Libby, and Helen. Each used a wheelchair and had severe learning disabilities. They were non-verbal, but three could vocalise when they wanted to draw another person's attention, when they were excited or upset. Helen arrived at one of the sessions, vocalising repeatedly in a manner that sounded like distress and even evoked a sense of desolation. She had been doing so for three days. Her support workers found it hard to interpret whether she was communicating pain, anger, or unhappiness or whether she was caught in a pattern of vocalisation that she couldn't interrupt. In the session, Helen began to respond through her vocalisations to the high register of the piano. North heard the pitch of her notes and echoed them vocally. She heard Helen sing and repeat a high tonic. She sang the tonic herself, increasing the intensity of the music with a vocal crescendo and a slide down during the climax to an octave below. The other group members, who had been silent, started to vocalise when Helen stopped. Mary and then Libby and Clare were drawn into vocalising too, while Jane offered a sound that seemed like a sigh to release tension. The intensity of Helen's sounds affected them all.

In the process of translational empathy, we are influencing each other's emotions. As Starcevic and Piontek (1997) explained, interpersonal interaction involves a flow of emotions and information in all directions. North (2014) speculated that Helen felt accompanied and journeyed with, rather than alone. Mary, Clare, and Libby began to offer new sounds and North introduced elements of a Spanish musical style to invite pauses and flourishes as she tried to reflect qualities of the unfolding drama. She found herself sitting on the edge of her seat, waiting eagerly to see what would unfold next. North explained that it was challenging to describe what she felt and what the other group members may have felt upon hearing Helen's initial sounds. As they continued and the others began to vocalise, a heightened sense of creating emerged. Each individual appeared to offer carefully placed sounds in relation to the whole. The support workers who were present in the room also listened intently to each woman as she communicated.

Simultaneously Being Together and Separate

As we engage in musicking, we build a sense of togetherness while honouring our separateness. Procter (2002) described therapeutic co-improvisation as a dynamic interplay, involving two-sided interaction where meeting, reflecting, and challenging can take place. In the context of musicking with persons with intellectual and physical disabilities, Carlson (2016) highlighted how music can call us to recognise and respond to another separate human being. Within the musical unfolding, all voices are present and need to be given their due. As the music develops, each participant "speaks" in their own distinct voice while the sonic dialogue exceeds the individual parts and creates a shared connection.

One way that music allows for both uniqueness and togetherness is through entrainment. When people make music together, they coordinate their behaviours in time, even as other musical facets, such as their melody and timbre, may differ. Usually, a regular pulse structure emerges and each player organises their contributions around it, enabling them to anticipate and orient their behaviours to those of others. There isn't typically one timekeeper to whom everyone else entrains. A multidirectional process unfolds that involves continuous reciprocal adaptation and collective convergence.

Entrainment is not always precise. In fact, within the "groove" of many genres of music, being slightly "out of time" is essential (Keil, 1994). Groove requires negotiation of balance between difference and sameness, collaboration, and discrepancy (in timing, timbre, and tuning). Through experiencing groove, clients may have opportunities to move past the barriers they encounter in their daily lives that reinforce social isolation (Ansdell, 2017).

Stige (2003) suggested a useful model of group musical relationships that seeks to accommodate both musical differences and shared musical participation. While communal musicking creates a shared focus of experience and expression, each member takes part in their own particular way. The meaning of the situation also differs for each participant. Stige contended that, in this way, communal musicking may have a powerful potential for realising the community ideal of unity beyond uniformity.

Difference is necessary for group growth and creativity. In a study with teenagers referred to group music therapy for aggression, I noted how joint musicking created opportunities for generative interactional blending: participants felt a sense of belonging while maintaining their individuality (dos Santos, 2018). They collaborated by fitting their unique ideas together to create a joint musical product. This kind of intermeshing is similar to Pavlicevic and Ansdell's (2009) notion of multisubjectivity, where an individual loses and retains subjectivity within a "collective 'I'" (p. 369). This is particularly important for teenagers (especially in Western formulations of this developmental stage). Teenagers are understood to be in the process of developing their capacity for "autonomous-relatedness" (Oudekerk et al., 2015, p. 472). While interpersonal disagreement is inevitable in life, the ability to balance asserting one's autonomy with confidence while also preserving closeness is essential for negotiating conflict within healthy relationships. If adolescents don't develop this, they may be at risk for using hostile conflict styles or undermining others' autonomy, as well as for experiencing loneliness and depression in their adult relationships (Miga et al., 2012; Soenens et al., 2008). If a person has a longing for sameness and to fit into others' expectations of them, otherness may feel frustrating and even threatening. Most human relations oscillate between sameness and otherness as complementary and affording co-creation. This tension can fuel healthy interpersonal processes (Berman, 2014).

Withness can be challenging when one does not have a clear sense of self. Drawing on Object Relations Theory to discuss the development of an individual sense of self, music therapist Sandra Brown (2002) described her work with David, a five-year-old boy with ASD. As a music therapist trained originally in the Nordoff-Robbins tradition, Brown's approach was an improvisational one in which she hoped to work to enable the freeing and development of David's potential. David, however, increasingly limited her musical contribution. He decided when she could play, what she should play, and then stopped her from playing altogether. Brown felt increasingly that they were dealing predominantly with his anxieties about maintaining control over his world and she wanted to allow him space to explore this and build his trust in her. As their work progressed, David grew increasingly able to allow her to participate musically, although still strictly controlling what or when she could play. When they did play together, though, David's playing showed no apparent awareness or influence of hers and a mutually interactive relationship was not possible.

The initial moment of change occurred in a session when David stopped playing a few beats after Brown did. His playing then gradually became more affected by what she did (his pulse becoming erratic when she altered the tempo, for example). One day, David strained to move out of the "cage" of the piano stool, insisting that Brown hold his feet and not let go. After repeating this again in the next session, he then launched himself away, shouted "I'm gone!" and rushed over to play the piano loudly and triumphantly. Brown joined in, drumming on the back of the piano and singing as she celebrated with him. She interpreted this as David asserting himself and being "born" as a separate being. Before this, it seemed he had two ways of being with her: omnipotently controlling or disregarding her existence altogether. Brown drew on Winnicott's (1965) notion of the shift from an infant's merger with the mother to being separate from her and being able to relate to her as "not-me" (p. 45). As their work progressed further, David became able to listen and respond to Brown's musical changes. He introduced songs he knew for them to sing and play, also allowing Brown to choose songs. David and Brown enjoyed exploring ways of being separate and together, playing on instruments individually and simultaneously, exploring dynamic levels, tempi, and registers, widening David's musical resources, cognitive awareness, and ability to enter creative, shared play. Brown described how David's growing awareness of her as separate from himself had allowed for the beginning of a true coming together for them.

References

Ahmed, S. (2014). *The cultural politics of emotion* (2nd ed.). Edinburgh University Press.

Aigen, K. S. (2012). Community music therapy. In G. E. McPherson & G. F. Welch (Eds.), *The Oxford handbook of music education* (Vol. 2, pp. 138–154). Oxford University Press.

Ansdell, G. (2014). *How music helps in music therapy and everyday life*. Ashgate.

Ansdell, G. (2017). Reflection belonging through musicing: Explorations of musical community. In B. Stige, G. Ansdell, C. Elefant, & M. Pavlicevic (Eds.), *Where music helps: Community music therapy in action and reflection* (pp. 41–62). Routledge.

Benjamin, J. (1988). *The bonds of love: Psychoanalysis, feminism, and the problem of domination*. Pantheon.

Bensimon, M. (2020). Relational needs in music therapy with trauma victims: The perspective of music therapists. *Nordic Journal of Music Therapy, 29*(3), 240–254.

Berman, A. (2014). Post-traumatic victimhood and group analytic therapy: Intersubjectivity, empathic witnessing and otherness. *Group Analysis, 47*(3), 242–256.

Black, S., Bartel, L., & Rodin, G. (2020). Exit music: The experience of music therapy within medical assistance in dying. *Healthcare, 8*(3), 331–342.

Brown, J. M. (2002, March). Towards a culturally centered music therapy practice. *Voices: A world forum for music therapy, 2*(1). https://voices.no/index.php/voices/article/view/1601

Bunt, L., & Hoskyns, S. (2002). Practicalities principles of and music basic therapy. In L. Bunt, S. Hoskyns, & S. Swami (Eds.), *The handbook of music therapy* (pp. 19–45). Taylor & Francis.

Carlson, L. (2016). Encounters with musical others. In L. Carlson & P. Costello (Eds.), *Phenomenology and the arts* (pp. 235–252). Lexington Books.

Cross, I. (2014). Music and communication in music psychology. *Psychology of Music, 42*(6), 809–819.

Cummings, L. (2016). *Empathy as dialogue in theatre and performance*. Palgrave Macmillan.

dos Santos, A. (2018). *Empathy and aggression in group music therapy with adolescents: Comparing the affordances of two paradigms* [Doctoral dissertation, University of Pretoria].

Gottesman, S. (2017). Hear and be heard: Learning with and through music as a dialogical space for co-creating youth led conflict transformation. *Voices: A World Forum for Music Therapy, 17*(1).

Herman, J. (1997). *Trauma and recovery: The aftermath of violence: from domestic abuse to political terror.* Basic Books.

Kaminer, D. (2006). Healing processes in trauma narratives: A review. *South African Journal of Psychology, 36*(3), 481–499.

Keil, C. (1994). Participatory discrepancies and the power of music. In C. Keil & S. Feld (Eds.), *Music grooves* (pp. 275–283). University of Chicago Press.

Kopf, M. (2010). Trauma, narrative and the art of witnessing. In B. Haehnel & M. Ulz (Eds.), *Slavery in art and literature* (pp. 41–58). Frank & Timme.

Laub, D. (2013). Bearing witness, or the vicissitudes of listening. In S. Felman & D. Laub (Eds.), *Testimony* (pp. 77–94). Routledge.

Lipari, L. (2010). Listening, thinking, being. *Communication Theory, 20*, 348–362.

Miga, E., Gdula, J. A., & Allen, J. (2012). Fighting fair: Adaptive marital conflict strategies as predictors of future adolescent peer and romantic relationship quality. *Social Development, 21*, 443–460.

Nelligan, S., Hayes, T., & McCaffrey, T. (2020). A personal recovery narrative through Rap music in music therapy. In A. Hargreaves & A. Maguire (Eds.), *Schizophrenia: Triggers and treatments* (pp. 233–268). Nova Science.

North, F. (2014). Music, communication, relationship: A dual practitioner perspective from music therapy/speech and language therapy. *Psychology of Music, 42*(6), 776–790.

Orange, D. (2003). *The post-Cartesian witness and the psychoanalytic profession.* Unpublished manuscript.

Oudekerk, B. A., Allen, J. P., Hessel, E. T., & Molloy, L. E. (2015). The cascading development of autonomy and relatedness from adolescence to adulthood. *Child Development, 86*(2), 472–485.

Pavlicevic, M. (1997). *Music therapy in context.* Jessica Kingsley.

Pavlicevic, M., & Ansdell, G. (2009). Between communicative musicality and collaborative musicing: A perspective from community music therapy. In S. Malloch & C. Trevarthen (Eds.), *Communicative musicality: Exploring the basis of human companionship* (pp. 357–376). Oxford University Press.

Poland, W. (2000). The analyst's witnessing and otherness. *Journal of American Psychoanalytic Association, 48*(1), 17–34.

Procter, S. (2002). The therapeutic, musical relationship: A two-sided affair? *Voices: A World Forum for Music Therapy, 2*(3).

Rolvsjord, R. (2016). Five episodes of clients' contributions to the therapeutic relationship: A qualitative study in adult mental health care. *Nordic Journal of Music Therapy, 25*(2), 159–184.

Scrine, E., & McFerran, K. (2018). The role of a music therapist exploring gender and power with young people: Articulating an emerging anti-oppressive practice. *The Arts in Psychotherapy, 59*, 54–64.

Shay, J. (1994). *Achilles in Vietnam: Combat trauma and the undoing of character.* Atheneum.

Soenens, B., Vansteenkiste, M., Goossens, L., Duriez, B., & Niemiec, P. (2008). The intervening role of relational aggression between psychological control and friendship quality. *Social Development, 17*, 661–681.

Starcevic, V., & Piontek, C. M. (1997). Empathic understanding revisited: Conceptualization, controversies, and limitations. *American Journal of Psychotherapy, 51*(3), 317–328.

Stern, D. (2010). The issue of vitality. *Nordic Journal of Music Therapy, 19*(2), 88–102, 570.

Stige, B. (2003). *Elaborations towards a notion of community music therapy.* Unipub Forlog.

Storsve, V., Westbye, I. A., & Ruud, E. (2010). Hope and recognition: A music project among youth in a Palestinian refugee camp. *Voices: A World Forum for Music Therapy, 10*(1).

Swaney, M. (2020). Four relational experiences in music therapy with adults with severe and profound intellectual disability. *Music Therapy Perspectives, 38*(1), 69–79.

Ullman, C. (2006). Bearing witness: Across the barriers in society and in the clinic. *Psychoanalytic Dialogues, 16*(2), 181–198.

Verney, R., & Ansdell, G. (2010). *Conversations on Nordoff-Robbins music therapy* (Vol. 5). Barcelona Publishers.

Weine, S. M. (1996). The witnessing imagination: Social trauma, creative artists, and witnessing professionals. *Literature and Medicine, 15*(2), 167–182.

Winnicott, D. (1965). *The maturational processes and the facilitating environment.* International Universities Press.

Zharinova-Sanderson, O. (2004). Promoting integration and socio-cultural change: Community music therapy with traumatised refugees in Berlin. In M. Pavlicevic & G. Ansdell (Eds.), *Community music therapy* (pp. 233–248). Jessica Kingsley.

10

Negotiating Meanings of Emotions

When engaging in translational empathy, parties express their emotions (in a manner that they choose to or feel compelled to in the moment) and respond to each other's expressions, thereby developing emotional meaning(s) through their interaction. This is how translational empathy is productive. A conversation (verbal, emotional, musical, or otherwise) need not depend on shared meanings but on the shared exploration of meanings. A clash or gap between various meanings can provide an ideal space for mutual learning.

Clark (2000) examined empathy through the lens of ethnographic approaches to cross-cultural social work. Ethnographers are trained to notice "translation competence," where people translate their experience into a form they think outsiders will better understand. Ethnographers attempt to elicit "untranslated" speech and meaning by encouraging participants to tell their stories in their own words. They use strategies such as incorporating and re-stating key terms and phrases used by the respondent to foreground the respondents' voice in front of the professional's voice. Similarly, as therapists, we aim to hear a client's story openly, without attempting to force sense onto it based on our assumptions. We hold emerging understandings as tentative hypotheses that can

A. dos Santos, *Empathy Pathways*,
https://doi.org/10.1007/978-3-031-08556-7_10

be supported, changed, or disconfirmed by clients in an iterative and reciprocal exchange. It is at the end of the process that ethnographers examine how inductively derived understandings cohere with relevant theoretical materials. Meanings of experiences are "named" through collaborative processes. Ethnographic knowing is conversational, interactive, and invites further dialogue. Collaborative music-making in music therapy sessions offers parallels. We use open listening and create "sense" through a reciprocal, iterative, and negotiated exchange that invites continued musical conversations. Through this process, we produce the meanings of emotions in sessions.

Reflexivity

In addition to an ongoing exploration of the Other, translational empathy requires that I ask questions of myself. How am I situated within the emotional interaction and what are the implications of that? When engaging in the emotional moves of translational empathy, we benefit from reflexive awareness of how ongoing and multiple processes of cultures, languages, and time shape our interaction and how we are each influencing the potential creative unfolding.

The term "reflection" comes from the Latin *re-* meaning "back" and *flectere* meaning "to bend." It was first used to explain light bending back off reflective surfaces (Dallos & Stedmon, 2009). The term "reflective practice" has become prominent in contemporary clinical practice. It's a core component of training and best practice for many healthcare professionals. We can't stand outside the therapy act. Instead, from the "inside," we engage in personal reflection as an immediate and spontaneous act in the moment. We engage in this reflection-in-action not only during therapy sessions but also in supervision, teaching, and learning. Personal reflection usually includes self-awareness of emotions, bodily sensations, and any thoughts, memories, and experiences evoked in the moment.

Personal reflexivity is the act of reflecting on and looking back over action. This is primarily an intentional cognitive process during which we apply knowledge and theory to make sense of the process that has occurred. Reflexivity can also be a playful, creative, and artistic activity.

According to Etherington (2004), reflexivity refers to the ability to notice our responses to the world around us, other people and events, and to allow that knowledge to inform our actions, understandings, and communications. Reflexivity requires us to be aware of our responses and make intentional choices about using them. Multiple sources of prior knowledge are relevant to include, such as model-specific theories, and theorised understanding of our own situation and status in terms of class, gender, ethnicity, and autobiography. Chinn (2007) used the term "personal reflexivity" to refer to how a therapist "acknowledges how her own agendas, experiences, motivations and political stance contribute to what goes on in work with clients" (p. 13). A therapist's perceptions of power, privilege, and inequality are crucial aspects of both reflective and reflexive processes (Dallos & Stedmon, 2009).

Besides the importance of reflexivity in translational empathy for therapy, it's also vital for empathic researchers. Research is not a "view from nowhere" (Hesse-Biber, 2016, p. 19). Researchers have also been socialised into specific socio-psychological, socio-economic, and socio-cultural ways of being. They carry resultant values and biases, interpreting meaning through those lenses. Reflexivity is a form of critical thinking that enables researchers to notice, explore, and account for the contexts that shape research processes and the knowledge they produce. The control or elimination of bias is neither possible nor desirable. We can carefully incorporate subjectivity into research by making our biases explicit and acknowledging how they dictate our questions and interpretations (Clark, 2000). Emotional intensities can provide rich insight (Lazard & McAvoy, 2020).

Using Imagination and Creativity to Create Emotional Meaning

We use a combination of imagination-based and experience-based information to understand both individuals and contexts (and individuals within contexts) (Redmond, 2018). Through empathy, one imaginatively situates experiences that the other expresses within the context of their life and against the backdrop of their aspirations, hopes, commitments, projects, loves, fears, disappointments, successes, and vulnerabilities (Ratcliffe, 2017).

Knowledge of the other and imagination-based information have a symbiotic relationship as experience grounds imagination. The more information we gain about another person through our experiences with them, the greater our potential for generating a useful imaginative account. The less information we have, though, the greater the need to engage our imagination. We can also use our imagination to critique and reconsider the beliefs we've generated by interpreting experience-based information. For example, suppose we perceive another person expressing emotion in a way that we would not have imagined they would. In that case, this can motivate us to gather more information or expand our imagination.

Arts practices produce possibilities for creative and imaginative translation (Burnard et al., 2015). Halpern (2001) argued that the principal role of imagination in empathy is not to replicate another person's experience, but to move towards an increasingly holistic and cohesive appreciation of their experience. We achieve this by making connections rather than attributing an emotion to a client and understanding an experience in isolation. Imagination serves to unify nuances and details of a client's life "into an integrated affective experience" (p. 88). Although we can't assume that imagination allows us to know the experience of another, it can enhance our awareness of otherness, helping us think about and consider alternative perspectives and realities. It invites us to step away from what we have simply taken for granted, setting aside familiar definitions and distinctions (Waite & Rees, 2014).

In Metzner's (2004) discussion of receptive music therapy, she highlighted the importance of the therapist assuming an empathetic attitude when trying to relate to a client's inner world and described this as an imaginative endeavour. Empathic imagination doesn't take place in a vacuum but is informed by the empathiser's emotional interactions with the other. This triggers a network of associated images, memories, and meanings that are mapped onto the perspectives and experiences of the other to facilitate evolving understanding (Hollan, 2008).

Imaginatively exploring the idea that "this could have been me" is a valuable tool. For example, we may work with clients who have committed grossly violent acts. Gaining some understanding of their life history, family processes, and sociocultural, political, and economic circumstances may not afford accurate "knowing" of their inner emotional states. Still, it may enable us to conceive mentally of a landscape within which we could creatively place ourselves. In this landscape, we could imaginatively move around and gain a sense of how we might respond if this had been our journey.

Working with Differences and Conflict

The emotion translation moves of expression and response within translational empathy can include agreement and synchrony, but they can also involve conflict, difference, and negotiation. The purpose of such negotiation may not be to reach a consensus. Translational empathy can hold ongoing paradox.

Intersubjectivity (The Version that Focusses on Coordination)

In the previous part of this book, we met one version of intersubjectivity (that referred to shared agreement about an objective reality). Along the path of translational empathy, another version of intersubjectivity that is also present in the literature will be more helpful. Here, the term refers to coordination (generating a sense that we're in this together). In *Intersubjectivity without agreement*, Matusov (1996) explained that intersubjectivity entails a process of participants coordinating contributions during joint activity. He argued that conventional formulations of intersubjectivity overemphasise agreement and minimise the value of disagreement. They focus exclusively on what the participants have in common and on consensus-seeking. Integrative activities are highlighted where disagreement is viewed as an initial state that should be resolved or as an obstacle or nuisance along the way. They understand the process of sharing subjectivities as all participants holding the same vision of an activity so they can engage in sync.

Suppose we view intersubjectivity as a process of coordination of individual participation rather than as a relationship of correspondence between individual actions. This would allow for consideration of conflict, disagreements, and fights, as well as consensus. Some activities are destroyed when the efforts of all the individuals perfectly align (think of an orchestra). Matusov called for valuing diversity, disharmony, divergent perspectives, opposition, and resistance to communication. These don't need to be viewed as failed intersubjectivity attempts, but as a particular form of intersubjectivity. Disagreements and misunderstandings are no less relevant than agreements and similarities. Disagreement can prompt communication and instigate change. In addition, underneath disagreement, we often find an agreement that grounds the argument. Agreement and disagreement exist in a dialectic relationship.

Negotiation

From the perspective of intercultural encounters, Pedersen et al. (2008) defined inclusive cultural empathy (ICE) as the ability of counsellors to "accurately understand and respond appropriately to the client's comprehensive cultural context, both in its similarities and differences, which may include confrontation and conflict" (p. 42). While we could well argue that the focus on "accurate understanding" places ICE within the realm of insightful empathy, the pervasive valuing of differences and conflict (whilst still welcoming similarity) within complexity and ambiguity resonates strongly with our current pathway of translational empathy. As Pedersen et al. explained, an inclusive approach that acknowledges and honours both similarities and differences contrasts sharply with the exclusive emphasis on dissonance reduction found in conventional (insightful) empathy. Here, the therapist and the client are instead encouraged to accept the therapeutic relationship as it is, in its complexity and ambiguity. ICE does not demand the establishment of common ground but enhances the therapeutic relationship by including both the therapist's and the client's diverse perspectives because these contribute to the dynamic relationship.

Specific psychoanalytic perspectives have also theorised empathy as ambivalent. According to Phillips and Taylor (2009), empathy, like kindness, cannot (and should not) be rooted out from conflict and aggression. For the British psychoanalyst Donald Winnicott and Freud before him, fellow feeling and genuine kindness allow for conflict and ambivalence. In contrast, magical or false kindness distorts our perceptions of other people to avoid conflict. Authentically feeling with others includes ambivalence and conflict.

There are a variety of approaches to working with conflict in interpersonal and broader socio-political contexts. Conflict management is used when conflict is considered inevitable, and containment is seen as the most realistic approach. Conflict resolution seeks to transcend conflict by assisting parties to reframe their positions and find creative solutions. Conflict transformation seeks more than reframing and win–win outcomes (Miall, 2004). Here, conflict is seen as a natural feature of relationships and has the potential to bring about growth (Lederach, 2015). "Transformation" refers to the end product as well as the process itself and relates to changes in actors, issues, rules, and structures (Mitchell, 2002). American Professor of International Peace-building, Lederach, defined conflict transformation as viewing and responding "to the ebb and flow of social conflict as life-giving opportunities for creating constructive change processes that reduce violence, increase justice in direct interaction and social structures, and respond to real-life problems in human relationships" (p. 24). No change is possible without conflict (Ramsbotham et al., 2011).

By its very nature, music deals with manageable layers of conflict (Stock, 2018) within which the aesthetic interest lies (Robertson, 2018). The arts hold paradoxes of predictability and unpredictability, destruction and creation, as equally valuable (Kenny, 2016). While a group of musicians needs a common objective to accomplish some form or order, having too many shared structures interferes with creating newness (Sawyer, 2010). Some destructiveness or disintegration of structure enables new ideas. The flexibility of a community music environment, in particular, can stimulate openness to dialogue, discovery, and disagreement, inviting participants to work in the moment and to approach others curiously as creative agents (Impey, 2013; Middleton, 2018).

Several music therapy studies have examined expressions of destructiveness through musicking. Odell-Miller (2002), for example, described how a client could use a music therapy group to work through destructive relational tendencies by sounding anger and omnipotence and drowning out others' music. As expressions of aggression are heard and validated in music therapy groups, participants start to explore new forms of engaging within their social world (dos Santos, 2019). Jackson (2010) found that working through anger doesn't necessarily involve order, but it can deepen group relations.

Music therapy groups can hold multifaceted connections and conflicts between and within individuals, groups and subgroups, the therapist, music, and broader environments. Oosthuizen and McFerran (2021) discussed how music therapy groups in South Africa with young people of different languages, genders, and race groups became a microcosm for practising the kind of chaotic, uncomfortable connectedness necessary for developing shared hospitality in everyday life. If acknowledged and given space through healthy expression, chaos in group relationships could awaken the potential for "resilient adaption" (Arndt & Naudé, 2016, p. 267) and essential disruption of inhospitable and oppressive systems. Welcoming, or even inviting chaos into a group could be necessary for group transformation. Playing with the disorder through music can offer youth who have committed offences a way to negotiate contradictory identities of being perpetrators and victims, feeling powerful as well as vulnerable, seeking to develop healthy relationships within frequently dysfunctional contexts, or trying to move forward while processing remorse (Oosthuizen, 2019).

Sometimes clients are more familiar and at ease with chaos than music therapists. It can then be counter-productive to try to remove chaos from sessions to create ordered and highly structured groups because this is at odds with the contexts to which such clients will return. Nitsun (1996) described how the paradoxical interplay of creative and destructive forces in a group deepens acceptance and awareness of the natural flow of life with contradictions and frustrations. Through accepting paradox, multiple viewpoints and possibilities can be welcomed.

There is increasing interest in and application of embodied knowledge in the conflict transformation field. Our bodies play an integral role in the conflict. All decision-making processes have kinaesthetic elements that shape and are shaped by relational factors. LeBaron and Alexander (2012) showed how dance and movement could facilitate creativity, the ability to recognise and articulate deep feelings and needs, embrace new ways of knowing, awareness of habitual responses, and new connections. An increasing repertoire of possible conflict behaviours and new ways of being together can be developed. When we see relations between people in conflict as a dance, this removes simplistic, binary zero/sum notions that permeate much colloquial conflict language. Young and Schlie (2011) advocated the use of the dance as a metaphor for negotiation because it challenges dichotomies such as harmony versus war, fight versus flight, and partners versus adversaries. Dancing requires attunement to the environment and another's intentions, as well as sympathetic responses to the unexpected. Conventional conflict approaches that are often verbal, linear, deliberative, and disembodied are too limited to address the diversity of human modes of understanding. Understanding the self as multifaceted with multiple forms of intelligence is key to engaging diverse and complex individuals in conflict settings (LeBaron & Alexander, 2012).

Peace is a contested term (ironically). Howell (2018) emphasised that harmony requires scrutiny when drawn on in contexts of social divisions and conflict. It's important to explore what kind of harmony is being evoked, who is calling for it and why, who will enact it, who benefits and how, and who is excluded. When the intention is to deliver peace and justice rather than only surface harmony, these are crucial questions.

Peace-building is a dynamic process that focuses at the grassroots on the psychological, social, and economic environment to address underlying causes of violence and create peace built on equity, justice, and cooperation (Gawerc, 2006). Lederach (2005) described peace-building as a creative process, drawing on moral imagination, which is the ability to imagine something rooted in the struggles of the real world but capable of giving birth to something that doesn't yet exist. Peace-building is engagement with the unknown. The creative act of peace-building requires space. Lederach's process holds such space to engage with existing contestations and bring diverse voices, injustices, and constructive conflict to the table.

Lederach's peace-building ideas are central to community music and community music therapy practice. Musicking offers a form of dialogue across differences (Pruitt, 2011; Urbain, 2008). In contexts of peace-building, people from diverse and potentially hostile communities can begin to develop shared sonic vocabularies and engage in dialogue (Mullen, 2018). The flexibility and informality of community music environments can foster openness to new musical discoveries within participants and regarding others, such as viewing other participants as creative agents. Community music practice emphasises community participation, increasing access, and shared ownership, which resonates with the peace-building ideas of Lederach (2005). Transcending violence requires building webs of relationships that include the enemy, cultivating curiosity that can welcome complexity, pursuing the creative act, and being willing to embrace the risk of the unknown that lies beyond the violence.

References

Arndt, N., & Naudé, L. (2016). Contrast and contradiction: Being a black adolescent in contemporary South Africa. *Journal of Psychology in Africa, 26*(3), 267–275.

Burnard, P., Hasslen, L., Jong, O., & Murphy, L. (2015). The imperative of diverse and distinctive musical creativities as practices of social justice. In C. Benedict, P. Schmidt, G. Spruce & P. Woodford (Eds.), *The Oxford handbook of social justice in music education* (pp. 357–371). Oxford University Press.

Chinn, D. (2007, October). Reflection and reflexivity. *Clinical Psychology Forum, 178*.

Clark, J. (2000). *Beyond empathy: An ethnographic approach to cross-cultural social work practice.* Unpublished manuscript, Faculty of Social Work, University of Toronto.

Dallos, R., & Stedmon, J. (2009). Flying over the swampy lowlands: Reflective and reflexive practice. In J. Stedmon & R. Dallos (Eds.), *Reflective practice in psychotherapy and counselling* (pp. 1–22). Open University Press.

dos Santos, A. (2019). Empathy and aggression in group music therapy with teenagers: A descriptive phenomenological study. *Music Therapy Perspectives, 37*(1), 14–27.
Etherington, K. (2004). *Becoming a Reflexive Researcher.* Jessica Kingsley.
Gawerc, M. I. (2006). Peace-building: Theoretical and concrete perspectives. *Peace & Change, 31*(4), 435–478.
Halpern, J. (2001). *From detached concern to empathy*. Oxford University Press.
Hesse-Biber, S. N. (2016). *The Oxford handbook of multimethod and mixed methods inquiry*. Oxford University Press.
Hollan, D. (2008). Being there: On the imaginative aspects of understanding others and being understood. *Ethos, 36*(4), 475–489.
Howell, G. (2018). Harmony. *Music & Arts in Action, 6*(2), 45–58.
Impey, A. (2013). The poetics of transitional justice in Dinka songs in South Sudan. *UNISCI Discussion Papers, 33*, 57–77.
Jackson, N. A. (2010). Models of response to client anger in music therapy. *The Arts in Psychotherapy, 37*(1), 46–55.
Kenny, C. (2016). The field of play: A focus on energy and the ecology of being and playing. In J. Edwards (Ed.), *The Oxford handbook of music therapy* (pp. 472–482). Oxford University Press.
Lazard, L., & McAvoy, J. (2020). Doing reflexivity in psychological research: What's the point? What's the practice? *Qualitative Research in Psychology, 17*(2), 159–177.
LeBaron, M., & Alexander, N. M. (2012). Dancing to the rhythm of the role-play, applying dance intelligence to conflict resolution. *Applying Dance Intelligence to Conflict Resolution, 33*, 2.
Lederach, J. (2015). *Little book of conflict transformation.* Simon and Schuster.
Lederach, J. P. (2005). *The moral imagination.* Oxford University Press.
Matusov, E. (1996). Intersubjectivity without agreement. *Mind, Culture, and Activity, 3*(1), 25–45.
Metzner, S. (2004). Some thoughts on receptive music therapy from a psychoanalytic viewpoint. *Nordic Journal of Music Therapy, 13*(2), 143–150.
Miall, H. (2004). Conflict transformation: A multi-dimensional task. In A. Austin, M. Fischer, & N. Ropers (Eds.), *Transforming ethnopolitical conflict* (pp. 67–89). Springer Fachmedien Wiesbaden.
Middleton, I. (2018). Trust. *Music and Arts in Action, 6*(2). https://musicandartsinaction.net/index.php/maia/article/view/187
Mitchell, C. (2002). Beyond resolution: What does conflict transformation actually transform? *Peace and Conflict Studies, 9*(1), 1–23.
Mullen, P. (2018). Community music. *Music and Arts in Action, 6*(2), 4–15.

Nitsun, M. (1996). *The anti-group: Destructive forces in the group and their creative potential*. Routledge.
Odell-Miller, H. (2002). One man's journey and the importance of time: Music therapy in an NHS mental health day centre. In E. Richards & A. Davies (Eds.), *Music therapy and group work: Sound company* (pp. 63–76). Jessica Kingsley.
Oosthuizen, H. (2019). The potential of paradox: Chaos and order as interdependent resources within short-term music therapy groups with young offenders in South Africa. *Qualitative Inquiries in Music Therapy, 14*(2).
Oosthuizen, H., & McFerran, K. (2021). Playing with chaos: Broadening possibilities for how music therapist's consider chaos in group work with young people. *Music Therapy Perspectives, 39*(1), 2–10.
Pedersen, P., Crethar, H., & Carlson, J. (2008). *Inclusive cultural empathy: Making Relationships central in counseling and psychotherapy.* American Psychological Association.
Phillips, A., & Taylor, B. (2009). *On kindness.* Penguin.
Pruitt, L. (2011). Creating a musical dialogue for peace. *International Journal of Peace Studies, 16*(1), 81–103.
Ramsbotham, O., Miall, H., & Woodhouse, T. (2011). *Contemporary conflict resolution: The prevention, management and transformations of deadly conflicts* (3rd ed.). Polity Press.
Ratcliffe, M. (2017). Empathy without simulation. In T. Fuchs, M. Summa, & L. Vanzago (Eds.), *Imagination and social perspectives: Approaches from phenomenology and psychopathology* (pp. 199–220). Routledge.
Redmond, M. (2018). *Social decentering: A theory of other-orientation encompassing empathy and perspective taking.* De Bruyter.
Robertson, C. (2018). Musical processes as a metaphor for conflict transformation processes. *Music and Arts in Action, 6*(3), 31–46.
Sawyer, K. (2010). *Group creativity: Music, theatre, collaboration.* Routledge.
Stock, J. (2018). Violence. *Music & Arts in Action, 6*(2), 91–104.
Urbain, O. (2008). *Music and conflict transformation: Harmonies and dissonances in geopolitics.* I.B. Tauris.
Waite, S., & Rees, S. (2014). Practising empathy: Enacting alternative perspectives through imaginative play. *Cambridge Journal of Education, 44*(1), 1–18.
Young, M., & Schlie, E. (2011). The rhythm of the deal: Negotiation as a dance. *Negotiation Journal, 27*(2), 191–203.

11

Responding Through Accompaniment

In translational empathy, we take a stance of honouring opacity, including cultural humility and curiosity, and acknowledge that we can gain a sense of togetherness while still being separate individuals. We become aware of the emotion translation processes taking place in our interactions, exploring how we create meaning through our interaction, how parties may communicate their emotions in different ways, and that everyone involved plays a vital role. We responsively encounter each other's emotional expressions, collaboratively engaging in richly ambiguous communicative modalities such as musicking while recognising that social, political, and cultural dimensions impact how we express emotions and interpret each other's emotions in incomplete ways. As we navigate meanings in our unfolding emotional interaction, we realise the need for reflexivity as we critically negotiate meaning together, valuing conflict and the potential for conflict transformation as a life-giving process. Not all participants in an encounter may embrace (all of) these empathic facets. Translational empathy may only unfold partially. Also, the consequences are unpredictable as we mutually navigate our way through the process with neither party fully controlling the

A. dos Santos, *Empathy Pathways*,
https://doi.org/10.1007/978-3-031-08556-7_11

outcome. As we tie up this pathway by exploring the stage of "response," we'll examine the idea of accompaniment, which includes processes of affirmation, allyship, and accountability.

Companionship and Psychosocial Accompaniment

Musicality serves our need for companionship (Malloch & Trevarthen, 2009; Nordoff & Robbins, 1971, 2004). This begins from the start of life between caregiver and infant, who draw on musical qualities to engage with each other. Through the parameters of communicative musicality (pulse, quality, and narrative), human interactions become "a music-like composition, an improvised song or dance of companionship with someone we trust" (Trevarthen, 2002, p. 35). Trevarthen defined communicative musicality as "the dynamic sympathetic state of a human person that allows coordinated companionship to arise" (p. 21). The craft of a music therapist is to support and mobilise a client's communicative musicality in the service of finding and developing relational connection (Rolvsjord, 2006). Musical interaction in music therapy offers experiences of mutuality (Rolvsjord, 2004; Ruud, 1998).

Ansdell (2014) presented the example of Felicity, a music therapist working with an infant called Julia, who had significant developmental problems. As the child cried loudly, averting her gaze and arching away, Felicity didn't try to calm her down but cradled her and sang "Ju-li-a" in a short downwards phrase, matching the brief bursts of her crying, synchronising with her rocking and reflecting both the shape of her phrases as well as some clear pitches. Sometimes Felicity sang with Julia, matching the length of the bouts of crying; other times, she fitted in "Ju-li-a" as the child breathed between phrases. Ansdell heard this as a musical duet. They seemed to experiment with their duet for a while, and it became a shared, mutual interplay. Suddenly, they both paused. Julia stopped crying and turned her head to look Felicity directly in the eye. This is a moment of withness. Once the "spark" of withness presents itself, an unfolding companionship can become possible.

Moments of withness lay out stepping stones for a journey of more sustained accompaniment.

Ansdell (2014) also described moments in music therapy between Richard and a woman with advanced dementia (Ansdell didn't offer her name or a pseudonym for her). She had no language left, just a series of gestures. Within the care home, she was often distressed and isolated. She sat in a wheelchair, and Richard faced her. He called to her with open vowels in a rising melodic phrase. She sat up, moved her upper body towards him, and smiled broadly. Here, again, is withness. Richard mirrored her movement as he leaned towards her while repeating the phrase. She seemed to match her breathing to Richard's as he did so, tracking his actions by looking at and listening to him and synchronising her movements as the tones ascended and descended. In the next few minutes, each musical and physical gesture was spontaneously and exquisitely choreographed between them. There was synchrony between movement and sound. Richard matched the timing of his phrases to the pace she could manage. Her mouth was open as she wordlessly mirrored Richard's voice, her eyes shining. At the peak of the harmonic tension, she reached forward, clasped Richard's hand, and uttered a single, placed "Ah!" with a bowing gesture, acknowledging their experience of companionship.

In each of these cases, the music therapist helped to mobilise communicative musicality by hearing the person musically. Felicity and Richard picked up the precise musical dimensions of the gestures and sounds of their client. They treated everything their client did as having potential for musical communication. However, they did not lose their own musical voice. Their contributions helped to extend the musical conversations. Julia and the elderly lady emerged from their isolation in these moments as they entered companionship with Felicity and Richard. The two music therapists facilitated musical communication and interpersonal intimacy with their clients in contexts where isolation and non-communication predominated. Ansdell (2014) explained that the origin of the word "companionship," as breaking bread together, highlights the qualities of solidarity and mutuality through something being shared. As musicians, we also have the related notion of accompanying, as one musician supportively travels together through the music with another.

A notion of accompaniment appeared in liberation theology in Latin America and was absorbed into liberation psychology as "psychosocial accompaniment" (Watkins, 2015). Hunter (2020), a social worker who practised in El Salvador, explained how this concept of listening presence means to "be with" and to be present. It entails showing up in person and communicating: "I am with you in your pain." This kind of bearing witness can be a powerful act as it's a way of honouring people's realities and showing support. Psychosocial accompaniment involves a particular type of presence. Rather than attempting to "fix" or "intervene," re-inscribing colonial hierarchies of values and power, accompaniment involves listening, witnessing, advocacy, and critical inquiry. Sacipa-Rodríguez et al. (2009) explained that accompaniment should be directed towards affirming people as the lead in their own stories and reconstructing the social fabric of communities. We need to refrain from suggesting solutions to their problems on their behalf; instead, we intently listen to their own strategies. Liberation perspectives hold a commitment to treat the interests and perspectives of impoverished and oppressed communities as the primary source of insight about daily truth and social reality (Adams et al., 2015). Just as a musical accompanist listens acutely to the unfolding song, for example, the psychosocial accompanier takes a supportive role rather than occupying the limelight (Watkins, 2015). Australian Aborigine artist and elder Lilla Watson cautioned: "If you have come here to help me, you are wasting your time. If you come because your liberation is bound up with mine, then let us work together" (Indigenous Action, 2014).

Accompaniment entails walking "in the company of others" (Fanon, 2004, p. 238) who desire witnessing, collective imagination, advocacy, and action. An accompanier waits patiently for a clear invitation to be presented and withdraws if this is not offered. The accompanier needs to be honest and transparent about the function of their involvement, who will profit from it, and how. Those offering an invitation should have the fullest possible knowledge of who seeks to come and what their motivations are. Those within the inviting community are free to take part or not and are welcome to ask them to leave (Watkins, 2015).

Recognition and Affirmation of Empowerment and Expertise

Within the recovery approach outlined by McCaffrey et al. (2018), the starting point is the needs, preferences, and objectives of the person concerned. Treatment goals and approaches are determined collaboratively by the client and music therapist as far as possible. This relationship is built as a respectful and mutually trusting one (Slade, 2009). While expertise may be gained through training and skill development, it's also gathered through lived experience. This source of wisdom that the client carries is important for directing decisions about care.

Stige and Aarø (2011) offered the acronym PREPARE. At its core, community music therapy is participatory, resource-oriented, ecological, performative, activist, reflective, and ethics-driven. Participation occurs collaboratively; all who take part have the right to be welcomed and heard. The emphasis lies on empowerment and enabling through music-making, rather than pathologising people with disabilities and mental health issues (Procter, 2001). The facilitator steps back, and the participants influence the decision-making, taking responsibility for constructing their own boundaries and guiding their own processes. For example, Pavlicevic (2010) described a case study of a choir for children facilitated by two South African music therapists. The group members spent a session brainstorming norms together, which they continued to implement over the process with little input from the music therapists. Community music therapy programmes are not characterised by a lack of structure, but by a structure that emerges from the group itself.

As empowerment is a politically loaded concept, Rolvsjord (2006) highlighted the distinction between "power to" and "power over." "Power to" (make decisions, act, change) is grounded in values to do with respect, collaboration, and mutuality. "Power over" refers to traditional patriarchal forms of power. When conventional forms of empathy situate one individual in the powerful position of empathiser and another in the place of one who requires empathy, and assumes that the empathiser can fully know their inner emotional experiences, this can be a disempowering process. Hierarchical roles and the assumptions of those with more power can be re-inscribed. Translational empathy purposively troubles this power dynamic by seeking a more collaborative emotional encounter.

Being an Ally

The origin of the word "ally" dates back to c. 1300, from the Middle English term *allien*, the AngloFrench word *alier*, and the Latin *alligare*, meaning "to bind to" (Merriam Webster, 2014). Synonyms for "ally" include accessory, accomplice, associate, aide, assistant, colleague, co-worker, collaborator, coadjutor, confederate, helper, friend, and partner. A critical part of being an ally in social justice movements entails examining issues of privilege and oppression and working towards equality (Oswanski & Donnenwerth, 2016). Ayvazian (2010) described an ally as a person who is part of a dominant group in society and who works to dismantle any form of persecution from which they get an advantage. As a white, able-bodied, heterosexual music therapist, I have significantly benefitted from systems that advantage me and disadvantage many of the clients I am working with. My task as an ally extends beyond the walls of the therapy room. Oswanski and Donnenwerth (2016) argued that being an ally is not optional as a music therapist; it is an ethical requirement.

Bishop (2002) outlined distinguishing characteristics and critical attributes of allies, which Oswanski and Donnenwerth (2016) translated into music therapy practice. I would add that these characteristics connect strongly with the main components of translational empathy. Music therapy allies don't hold power or control over others but stand with others. Music therapy allies gather knowledge about their personal history, the histories of the groups they are part of, and how these intersect with the histories of other groups. They strive to explore and appreciate differences within community groups and organisations and try to understand systems of power and privilege and the reality of the struggle that a non-dominant person or group experiences. However, they are also acutely and honestly aware of their limitations and blind spots. Music therapy allies thoughtfully assess their privilege, influence, and identity markers and practice with respect for others. They know that their work is a series of progressive adjustments and alterations. Allies strive to create a space, place, and time inhabited by justice, empowerment, and equality for those who have been oppressed.

Accountability

Compassion is a term that can denote privilege. I—the compassionate one—have a resource that could alleviate the pain of the struggling sufferer (Berlant, 2004). In most Eurocentric notions of empathy, a subject with class, gender, race, and political privilege is seen to occupy the centre, comes across difference, and then elects whether to show empathy and compassion (Pedwell, 2016). Deciding to offer empathy can itself be a display of power. If empathy is understood as an experience of "co-feeling," problematic projections or appropriations on the part of privileged subjects can be invited, and complicity in the broader power relations through which suffering, oppression, and marginalisation occur can be obscured.

Efforts to cultivate empathic experiences by extracting individuals from structural relations are risky. Empathy is not necessarily about trying to "feel" or "know" another's emotions. It can be about "seeking to understand the structures of feeling and the feelings of structure that produce and mediate us differentially as subjects and communities who feel" (Pedwell, 2012, p. 294). Decolonising empathy calls for understanding conditions like structural inequalities and poverty that generate suffering, creating realistic opportunities for solidarity, engaging in specific actions to address the suffering of people experiencing the consequences of colonisation and oppression, and critically and honestly evaluating the effects of these actions (Zembylas, 2018).

Pedwell (2014) argued that if those who have played a role in perpetuating the political and social status quo want to take part in disrupting existing power relations, complicity must be acknowledged. "The past" remains alive in the present. This kind of empathy refuses to deny anger, shame, and sadness. Instead of being a "positive" emotion that is cultivated to pacify "negative" emotions, empathy then entails "a critical receptivity to being affected by ways of seeing, being and feeling that does not simply confirm what we think we already know" (p. 36). It's likely that recognising one's complicity in the suffering of others can elicit feelings of shame (Zembylas, 2018). Acknowledging shame, while difficult, can have transformative potential. Shame can sensitise an individual to transform that which has elicited shame in the first place. When empathy involves an ongoing and complex process of confrontation, attunement, and negotiation practices (Pedwell, 2016), it can become transformative, opening spaces for mutual influence.

References

Adams, G., Dobles, I., Gómez, L. H., Kurtiş, T., & Molina, L. E. (2015). Decolonizing psychological science: Introduction to the special thematic section. *Journal of Social and Political Psychology, 3*(1), 213–238.

Ansdell, G. (2014). *How music helps in music therapy and everyday life*. Ashgate.

Ayvazian, A. (2010). Interrupting the cycle: The role of allies as agents of change. In M. Adams, W. J. Blumenfeld, C. R. Casta-ñeda, H. W. Hackman, M. L. Peters, & X. Zúñiga (Eds.), *Readings for diversity and social justice* (2nd ed., pp. 625–628). Routledge.

Berlant, L. (2004). Introduction: Compassion (and withholding). In L. Berlant (Ed.), *Compassion: The culture and politics of an emotion* (pp. 1–14). Routledge.

Bishop, A. (2002). *Becoming an ally: Breaking the cycle of oppression in people* (2nd ed.). Fernwood.

Fanon, F. (2004). *The wretched of the earth*. Grove Press.

Hunter, C. (2020). I am with you in your pain: Privilege, humanity, and cultural humility in social work. *Reflections: Narratives of Professional Helping, 26*(2), 89–100.

Indigenous Action. (2014). *Accomplices not allies: Abolishing the ally industrial network*. http://www.indigenousaction.org/accomplices-not-allies-abolishing-the-ally-industrial-complex/

Malloch, S., & Trevarthen, C. (2009). Musicality: Communicating the vitality and interests of life. In S. Malloch & C. Trevarthen (Eds.), *Communicative musicality* (pp. 1–10). Oxford University Press.

McCaffrey, T., Carr, C., Solli, H. P., & Hense, C. (2018). Music therapy and recovery in mental health: Seeking a way forward. *Voices: A World Forum for Music Therapy, 18*(1). https://doi.org/10.15845/voices.v18i1.918
Merriam Webster. (2014). *Merriam Webster online.* https://www.merriam-webster.com
Nordoff, P., & Robbins, C. (1971/2004). *Therapy in music for handicapped children.* Barcelona Publishers.
Oswanski, L., & Donnenwerth, A. (2016). Allies for social justice. In A. Whitehead-Pleaux & X. Tan (Eds.), *Cultural intersections in music therapy: Music, health, and the person* (pp. 257–270). Barcelona Publishers.
Pavlicevic, M. (2010). Action: Because it's cool. Community music therapy in Heideveld, South Africa. In B. Stige, G. Ansdell, C. Elefant, & M. Pavlicevic (Eds.), *Where music helps: Community music therapy in action and reflection* (pp. 93–101). Ashgate.
Pedwell, C. (2012). Economies of empathy: Obama, neoliberalism and social justice. *Environment and Planning d: Society and Space, 30*(2), 280–297.
Pedwell, C. (2014). *Affective relations: The transnational politics of empathy.* Palgrave Macmilllan.
Pedwell, C. (2016). Decolonizing empathy: Thinking affect transnationally. *Samyukta: A Journal of Women's Studies, XVI*(1), 27–49.
Procter, S. (2001). Empowering and enabling. *Voices: A World Forum for Music Therapy, 1*(2).
Rolvsjord, R. (2004). Therapy as empowerment. *Nordic Journal of Music Therapy, 13*(2), 99–111.
Rolvsjord, R. (2006). Therapy as empowerment: Clinical and political implications of empowerment philosophy in mental health practises of music therapy. *Voices: A World Forum for Music Therapy, 6*(3).
Ruud, E. (1998). *Music therapy: Improvisation, communication and culture.*
Sacipa-Rodríguez, S. S., Guerra, C. T., Villarreal, L. F. G., & Bohórquez, R. V. (2009). Psychological accompaniment: Construction of cultures of peace among a community affected by war. In C. Sonn & M. Montero (Eds.), *Psychology of liberation* (pp. 221–235). Springer.
Slade, M. (2009). *100 ways to support recovery.* Rethink.
Stige, B., & Aarø, L. E. (2011). *Invitation to community music therapy.* Routledge.
Trevarthen, C. (2002). Origins of musical identity: Evidence from infancy for musical social awareness. In R. A. R. MacDonald, D. J. Hargreaves, & D. Miell (Eds.), *Musical identities* (pp. 21–38). Oxford University Press.

Watkins, M. (2015). Psychosocial accompaniment. *Journal of Social and Political Psychology, 3*(1), 324–341.
Zembylas, M. (2018). Reinventing critical pedagogy as decolonizing pedagogy: The education of empathy. *Review of Education, Pedagogy, and Cultural Studies, 40*(5), 404–421.

Part III

An Empathising Assemblage

Territorialising an assemblage as mutually affectively response-able.

12

A Stance of Acknowledging That We Are Always Already Entangled

As we leave the previous pathway and step onto a new one, we're turning from the radical other towards the radically relational. Instead of considering the starting point to be individuals (who then enter relationships with one another), we will perform an ontological flip and begin with relationships that then generate certain expressions of selves. Here, empathy won't entail getting "inside the skin" of a client but being "inside the skin of the relationship" (O'Hara, 1997, p. 18).

Assuming a relational ontology has significant implications. The idea of empathy can no longer explain how one unitary self shares in and understands the emotions of another; or how emotions can be translated between opaque individuals. Therefore, we need to re-conceive empathy in relational terms. You may wonder whether we're stretching the concept so far that it will become unrecognisable. I hope to make the case that all the core components of empathy are still present in this pathway. As we immerse ourselves in a relational ontology, it becomes helpful to think of emotion in broader terms as part of affect; as relationships are ontological, the empathic process takes place in the "between" (of assemblages) rather than "inside" individuals; and we can think of sharing,

A. dos Santos, *Empathy Pathways*,
https://doi.org/10.1007/978-3-031-08556-7_12

understanding, and being moved by the affect flowing between us as our capacity to respond, in other words, we are seeking to be "response-able."

In this chapter, we delve into a stance of acknowledging that we're always already relationally entangled. In the previous two empathy pathways (insightful and translational) we followed up the chapters on stance by concentrating on what we become aware of when empathising. Emotions and meaning subsequently followed. In our current empathy pathway, however, we'll follow up an examination of stance with a chapter on affective attunement, where we'll integrate a discussion on emotions and meaning. Only then will we explore what we become aware of, as all the necessary theoretical concepts will then be in place. This part of the book ends by investigating how we can make our practices more empathically response-able.

In this present chapter we'll draw on some ideas from a range of locations and people groups. While scholars should acknowledge diverse traditions for their role in articulating concepts that are frequently considered "new" in Western scholarship, we can't do so without recognising the potential for cultural appropriation. It's problematic to draw on Indigenous stories without Indigenous people present to inform their application, while it is also problematic for a Western thinker/writer to omit and erase Indigenous knowledge (Todd, 2016). It's with clear and intentional hesitance, therefore, that Indigenous Knowledge Systems should be represented in texts written by authors outside of these traditions (Wu et al., 2018).

Sundberg (2013, 2014) and Watts (2013) offered Euro-Western scholars practical tools for carefully and respectfully incorporating Indigenous ontologies into their work. Knowledge comes from somewhere, and, as a result, is always bound up in power relations. Situating self in relation to place and community affiliation is an important step towards avoiding essentialist conceptions of pan-Indigenous epistemology. Watts used the idea of "Place-Thought." Writing involves writing as part of a place and explicitly acknowledging how one is plugging in and out of the writing of others who are also engaged in "place-thought."

Notably, any use of "Indigenous" as an umbrella term covers an enormous range of sociocultural beliefs and practices. In fact, both

"Indigenous" and "Western" are polyvocal and polyvalent domains of discourse. A simple binary division between them reflects a Western Cartesian split that doesn't exist in many Indigenous cultures. Also, a significant number of indigenous people live within globalising Western cultures, experiencing and expressing selective appropriation, multiple and fluid identities, negotiation, and resistance (Shaw et al., 2006).

Strong Relationality

Relationality has weak and strong versions. From the perspective of weak relationality, people, things, places, and practices are considered self-contained units, and relationships involve reciprocal exchanges of information between them. In contrast, strong relationality is a kind of ontological relationality. Here, "relationships are relational 'all the way down'" (Slife, 2004, p. 159). Relationships come first. As De Quincey (2005) wrote, "we don't form relationships, they form us" (p. 182). Every entity, including each human being, is a nexus of relations. No element of the relation precedes the relation, and the property, quality, and identity of anything depend on how it is related (Selg & Ventsel, 2020).

Slife (2004) used the example of a stick figure. The circle at the top is only a "head" because of its relationships with the rest of the figure. If we erase the figure's legs, the image becomes the symbol used for females. A hammer and nail become what they are through their relationship. A person's spoken accent is only recognised as such due to its relation with the accent of other groups. Humour, as a further example, is a dynamic process that (re)constitutes the involved actors (for instance, as jokers, audience, laughers, and so on) within the dynamic relation they are part of. Also, jokes work only when everyone involved in the transaction knows the socio-political, historical, and cultural contexts that a joke taps into (Selg & Ventsel, 2020). In a piece of music, we experience a note in a specific way because of its relationship to the preceding and following notes (Levitin, 2006). Even silence in music gains meaning relationally. Silences, as Sutton (2005) argued, can be "a lot of different kinds of 'somethings'" (p. 378). Silence takes shape through relationships with context, musicians, listeners, notes, places, emotions, and so on, as, for example, vulnerable silence, unbearable silence, suspended silence, anticipatory silence, heavy silence, spacious silence, dead silence, ambivalent silence, or peaceful silence. When investigating relationships, the primary unit of analysis is the unfolding, dynamic process, not the constituent elements (Emirbayer, 1997).

Valuing Humans and Non-humans in Relational Webs

In *The self delusion*, Oliver (2020) highlighted how many of us see ourselves as discrete entities who are distinguishable and separate from our surroundings. Yet, from our physical bodies to our minds, we are inextricably linked to the world around us. The idea of an "I" that we affirm, protect, and nurture is merely an illusion. Our identity is contingent on who we are with, the time of day, and the qualities of the environment. Atoms "from the furthest reaches of the universe are gathered in our bodies" (p. 23). As walking ecosystems, we host and dynamically interconnect with thousands of species of fungi, bacteria, protozoa, and viruses. The boundary (if one can even call it that) between our body and the rest of the world is fuzzy indeed.

Relationships involve complex webs of humans and non-humans. This inclusive focus is by no means new. In Taoism, one of the three major religious and philosophical traditions in China, for example, humans and non-humans are considered intricately connected, mutually constitutive, dependent, and indivisible. Oneness is not sought; oneness is recognised (Wu et al., 2018).

Many Indigenous people groups' philosophical, cultural, and legal traditions are situated in profoundly relational ontologies (Long Soldier, 2017; Reddekop, 2014) and epistemologies (Lee Soon, 2016). Selves-in-relation are viewed as connected to each other and all living and non-living things as parts of a web of life (Kovach, 2012; Waterfall et al., 2017; Wilson, 2008).

Animism was coined by E. B. Tylor in 1871 as one of the earliest concepts in anthropology. Encyclopaedias of anthropology often present animism as a form of religious belief in which life or divinity is attributed to natural phenomena, such as thunder or trees. Tylor's stance was that "animists" view the world erroneously and childishly. Authors such as Bird-David (1999) and Harvey (2013) revisited the concept, challenging this singular and flawed understanding. They described animisms as plural, local, and specific. Instead of focussing on life "within," animism highlights the life "in-between" in terms of how persons (often of different species) relate together.

In the 1930s, Irving Hallowell conducted an ethnography of the Ojibwa in the Lake Winnipeg region of northern Canada. He observed that the Ojibwa used the overarching category "person" to include subcategories of "human person," "wind person," "animal person," and so on. It's considered possible and desirable to form a range of social relationships involving exchange, obligation, nurture, and empathy with many of these persons (Reddekop, 2014). To refer to a particular thing as animate is to describe its positioning within a relational field. The animacy of a rosebush can be understood in its having thorns, and therefore the capacity for self-protection. Spoons are categorised as animate, as they give shape to and contain the liquid they hold. We see clearly in this case how animacy functions as a relational concept: the spoon's animacy becomes apparent through its relationship with the liquid (Bird-David, 1999).

As Reddekop (2014) described, for Quichua-speaking people living in the Amazonian lowlands of Ecuador, humans, non-humans, and places become inseparable at many levels. The land is embedded in people, and people are embedded in the land. Quichua-speaking people refer to the notion of *yakichina*, a translation of which could be "causing another to feel empathy or compassion towards you" (p. 137). *Yakichina* emerges in interactional relations and cooperation. It tends to involve noticing the aloneness and helplessness of others who should be multiple, in other words, connected to supporting relatives. Responding to the contingencies and uncertainties of life to become a successful person, one needs to develop bonds of *yakichina* with a plethora of human and non-human others. One can then gain strength and attractiveness, medicines, do the tasks of gardening or hunting, and make and do beautiful things.

In Lowland South America, much ritual music is directed to and/or received from non-human beings such as animals, plants, or spirits

(Brabec de Mori & Seeger, 2013). Keller (2012) argued that when musical scholars exclude non-human animals, they can't fully describe "how musical is man," referencing John Blacking's (1973, p. 3) definition. Indigenous people groups have developed methods for listening to the music of non-human beings, studying their potentials and behaviours, translating this understanding, and bringing this into their communal, musical life.

Shifting our attention to the African continent, in former South African president Thabo Mbeki's (1996) speech "I am an African," he began by saying, "I owe my being to the hills and the valleys, the mountains and the glades, the rivers, the deserts, the trees, the flowers, the seas and the ever-changing seasons that define the face of our native land" (http://www.info.gov.za/speeches/1996/960819_23196.htm). Within African Indigenous perspectives, social reality is understood in terms of the self-in-relation to others, the environment (Murove, 2005; Prozesky, 2001), and the spirits (Chilisa et al., 2017; Nzewi, 2002). In African Cosmology, the idea of holism means that everything exists interdependently with everything else (Semenya & Mokwena, 2012).

At the heart of *ubuntu*, a notion shared by many Southern African people, is a relational, animistic epistemology (Taringa, 2020). *Ubuntu* is a multifaceted ontology, epistemology, and practice, described as a philosophy, a value system, a moral notion, a capacity, a manifestation of community (Mawere & Mubaya, 2016), and an obligation to promote harmonious relationships (Metz, 2007). Swanson (2012) explained *ubuntu* as humble togetherness, capturing caring, communality, and generative relating. Everything is in relation with every other being or thing and personal identity can't be separated from the environment as a whole (Church, 2012). Rather than asking "Who are you?", the question is "Who are we?" (Forster, 2010). According to Mbiti (1969), who wrote the seminal text *African religions and philosophy*, "whatever happens to the individual happens to the whole group, and whatever happens to the whole group happens to the individual. The individual can only say: 'I am because we are; and since we are, therefore I am'" (p. 108). Edwards (2010) added that *ubuntu*, in theory and practice, constitutes social empathic interaction through joint cultural dialogue, deepening communication and social healing. It is only through relations that meaning becomes possible.

Musicking is an embodiment of *ubuntu*, as it can create possibilities for kindness, collaboration, generosity, and support, which can extend to improving living conditions (Kagumba, 2021). *Ubuntu* supports Small's (1998) notion of musicking as a social process because individuals are seen as always situated in collective existence.

Drawing on indigenous ways of knowing within her "Field of Play" approach, Kenny (2016) identified an "ecology of being" for music therapists that included both the condition of the music therapy participant and the condition of the earth itself. In Kenny's view, there is a continual, dynamic interplay between the earth and ourselves within all situations, including music therapy. The earth gives energy to the people who take part in music therapy, just as they give energy back to the earth.

Western posthuman ways of knowing, doing, and be(com)ing are closely related to but not entirely synonymous with relational approaches in Indigenous Knowledge Systems (Sundberg, 2014; Tuck & Yang, 2014; Watson & Huntington, 2008; Zapata et al., 2018). It could be said that posthumanism spins itself on the back of Indigenous Knowledge Systems (Todd, 2016). "Posthuman" has become an umbrella term that includes Posthumanism, Transhumanism, New Materialisms, Antihumanism, Object-Oriented Ontology, Posthumanities, and Metahumanities (Ferrando, 2019). The posthuman emphasis on the importance of matter in cultural and social practice is connected with indigenous philosophies, which, as discussed above, vitally attune to matter (McPhie, 2019).

The post-anthropocentrism that is key within posthumanism is a response to the realisation that humans have been considered superior to non-human nature in Western thought, with catastrophic consequences for the environment. In posthumanism, humans are decentred, and humans and other-than-humans are granted equivalent ontological status (McPhie, 2019). The goal of posthumanism is not to get rid of humans but to see humans as part of relations with non-human and more-than-human subjects (Bogost, 2012). A posthuman approach to musicking, for example, engages "a broad spectrum of nonhuman and agentic actors in the process of developing new forms of meaning through music" (Woods, 2020, p. 18).

Even human subjectivity is inseparable from relationality with other people and things. Feelings are generated through interactions between self and world, not as properties of the self (Labanyi, 2010). As Braidotti (2011) wrote, subjectivity doesn't just involve an individual person; it's an open frame that includes the human and the non-human and even the planet as a whole (we see this in examples like the cyber-self; Gmail, online identities, or pacemakers).

Within New Materialisms, no division exists between language and matter: culture is constructed materialistically, and biology is mediated culturally. Matter is not understood as static, passive, or fixed, waiting to be crafted by an external force. Matter is instead an ongoing "process of materialisation" (Butler, 1993, p. 9). (A drum is being a drum; a bird is "birding.") When explaining her idea of "vital materiality," Bennett (2010) wrote that vital materialists seek to "linger in those moments during which they find themselves fascinated by objects, taking them as clues to the material vitality that they share with them" (p. 18).

McPhie (2018), a specialist in outdoor and environmental learning, taught an environmental ethics module where he asked his students to interview a building for which, he wrote, "I received funny looks" (p. 307). He continued to explain how, gradually, the students ventured into the world of inorganic life. Some decided to investigate the buildings' backstory; others tried to listen attentively to what the building said to them through its uniqueness. Afterwards, they made comments such as "A building (and lots of other things) can tell their stories in ways other than words…"; "It developed an atmosphere that was and still is friendly, welcoming, happy and supposedly deeply connected with its surroundings" (p. 308).

Deleuze and Guattari (1987) were highly influential in developing a philosophical posthumanist ontology. According to Duff (2014), a Deleuzian approach to studying health and development acknowledges a spectrum of material, non-human, and inorganic elements, which up to this point were mostly demoted to a nominal status of "context." Rather than being mere "props," non-humans are relationally intertwined with humans. According to Ansdell and DeNora (2016), music therapy research has tended to neglect the role of the material culture of social musicking. They described their project SMART (St. Mary Abbotts Rehabilitation and Training), involving community music therapy conducted at a community centre in London with people experiencing mental health difficulties. The direction and success of the musicking that took place were highly mediated by various musical instruments and their associated aspects, such as scores and texts, mallets, CDs, recordings, lyrics, costumes, microphones, amplifiers, and so on. Ansdell and DeNora couldn't gloss over these aspects as they tried to understand the "real music."

Interobjectivity (The Version of Intersubjectivity That Includes Interactions with Objects Within Assemblages)

Although intersubjectivity speaks to relationships, it still has an individualistic bias that arises from Western culture. Cartesianism remains at work. In our current empathy pathway, we understand living beings not as separate, self-contained entities but as part of ongoing, dynamic relations with their surroundings. Human relations, and human intersubjectivity, are framed, structured, and mediated by the use of objects. In Latour's (1996) view, objects are implicated in social reality as much as subjects. Human intention is lodged in objects. Think of a fence that keeps pedestrians off your lawn. The fence is in interaction with the person who encounters it. The fence is acting in its own right. We see this idea in music therapy all the time. Our intention develops in sessions through our interactions with instruments we play, the space we move in, and the paper and pastels we draw with. Therefore, we shift from intersubjectivity towards the concept of interobjectivity as we recognise not a world of separate things but a world of inseparable flowing processes (Shotter, 2013).

Extended Minds

In 1998, Clark and Chalmers co-wrote the now-famous paper entitled *The Extended Mind*, which began with a seemingly simple question: "Where does the mind stop and the rest of the world begin?" (p. 7). A family of (sometimes conflicting [Shapiro, 2010]) approaches have developed since then, which fall under the term "4E cognition." These involve the ideas that mind is embodied (thought is a "full-body" experience); embedded (mind is situated in a dynamic socio-material milieu); enactive (involving an autonomous and active process of "bringing forth" a world of meaning); and extended (involving technologies, objects, and other agents that form a cognitive ecology) (Van der Schyff & Krueger, 2019). The mind is not considered to entail processes that only reside in an individual's head (Gallagher, 2017). Instead, perception and cognition are ecologically situated actions (van der Schyff & Krueger, 2019). One's body extends into one's environment, just as the environment extends into one's body (Reynolds, 2018). Noë (2009) claimed that consciousness, like life itself, is a dynamic, world-involving process. Consciousness is achieved by a whole animal in its environment.

Elements of the world "outside" our heads can act effectively as "extensions" of our mind to enable us to think in ways our brains couldn't manage alone. Initially, Clark and Chalmers (1998) focused their attention on the extension of the mind through technology (an idea that originally received some criticism until the development of smartphones made their argument very hard to deny). In principle, there is no reason not to consider pen and paper, a wristwatch, musical instruments, landmarks, or a linguistic community as belonging to my mind. Why can't a caregiver be part of the mind of a person with Alzheimer's, for example?

Crossley (2017) discussed symbiosis between musicians and instruments that develops through the growing integration of instrumental and interpretative movements, forming a coherent whole. The instrument is increasingly absorbed into the player's reality, becoming an integral component of the extended body and the extended mind (Magnusson, 2009).

Music bridges the dualisms of "inner" (mind, body) and "outer" (the world). So-called higher cognitive processes emerge from physical interactions with the environment when musicking, without dualistic gaps. Already in the womb, we feel the relationship between movement, sound, and emotional states. Musical parameters such as pitch, tempo, rhythm, and melody are meaningful to us because they are compatible with the dynamic forms of our emotional, sensuous, and mental life (van der Schyff, 2013).

A child learns to restore their sense of calm when comforted by a caregiver. The caregiver facilitates the child's basic physiological processes, such as burping, by manipulating the child's posture. Caregivers also direct children's attention to particular things. The child only emerges gradually as an individual from this baby-caregiver relationship. For many of us, this separation is only partial. We don't completely detach from our community and from larger environmental situations and structures in relation with which we first emerged as a self. We are tangled up with the places we find ourselves as we are *of* these places and situations. When we experience a radical change to our environment, such as losing a spouse or immigrating to another country, we find these enormously impactful, perhaps even devastating. When we lose a feature of our environment in which our daily activities are intimately interwoven, we feel as if we've lost part of our self, because we have (Noë, 2009).

Rietveld et al. (2018) argued that a critical "E" is missing in the "4E cognition" model and called for the addition of "ecological." They named their framework the "skilled intentionality framework" (SIF), defining skilled intentionality as engaging selectively with a range of affordances in a specific, concrete situation. They used the term "affordances" to refer to possibilities for action that are present in an environment. Affordance entails the relationship between the particular features of the environment and things in it and the sensorimotor capacities that the perceiver uses to perceive and respond to these features.

Music listening, for example, affords movement, particularly in relation to entrainment and affective synchrony. When individuals entrain their movements with each other (getting into the "groove" together), they can offload some of their organisational and regulative processes onto the music and let it do a portion of this work for them. During episodes of deep listening, we may feel experientially consumed and transported by the music as we allow ourselves to let go and be taken on an affective, sonic journey (Krueger, 2014). When a person listens to a playlist to work through grief or anger, for example, they access an expanded spectrum of expressive capacities. Through this process, they may also learn something new about the character and causes of their experiences, as well as how to deal with them (Van der Schyff & Krueger, 2019). Music can sustain and sharpen our focus of attention. Listening to music while running can create greater ease of movement. Not only may we become distracted from unpleasant sensations, but music can act as a kind of surrogate agency in these situations as it assumes an external regulative function (Krueger, 2014).

When musicking together with others—in a jazz trio, string quartet, or ritual drumming and dancing ceremony—participants depend on each other to scaffold the extended musical environment as they take on and offload specific tasks such as entraining with a beat provided by the drummer. As Van der Schyff and Krueger (2019) emphasised, musical cognition can only be understood when we take into account the many ways we engage as extended beings with rich networks of bodily, emotional, material, social, and cultural scaffolding.

Immanence

Let's turn our attention back to posthuman perspectives, where we find the notion of immanence. While transcendence refers to being "on" or "outside" of the earth, immanence refers to being "of" the earth. A transitive cause produces an effect that is entirely distinct from itself. An example that is often used to explain this is the Christian notion of a creator God who is distinct from creation itself (even during the act of creating). An emanative cause is one in which an effect is produced, but the effect is not ontologically distant from the cause. A light source, for example, produces the effect of light but doesn't go "outside itself" to accomplish this. Light emanates from the source. We could think similarly of the relation between a mind and an idea (Adkins, 2018).

Deleuze's (2001) transcendental empiricism offered a way to explore life as an immanent process. ("Transcendental" and "transcendent" are quite different notions: the "transcendental field" refers to one that exceeds the bodies, ideas and entities that comprise it, while "the transcendent" refers to external causality.) Rather than examining "what is," Deleuze's empiricism seeks to explore how a subject is "becoming" (we will discuss this further in Chapter 13). The focus of transcendental empiricism is life in its becomings. The phrase "humans' relationship to nature" refers to an epistemology we would find in transcendent ontology. Within transcendental empiricism, we acknowledge that we are "always already of nature" (McPhie, 2019, p. 117). Therefore, it would be impossible to have a relationship *with* nature. We know the world because we are *of* the world; we are part of the world in its becoming (Barad, 2007).

We may choose to think here of a musical improvisation or music listening being *of* the environment, not taking place *in* an environment. In his article *Listening in the ambient mode*, Viega (2014) described his experience of listening to (becoming with) Keith Fullerton Whitman's (2002) album *Playthroughs* while walking through Central Park in New York City. In the ambient mode of being, one becomes immersed in raw materials of soundscapes (sonic environments) and shifts awareness nomadically across the environmental terrains, experiencing being in a liminal space while simultaneously being grounded in the here-and-now. Viega described weaving in and out of wooded paths surrounding a pond in the park's centre as he grew engulfed in the processed guitar drones. The urban landscape blurred almost seamlessly with the pond and pre-storm environment. As a low bass tone arose from the music, he felt newly and firmly grounded in the space in the here-and-now. Viega described feeling physically present to himself, to the vitality of the heartbeat of a whole city, and the zeitgeist of the era. He experienced fully living his own life through being in the presence of all that came before. The first falling snowflake called him into renewed motion as he continued walking his path.

What Are Assemblages?

We (humans and non-humans) are all immanently intertwined. One way of thinking about intertwining is through Deleuze and Guattari's (1987) use of the French term "agencement," which carries a range of meanings, including combining, arranging, ordering, and putting together. While a single English equivalent is insufficient, the term has been translated into "assemblage," understood both as a verb and a noun (Law, 2004). However, defining what an assemblage "is" doesn't really make sense, as the concept refers to an ongoing process of becoming (Lancione, 2013). One might think of an assemblage as a "web-in-motion."

An assemblage is composed of heterogeneous elements (human and/or non-human, organic and/or non-organic) that are lines of becoming. Still, it can't be reduced to the sum of the elements or disassembled to reveal the constituent elements. What matters is what unfolds between the parts of an assemblage (Deleuze & Parnet, 1987). As Deleuze (1995) famously wrote, "I don't like points" (p. 161). For Deleuze and Guattari (1987), "lines of becoming" weren't defined by their connected points. A becoming refers to the in-between.

We could think about a "musicking assemblage," comprising of at least hand—voice—instrument—listening; an "eating assemblage" of at least mouth—appetite—food–energy; a healthcare assemblage entailing at least pathology—healthcare professional—therapy—consulting room, and so on. Each assembled element transforms in its relations with the others so it makes little sense to refer to "constituent parts" (McPhie, 2019). Assemblages are "a temporary grouping of relations" (Coleman & Ringrose, 2013, p. 9) that are constantly moving and changing as an uncertain, unfolding, and hesitant processes. Assemblages are productive (Fox & Alldred, 2013), and so the question we ask is: What does this assemblage *do*? As a process of many, often unexpected connections, an assemblage creates ways of functioning. A productive assemblage generates new means of expression, new institutions, new spaces, new behaviours, and new realisations (Mazzei & Jackson, 2017).

Barad (2003) used the term "intra-action" to replace interaction. Interaction implies two pre-existing bodies entering participation with one another. Instead, for Barad, bodies and environments emerge through intra-actions. "Things" are always inseparably entangled, and assemblages are mutually constitutive becomings of intra-relations. We see this also in Indigenous ontologies, where the identity of "things" are not understood as being independent or discrete but emerging through (and as) relations with everything else. For example, a term for Maori people is *tāngata whenua*, which literally refers to land-earth-placenta-human. Each are seen to form the other (McPhie, 2019).

In Small's (1998) explanation of musicking, he showed how musical meaning isn't in the sound structures themselves but in the relationships created in and by joint music-making and reception. Ansdell (2014) emphasised how the purpose of music therapy is to form and grow interdependent ecological relationships that unfold in complex ways between people and sounds, people and people, sounds, people and places, and so on. Even apparently non-musical things (like emotions, movements, identities, communications, and events) "become" in relation to apparently musical things (like sound forms). For DeNora (2003), this is the realm of the para-musical (the things that go with, become alongside, and connect through or with music), including categories of mental, physical, individual, relational, social, and political phenomena. Para-musical things are incorporated into musick(ing), and the musical affords and constitutes the para-musical. Music acts in concert with these phenomena and the environments in which it is located, such that the musical and para-musical can rarely be separated. Ansdell urged us, therefore, to explore where music is perceived, experienced, performed, acted, and reflected upon to understand how it functions as a resource for us through how it gathers together musical people, things, and events.

As an example of assemblage thinking, let's look at Lancione's (2013) exploration of homelessness. People without homes (just like people with homes) are (becoming) assemblages, made of flesh and bones, emotions, thoughts, and desires, who move, change, and relate. The city is also an assemblage "where homelessness becomes a heterogeneous lived experience, and where the subjectivity of each individual is relationally constituted" (p. 359). In assemblage thinking, people without homes and the city are not considered as two separate categories. The people and the city continuously entangle, co-constitute, and co-affect each other. Processes of assembling are not neutral, however. In visiting a shelter or a soup kitchen, a person without a home will encounter practices, artefacts, and discourses that carry relational power.

Assemblages are made of other assemblages, and processes of plugging in and out distinguish which ones come into play at any given time. Parts of the assemblage can be disconnected and plugged into another assemblage, where interactions are different (McPhie, 2019). I am already an assemblage, and when I improvise in a session with a particular client, I am plugging into a different assemblage, with the capacity to affect and be affected in many other ways.

Deleuze and Guattari (1987) used the term "rhizome" (p. 12) to articulate the connections that occur between similar and diverse assemblages. One can think of a rhizome as a tuber or bulb compared to an arborescent, which is a taproot. A rhizome is flat and has many different entry points. Rhizomatic thinking is ceaselessly and unpredictably emergent (Valente & Boldt, 2015). The internet is an excellent example of a rhizome. It has no beginning or end; you can start anywhere and follow infinite trails that branch off in any number of ways.

Anthropologist Ingold's (2007, 2008, 2010) idea of "meshwork" relates closely to the notion of assemblages (and can be difficult to distinguish [Rival, 2014]). Ingold (2000) concentrated on the practices that people engage in as they inhabit their "landscapes" and "taskscapes." He highlighted how human bodies and environments intertwine and how social relations and cultural meaning emerge through embodied experiences shared with others in particular places.

In her exploration of music therapy with children and parents, Flower (2019) drew on Ingold's idea of meshwork. Flower wondered how she could make sense of the intertwining character of music therapy as a meshwork of interlinking trails. She argued for a radical realignment away from the conventional dyadic perception of music therapy practice (a therapist intervenes for the benefit of a client) towards an ecological attitude that considers music therapy as "a meshwork of interweaving lines of musicking, expertise, and emergence, within and beyond the therapy room" (p. 3). Music therapy is performed through the fluid relational interplay between children, parents, therapists, and others. If we understand music therapy as interacting contexts, it can be viewed as not only taking place within the therapy room. When borders between therapy and everyday life are considered active, then musicking travels freely. A family's musical life can find its way fluidly into sessions, and things and events of sessions flow out.

Distributed Agency

Agency is a notion that is conventionally described in Western thinking as being associated with an individual actor and as related to individual human will (Emirbayer, 1997). Residues of Enlightenment rationality hold that agency is exclusively reserved for humans, as we make autonomous choices. As a humanistic belief, conscious agency aligns with subjectivity (McPhie, 2018).

Enfield (2017) described agency (in terms of generating goal-directed controlled behaviour) as comprising both flexibility and accountability. Agency as flexibility involves being able to select a behaviour, its function, and its execution, and anticipating how others might view and react to the behaviour. Your flexibility depends on how much freedom you have to determine a course of behaviour and its outcomes. Agents have accountability in that they may be subject to being evaluated for their behaviour, they may be assessed in relation to whether they were entitled to carry out the behaviour, and they may also be perceived as having a degree of obligation to behave in a particular way. Your accountability depends on the degree to which others can be expected to interpret what you do in specific ways and how constrained you may be by natural laws.

In reality, "an agent" is seldom, if ever, an individual. The elements of agency related to a single course of action are usually divided up and shared among many people. When I ask you to pass the tambourine, I plan the behaviour, but you execute it. One person may provide the flexibility needed for meeting another person's ends (to carry out a leader's wishes, for example), or multiple people may be involved in meeting shared ends (such as in a jazz ensemble). One person may bear the accountability of another person's flexibility. The locus of agency may be a social unit, which is not confined to individual bodies. People often merge to form units of motivation, flexibility, and accountability. If you and I decide to write a song, we commit to a common course of action with joint reasons for that action. We may carry out different subcomponents of the job—you write the lyrics; I compose the melody—but we act as one, and in the end, we may agree to share the praise or criticism that we receive (Enfield, 2017). We see how agency can be distributed instead of bound within an essentialised, isolated, and subjective self who determines their own destiny (McPhie, 2019).

Agents are also not always human. Natural selection can be understood as an agent. A music therapist is an agent, but so too is a musical instrument. An enzyme in the human body is an agent. Lightning bolts are agents. We have seen how Covid-19 has acted as an agent. A hurricane may topple buildings or even administrations, geophysical changes shaped colonial encounters, worms regenerate agricultural land, wind carves out the landscape, cigarette butts play a role in starting fires (Country et al., 2016). Most entities and events have multiple causes (and each of these causes is an entity or event with multiple causes, and so on). Agency, then, is always distributed in depth (as each cause of some effect is also the effect of some cause) and distributed in breadth (each effect has multiple causes and each cause has multiple effects) (Enfield, 2017).

Posthumanist theorists, in particular, tend to emphasise that agency is not attributable only to human beings (McPhie, 2019). Vital materialists like Braidotti (2013) conceived of matter as containing its own agentic vitality as part of the self-organising, dynamic force of life itself. New materialists, such as Bennett (2010), explicitly ascribed vitality to inanimate matter (she wrote of "lively matter"). Bennett defined vitality as the "capacity of things—edibles, commodities, storms, metals—not only to impede or block the will and designs of humans but also to as quasi agents or forces with trajectories, propensities, or tendencies of their own" (viii). Agency is therefore diffused across many entities and achieves its capacity within assemblages.

The notion of intra-action (Barad, 2003), as discussed above, also helps us understand how agency is distributed in an assemblage. Intra-action captures how agency emergences from relationships between components rather than emanating from components themselves. Empathy can be understood as an "affective, intra-active encounter" (Wilde & Evans, 2019, p. 797).

New materialist theorists clarify that deconstructing the humanist legacy of individual agency does not dissolve humans as identifiable agents altogether; it encourages an expanded sense of connection between self and human and non-human others. We don't lose all accountability. We just recognise that the story is complex, with multiple casual elements that intertwine. Every living thing is an entanglement: "To be is to be related" (Mol, 2006, p. 54). Humans and non-humans have always been involved in an intimate dance with each other. As Bennett (2010) wrote, human agency has never been "anything other than an interfolding network of humanity and nonhumanity" (p. 31). If we only value autonomy, self-governance, and rationality when explaining agency in therapy, we can miss the potential for change in the material environment (McPhie, 2019).

Accounting for human and non-human forms of agency enables consideration of new empowering capacities (Barad, 2003; Thrift, 2008). When we consider ill-being, for example, as an internal state, we might more readily blame individuals for their condition. Authors like Duff (2014), McLeod (2017), and McPhie (2019) shifted attention from the individual onto the collective body. Health is an assemblage of non-human and human forces. Recovery is sustained in particular territories. Is it, in fact, the assemblage that recovers.

In their Norwegian study on agency and dementia care, for example, Ursin and Lotherington (2018) understood care as a relational, world-forming process of collective action. Persons needing care were also seen as active agents. Care practices, as localised human-material relations, are social and political processes that form daily life for people with dementia, their caregivers, and the society in which they live. Urson and Lotherington examined links between multiple care relations, knowledge, and citizenship, demonstrating how agency emerges as an effect of care in practice. They described care-collectives that were malleable, heterogeneous networks shaped by and shaping relationships between the socio-material agents that constitute them.

What Is Meant by "Territorialising"?

Deleuze and Guattari (1987) described a child in the dark who is gripped with fear. He begins to sing under his breath to comfort himself. The song is calming stability within what feels scary and chaotic. The song establishes a space of comfort. What is essential is that the song has repeated elements within sound patterns. Through the singing of the song (through the repetition of simple phrase structures), a milieu is created because the song acts to mark the space as calm and stable. Deleuze and Guattari described a milieu as "a block of space–time constituted by the periodic repetition of the component" (p. 313). Through song, birds mark space (as a zone of influence) with sound. When a car drives by with a pounding, heavy bass playing, this shapes the space and temporarily changes the space's character. Some people turn towards the vehicle, others away from it, some bodies are energised, some tighten with tension, thoughts, and feelings bubble up (recognition, disdain). This resonant space that is created is a milieu. The street already had milieus before the car drove onto the road. These are changed by the new vehicle and its rhythm, but will reassert themselves once it has gone past. Space is continually in motion. Milieu effects are not just functional; they are expressive. It's not the object itself but the object's effect on the space that is most essential for the milieu (Wise, 2000).

Space is marked and shaped in many ways: physically (with walls and fences); by objects (a coat on a chair placed there to reserve the seat); by bodies (our posture invites interaction or closes off the space); with clothing styles (grunge, preppie, biker, words on a t-shirt); through graffiti and posters; by our spoken language and the accents we use, the ideas we articulate that affect the space around us (attracting or repelling others, drawing people together around a similar theme). Wise (2000) wrote about what makes "home." We create and experience the space "home" when we unpack and arrange our personal objects and when there are the presence, habits, and effects of parents, life partners, children, and companions. We may feel at home simply in a significant other's company.

An accumulation of milieu effects forms a territory (Deleuze & Guattari, 1987). For example, a collection of objects expressing comfort (milieu effects) is territorialising a space as home (as an expression of home). The territory is the resultant space. Deleuze and Guattari stated that "what defines the territory is the emergence of matters of expression (qualities). Territories are marked out by the expressive acts through which they are produced" (p. 346). Territory is not a place, but an act, namely territorialisation, which is the expression of a territory. Territory refers to how space "becomes" by being continuously made and remade through numerous systems of relations between humans and non-humans. We are constantly making our music therapy spaces. Exploring a territory entails micro-analysing the many relations through which it is being enacted (Tucker, 2010). A territory can be considered a "context of lived reality" (Stover, 2017, p. 20). A territory is also an event because the coming together of bodies to produce a context is always an active expression of territorialisation.

One may find oneself entangled within territories of freedom, patriarchy, authoritarianism, anxious competitiveness, criticism, creativity, or acceptance. Tucker (2010) explored territories of mental health service users in the United Kingdom, particularly home environments and drop-in day centres, in light of the emphasis on community care for persons with mental health challenges. Tucker described how Phil, who visited a day centre, took part in activities formed through relations between various human and non-human bodies that impacted each other in multiple ways. Phil described listening to music, drinking a cup of tea while having a chat and a good laugh. These functional activities are expressive of a territory that is not completely pathological. These activities are "normal" things to do, which is important considering the position frequently experienced by these service users as "abnormalised" in specific ways through the many types of discrimination they encounter.

Wise (2000) explained the subject as a repeated (or repeating) expression of territories. These repetitions that form subjects are habits. Habits involve the repetition of behaviour that reflects a process of learning that's no longer conscious. We are not who we are because of an essence that underpins our thoughts and motions, but through habitual repetition of thoughts and motions. Our identity is made of habits. Wise wrote, "there is no fixed self, only the habit of looking for one" (p. 303). If one walks into a setting that feels uncomfortable (maybe it is too crowded, noisy, or packed with people who seem very "different" to oneself), one has entered a place that isn't one's territory and that one can't territorialise. One's habits seem to have no place in this setting, and, as such, one feels as if one's "self" has no place here. It is a setting that others have territorialised.

Culture is a means of territorialising (Wise, 2000). We live cultures through the movements of our bodies, through signs, discourse, and meanings, through interacting with objects, technologies, and environments. Most of our habits are created through ongoing interactions. The notions of "internal" and "external" dissolve because, as Wise explained, we are "the result of our reactions to the world, and are as such an enfolding of the external…We are spoken by our spaces, by the effects of territorialisations" (p. 303).

Territories gain a certain form as stabilised assemblages (zones of ordering). To deterritorialise is then to erode, question, break away and escape such imposed order (Mulcahy, 2016). Deterritorialisation takes place when there is a form of rupture. As a simple example, Masny (2013) described walking down the corridor to your office at work when the smell of freshly brewed coffee suddenly wafts past your nose. Your trajectory has been disrupted; your plan has been deterritorialised. The rupture offers the desire for a coffee break, perhaps the wish to go out for coffee with friends, or even images of your upcoming holiday. We can't predict up front where this coffee aroma will lead. When a rupture occurs in a rhizome, it starts up again along a new line of growth, or what Deleuze and Guattari called a line of flight. Improvisational music is often used to generate lines of flight because it enables creative, relational connections. As Wood (2016) explained, music is an ideal example of rhizomatic behaviour because music always sends out lines of flight. Music is a creative system that is exquisitely capable of freeing itself from precharted power. Music re-creates, seeks newness, and reframes; it "affirms the power of becoming" (Deleuze & Guattari, 1987, p. 327).

Lines of flight escape the established territory and potentially generate new territories with different conditions. As "becomings," they branch out and create many rhizomatic connections, engaging playfully with limits (Jackson, 2003). Even as territories are established with traditions, conventions, structures, and discourses that restrict and control movements, they still hold possibilities for deterritorialisations and the creation of new territories (Johansson, 2016). People bend lines and seek fresh paths. Lines of flight are creative acts of resistance opening up new potentials for living (Winslade, 2009). Even a slight bend of a line can generate an enormously different trajectory (Hickey-Moody & Malins, 2007).

Imagine a music therapy session with teenagers referred for violent behaviours. One girl in the group, Aniya, is repeatedly in trouble at school for her brash and insulting verbal responses to teachers. Her teacher's anger grows, with her reactions to Aniya becoming ever more aggressive, too. This plays a role in territorialising the classroom space as a conflict zone. In the music therapy sessions, Aniya is invited to be the lead singer in a hip-hop song the group is creating. The group members develop the lyrics collaboratively, and Aniya has the freedom to craft her contribution within the bridge, where she will improvise her rap. As the drum track the group has selected on *Garageband* begins, the participants' bodies start to loosen and groove. Aniya sees the expectant faces of her peers and the open, receptive smile of the music therapist. These expressions are territorialising the space as receptive and warm. Aniya takes hold of the microphone, which contributes to marking the space as performative, as inviting "real" music. Even as she incorporates specific phrases into her rap that she sometimes finds herself directing at teachers in anger, these lyrics now take on different (creative, expressive) meanings within the space of the song as a line of flight. As she raps, the other members (the teenagers and the music therapist) click, stamp their feet rhythmically, and make vocal percussive sounds to support her. Aniya becomes as a musician, a driver of creative momentum, and as a supported team member.

References

Adkins, B. (2018). To have done with the transcendental: Deleuze, immanence Intensity. *The Journal of Speculative Philosophy, 32*(3), 533–543.

Ansdell, A. (2014). Yes, but, No, but: A contrarian response to Cross. *Psychology of Music, 42*(6), 820–825.

Ansdell, G., & DeNora, T. (2016). *Musical pathways in recovery: Community music therapy and mental wellbeing.* Routledge.

Barad, K. (2003). Posthumanist performativity: Toward an understanding of how matter comes to matter. *Signs: Journal of Women in Culture, 28*(3), 801–831.

Barad, K. (2007). *Meeting the universe halfway: Quantum physics and the entanglement of matter and meaning.* Duke University Press.

Bennett, J. (2010). *Vibrant matter: A political ecology of things.* Duke University Press.

Bird-David, N. (1999). "Animism" revisited: Personhood, environment, and relational epistemology. *Current Anthropology, 40*(S1), S67–S91.

Blacking, J. (1973). *How musical is man?* University of Washington Press.

Bogost, I. (2012). *Alien phenomenology, or, what it's like to be a thing*. University of Minnesota Press.

Brabec de Mori, B., & Seeger, A. (2013). Introduction: Considering music, humans, and non-humans. *Ethnomusicology Forum, 22*(3), 269–286.

Braidotti, R. (2011). *Nomadic theory: The portable Rosi Braidotti.* Columbia University Press.

Braidotti, R. (2013). *The posthuman.* Polity Press.

Butler, J. (1993). *Bodies that matter*. Routledge.

Chilisa, B., Major, T. E., & Khudu-Petersen, K. (2017). Community engagement with a postcolonial, African-based relational paradigm. *Qualitative Research, 17*(3), 326–339.

Church, J. (2012). Sustainable development and the culture of uBuntu. *De Jure Law Journal, 45*(3), 511–531.

Clark, A., & Chalmers, D. (1998). The extended mind. *Analysis, 58*, 7–19.

Coleman, R., & Ringrose, J. (2013). Introduction: Deleuze and research methodologies. In R. Coleman & J. Ringrose (Eds.), *Deleuze and Research Methodologies* (pp. 1–22). Edinburgh University Press.

Country, B., Wright, S., Suchet-Pearson, S., Lloyd, K., Burarrwanga, L., Ganambarr, R., Ganambarr-Stubbs, M., Ganambarr, B., Maymuru, D., &

Sweeney, J. (2016). Co-becoming Bawaka: Towards a relational understanding of place/space. *Progress in Human Geography, 40*(4), 455–475.
Crossley, J. (2017). *The cyber-guitar system: A study in technologically enabled performance practice* [Doctoral dissertation, University of the Witwatersrand].
De Quincey, C. (2005). *Radical knowing: Understanding consciousness through relationship*. Park Street Press.
Deleuze, G. (1995). *Negotiations: 1972–1990* (M. Joughin, Trans.). Columbia University Press.
Deleuze, G. (2001). *Pure immanence: Essays on a life* (A. Boyman, Trans.). Zone Books.
Deleuze, G., & Guattari, F. (1987). *A thousand plateaus: Capitalism and schizophrenia*. Athlone.
Deleuze, G., & Parnet, C. (1987). *Dialogues*. Athlone.
DeNora, T. (2003). *After Adorno: Rethinking music sociology*. Cambridge University Press.
Duff, C. (2014). *Assemblages of health: Deleuze's empiricism and the ethology of life*. Springer.
Edwards, S. (2010). A Rogerian perspective on empathic patterns in Southern African healing. *Journal of Psychology in Africa, 20*(2), 321–326.
Emirbayer, M. (1997). Manifesto for a relational sociology. *American Journal of Sociology, 103*(2), 281–317.
Enfield, N. (2017). Elements of agency. In N. Enfield & P. Kockelman (Eds.), *Distributed agency* (pp. 3–8). Oxford University Press.
Ferrando, F. (2019). *Philosophical posthumanism*. Bloomsbury Academic.
Flower, C. (2019). *Music therapy with children and parents: Toward an ecological attitude* [Doctoral dissertation, Goldsmiths, University of London].
Forster, D. A. (2010). A generous ontology: Identity as a process of intersubjective discovery—An African theological contribution. *HTS: Theological Studies, 66*(1), 1–12.
Fox, N. J., & Alldred, P. (2013). The sexuality-assemblage: Desire, affect, antihumanism. *The Sociological Review, 61*(4), 769–789.
Gallagher, S. (2017). *Enactivist interventions: Rethinking the mind*. Oxford University Press.
Harvey, G. (2013). Introduction. In G. Harvey (Ed.), *The handbook of contemporary animism* (pp. 15–16). Routledge.
Hickey-Moody, A., & Malins, P. (2007). Introduction: Gilles Deleuze and four movements in social thought. In A. Hickey-Moody & P. Malins (Eds.), *Deleuzian encounters: Studies in contemporary social issues* (pp. 1–26). Palgrave MacMillan.

Ingold, T. (2000). *The perception of the environment: Essays on livelihood, dwelling and skill*. Routledge.
Ingold, T. (2007). *Lines: A brief history*. Routledge.
Ingold, T. (2008). Bindings against boundaries: Entanglements of life in an open world. *Environment and Planning A, 40*, 1796–1810.
Ingold, T. (2010). The textility of making. *Cambridge Journal of Economics, 34*(1), 91–102.
Jackson, A. (2003). Rhizovocality. *Qualitative Studies in Education, 16*(5), 693–710.
Johansson, L. (2016). Post-qualitative line of flight and the confabulative conversation: A methodological ethnography. *International Journal of Qualitative Studies in Education, 29*(4), 445–466.
Kagumba, A. K. (2021). *Orchestrating social competence: On the transformative work of musicking in two Ugandan NGOs (M-LISADA and Brass for Africa)* [Doctoral dissertation, Texas Tech University].
Keller, M. S. (2012). Zoomusicology and ethnomusicology: A marriage to celebrate in heaven. *Yearbook for Traditional Music, 44*, 166–183.
Kenny, C. (2016). The field of play: A focus on energy and the ecology of being and playing. In J. Edwards (Ed.), *The Oxford handbook of music therapy* (pp. 472–482). Oxford University Press.
Kovach, M. (2012). *Indigenous methods: Characteristics, conversation, and contexts*. University of Toronto Press.
Krueger, J. (2014). Affordances and the musically extended mind. *Frontiers in Psychology, 4*, 1003.
Labanyi, J. (2010). Doing things: Emotion, affect, and materiality. *Journal of Spanish Cultural Studies, 11*(3–4). Special Issue: Cultural/political reflection—Lines, routes, spaces
Lancione, M. (2013). Homeless people and the city of abstract machines: Assemblage thinking and the performative approach to homelessness. *Area, 45*(3), 358–364.
Latour, B. (1996). On interobjectivity. *Mind, Culture and Activity, 3*(4), 228–245.
Law J. (2004). *After method: Mess in social science research*. Routledge.
Lee Soon, R. (2016). Nohana i waena i na mo'olelo/living between the stories: Contextualising drama therapy within an indigenous Hawaiian epistemology. *Drama Therapy Review, 2*(2), 257–271.
Levitin, D. (2006). *This is your brain on music*. Dutton.
Long Soldier, L. (2017). *Whereas*. Graywolf Press.

Magnusson, T. (2009). Of epistemic tools: Musical instruments as cognitive extensions. *Organised Sound, 14*, 168–176.
Masny, D. (2013). Rhizoanalytic pathways in qualitative research. *Qualitative Inquiry, 19*(5), 339–348.
Mawere, M., & Mubaya, T. (2016). *African philosophy and thought systems: A search for a culture and philosophy of belonging.* Langaa Research and Publishing.
Mazzei, L., & Jackson, A. (2017). Voice in the agentic assemblage. *Educational Philosophy and Theory, 49*(11), 1090–1098.
Mbeki, T. (1996). *Statement on behalf of the African National Congress, on the occasion of the adoption by the Constitutional Assembly of 'The Republic of South Africa Constitutional Bill 1996.* Cape Town, 8 May—issued by the Office of the Deputy President. http://www.info.gov.za/speeches/1996/960819_23196.htm
Mbiti, J. (1969). *African religion and philosophy*. Heinemann.
McLeod, K. (2017). *Wellbeing machine: How health emerges from the assemblages of everyday life.* Carolina Academic Press.
Mcphie, J. (2018). I knock at the stone's front door: Performative pedagogies beyond the human story. *Parallax, 24*(3), 306–323.
McPhie, J. (2019). *Mental health and wellbeing in the Anthropocene: A posthuman inquiry.* Palgrave Macmillan.
Metz, T. (2007). Toward an African moral theory. *Journal of Political Philosophy, 15*(3), 321–341.
Mol, A. (2006). *The multiple body: Ontology in medical practice.* Duke University Press.
Mulcahy, D. (2016). Policy matters: De/re/territorialising spaces of learning in Victorian government schools. *Journal of Education Policy, 31*(1), 81–97.
Murove, M. (2005). *The theory of self-interest in modern economic discourse: A critical study in the light of African humanism and process philosophical anthropology* [Doctoral thesis, University of South Africa].
Noë, A. (2009). *Out of our heads.* Hill and Wang.
Nzewi, M. (2002). Backcloth to music and healing in traditional Africa society. *Voices, A World Forum for Music Therapy, 2*(2).
O'Hara, M. (1997). Relational empathy: Beyond modernist egocentricism to postmodern holistic contextualism. In A. C. Bohart & L. S. Greenberg (Eds.), *Empathy reconsidered* (pp. 295–319). American Psychological Association.
Oliver, T. (2020). *The self delusion.* Weidenfeld & Nicolson.

Prozesky, M. H. (2001). Well-fed animals and starving babies: Environmental and developmental challenges in process and African perspectives. In M. Murove (Ed.), *African ethics: An anthology of comparative and applied ethics* (pp. 298–307). University of KwaZulu-Natal Press.

Reddekop, J. (2014). *Thinking across worlds: Indigenous thought, relational ontology, and the politics of nature; Or, if only Nietzsche could meet a Yachaj* [Doctoral dissertation, University of Western Ontario]. Electronic Thesis and Dissertation Repository, 2082.

Reynolds, J. (2018). The extended body: On aging, disability, and well-being. *Hastings Center Report, 48*, S31–S36.

Rietveld, E., Denys, D., & Van Westen, M. (2018). Ecological-enactive cognition as engaging with a field of relevant affordances. In A. Newen, L. De Bruin, & S. Gallagher (Eds.), *The Oxford handbook of 4E cognition* (pp. 41–70). Oxford University Press.

Rival, L. (2014). The materiality of life: Revisiting the anthropology of nature in Amazonia. In G. Harvey (Ed.), *The handbook of contemporary animism* (pp. 92–100). Oxon.

Selg, P., & Ventsel, A. (2020). The "relational turn" in the social sciences. In P. Selg & A. Ventsel (Eds.), *Introducing relational political analysis* (pp. 15–40). Palgrave Macmillan.

Semenya, B., & Mokwena, M. (2012). African cosmology, psychology and community. In M. Visser & A. Moleko (Eds.), *Community psychology in South Africa* (pp. 71–84). Van Schaik Publishers.

Shapiro, L. (2010). *Embodied cognition.* Routledge.

Shaw, W. S., Herman, R. D. K., & Dobbs, G. R. (2006). Encountering indigeneity: Re-imagining and decolonizing geography. *Human Geography, 88*(3), 267–276.

Shotter, J. (2013). From inter-subjectivity, via inter-objectivity, to intra-objectivity: From a determinate world of separate things to an indeterminate world of inseparable flowing processes. In G. Sammut, P. Daanen, & F. Moghaddam (Eds.), *Understanding the self and others: Explorations in intersubjectivity and interobjectivity* (pp. 83–125). Routledge.

Slife, B. D. (2004). Taking practice seriously: Toward a relational ontology. *Journal of Theoretical and Philosophical Psychology, 24*(2), 157.

Small, C. (1998). *Musicking: The Meanings of Performing and Listening.* Wesleyan University Press.

Stover, C. (2017). Affect and improvising bodies. *Perspectives of New Music, 55*(2), 5–66.

Sundberg, J. (2013). Decolonizing posthumanist geographies. *Cultural Geographies, 21*(1), 33–47.
Sundberg, J. (2014). Decolonizing posthumanist geographies. *Cultural Geographies, 21*(1), 33–47.
Sutton, J. (2005). Hidden music—An exploration of silences in music and in music therapy. *Guidelines Article Formatting, 6*, 375.
Swanson, D. (2012). Ubuntu, African epistemology, and development. Contributions, contradictions, tensions, and possibilities. In H. K. Wright & A. A. Abdi (Eds.), *The dialectics of African education and Western discourses: Counterhegemonic perspectives* (pp. 27–52). Peter Lang.
Taringa, N. T. (2020). The Potential of Ubuntu Values for a Sustainable Ethic of the Environment and Development. In E. Chitando & M. R. Gunda (Eds.), *Religion and development in Africa* (pp. 387–399). University of Bamberg Press.
Thrift, N. (2008). *Non-representational theory: Space/politics/affect.* Routledge.
Todd, Z. (2016). An indigenous feminist's take on the ontological turn: 'Ontology' is just another word for colonialism. *Journal of Historical Sociology, 29*(1), 4–22.
Tuck, E., & Yang, K. W. (2014). R-words: Refusing research. In D. Paris & M. T. Winn (Eds.), *Humanizing research: Decolonizing qualitative inquiry with youth and communities* (pp. 223–247). Sage.
Tucker, I. (2010). Mental health service user territories: Enacting 'safe spaces' in the community. *Health, 14*(4), 434–448.
Ursin, G., & Lotherington, T. A. (2018). Citizenship as distributed achievement: Shaping new conditions for an everyday life with dementia. *Scandinavian Journal of Disability Research, 20*(1), 62–71.
Valente, J., & Boldt, G. (2015). The rhizome of the deaf child. *Qualitative Inquiry, 21*(6), 562–574.
van der Schyff, D. (2013). Emotion, embodied mind and the therapeutic aspects of musical experience in everyday life. *Approaches: Music Therapy and Special Music Education, 5*(1).
Van der Schyff, D., & Krueger, J. (2019). *Musical empathy, from simulation to 4E interaction. Music, sound, and mind.* Brazilian Association of Music Cognition.
Viega, M. (2014). Listening in the ambient mode: Implications for music therapy practice and theory. *Voices: A World Forum for Music Therapy, 14*(2).
Waterfall, B., Smoke, D., & Smoke, M. (2017). Reclaiming grassroots traditional indigenous healing ways and practices within urban indigenous community contexts. In S. L. Stewart, R. Moodley, & A. Hyatt (Eds.),

Indigenous cultures and mental health counselling: Four directions for integration with counselling psychology (pp. 3–16). Routledge.
Watson, A., & Huntington, O. (2008). They're here—I can feel them: The epistemic spaces of Indigenous and Western knowledges. *Social and Cultural Geography, 9*, 257–281.
Watts, V. (2013). Indigenous place-thought and agency amongst humans and non-humans (First woman and sky woman go on a European tour!). *DIES: Decolonization, Indigeneity, Education and Society, 2*(1), 20–34.
Whitman, K. (2002). *Playthroughs.* Kranky.
Wilde, P., & Evans, A. (2019). Empathy at play: Embodying posthuman subjectivities in gaming. *Convergence, 25*(5–6), 791–806.
Wilson, S. (2008). *Research is ceremony: Indigenous research methods.* Fernwood Publishing.
Winslade, J. (2009). Tracing lines of flight: Implications of the work of Gilles Deleuze for narrative practice. *Family Process, 48*(3), 332–346.
Wise, J. (2000). Home: Territory and identity. *Cultural Studies, 14*(2), 295–310.
Wood, J. (2016). *Interpersonal communication: Everyday encounters* (8th ed.). Cengage Learning.
Wood, S. (2016). *A matrix for community music therapy practice.* Barcelona Publishers.
Woods, P. (2020). Reimagining collaboration through the lens of the posthuman: Uncovering embodied learning in noise music. *Journal of Curriculum and Pedagogy,* 1–21.
Wu, J., Eaton, P. W., Robinson-Morris, D. W., Wallace, M. F., & Han, S. (2018). Perturbing possibilities in the postqualitative turn: Lessons from Taoism (道) and Ubuntu. *International Journal of Qualitative Studies in Education, 31*(6), 504–519.
Zapata, A., Kuby, C., & Thiel, J. (2018). Encounters with writing: Becoming-with posthumanist ethics. *Journal of Literacy Research, 50*(4), 478–501.

13
Affective Attunement

As we broaden our scope from singular humans who encounter one another to relational webs, the concept of emotions—as particularly human subjective experiences and interpretations—is insufficient. The best notion we can use here then is affect. Affects don't "belong" to anyone but emerge in encounters and interactions between bodies. They affect the bodies involved and mark them as they transform (Seyfert, 2012).

Affects as Intensities

We can first understand affect as a kind of sensation or intensity (Wetherell, 2012). For example, suppose we encounter something unexpected, like musicians playing out of key, the smell of soured milk, or a shocking image. Our body may experience a particular force or sensation before we make sense of or articulate the experience of aversion (Hickey-Moody & Malins, 2007). The affect is first felt viscerally before it is thought about and given emotive or subjective meaning. Our initial reaction to music often entails a sensation that is yet to have a meaning

A. dos Santos, *Empathy Pathways*,
https://doi.org/10.1007/978-3-031-08556-7_13

or an emotion label assigned to it (Meelberg, 2009). The flow of affective activity when experiencing music could take the form of chills running down our spine perhaps, a sense of stillness, or a sensation of quickening.

Affective practices become organised and unfurl as patterns and sequences. Wetherell (2012) described how self-pity, for example, can flare up, crescendo, and diminish along with the articulation and interpretation of the experience. Affective phenomena characterised by intense bodily surges (such as panic attacks) usually occur in relatively brief episodes or bursts. Others may entail a fairly continuous set of background feelings that shift in and out of focus throughout every day.

The daily affects we experience are shaped by what Daniel Stern (2010) referred to as "forms of vitality." In Chapter 9, we encountered vitality affects in relation to translational empathy (as a way of being with a client through "bursting" in the music together, for example). We also discussed how vitality affects are multi-modal: a river can flow, a dancer's body can flow, a melody line can flow. In that pathway, we were still thinking of the quality of "bursting" as being "in" the client (or the music or the body movement, etc.). We thought of it as being communicated through the client's expression, and as the therapist, we then responded through expressions that matched that vitality affect.

In this current pathway, we are thinking about how vitality affects can be experienced as sensations and intensities in bodies and between (human and non-human) bodies. They are part of the flow of connections between all elements in an assemblage. The flow of affective activity can take on the form of a surge in the momentum between all the different human and non-human elements within an improvisation, for example.

Our energies are not self-contained, and there is no neat distinction between the individual and the environment. However, as Brennan (2004) explained, acknowledging this transmission doesn't imply that a person's emotional experience is irrelevant. Affects are not encountered and perceived in a vacuum. If I'm feeling anxious when I enter a session, that will shape the affects (as impressions) that I receive in the group. Suppose I perceive flat, sinking, dull affects in an interaction with a client. This may relate to the present moment or I may make sense of this in terms of my personal history (perhaps an event that happened with a client in a previous session that I'm still thinking about). The client may interpret this affect in relation to their personal embodied process in light of a recent loss they have experienced.

Importantly, as Brennan argued, affects are not necessarily generated by thoughts (thinking of an event and then experiencing a surge of associated affect). Affects can be present in the spaces we're in, and then we seek ways of interpreting them by "glueing" thoughts and meanings onto them (which, as we have just discussed, different parties could be doing in various ways within the same encounter). Brennan wrote that "...affects may, at least in some instances, find thoughts that suit them, not the other way around" (p. 7). We may have an underlying sense of gnawing guilt, and we think it relates to a particular relational conflict, but then when that conflict is solved, the sensation remains, and we find a new explanation for it. Affects can be described as having more of a "life of their own" in this empathy pathway, as we are territorialising assemblages as mutually response-able.

Being Affected and Affecting

The second prominent use of the term "affect" refers to "being affected" by an event (even when it's unclear what the impact may be). In the introduction to Deleuze and Guattari's (1987) *A Thousand Plateaus*, Brian Massumi wrote that affect is "an ability to affect and be affected" (p. xvi). This refers to the capacity of bodies and things to affect one another. Again, it's important to keep in mind that not only humans affect and are affected. The sun affects the moon; waves affect the shape of the coastline; sound waves played through a plastic sheet affect salt that we've sprinkled on it, creating intricate patterns. Here, the term "affect" refers to something like a force relation in that affect is the change that takes place when bodies come into contact (Kofoed & Ringrose, 2012). Affect can be understood as a potential for action. As catalysts, affects can provoke us to look differently, interpret in new ways, relate differently, move our bodies, or extricate ourselves from habitual ways of thinking and behaving (Hickey-Moody & Malins, 2007).

Deleuze (1992) drew strongly on the thinking of seventeenth Century Dutch philosopher Spinoza, who distinguished between two definitions of affect through semantic differentiations within Latin. Spinoza referred first to "affectio," as the lived experience (in the body and the mind) of a feeling state. This implies the presence of something "doing" the "affecting." Second, "affectus" refers to the transition from one affective state to another, involving feelings in the body and impressions or ideas in the mind. For this reason, while separate chapters on emotions and meaning were presented in parts one and two of this book, they need to be integrated here in part three. Gregg and Seigworth (2010) drew on both the dimensions of affectio and affectus in their explanation of affect as "an impingement or extrusion of a momentary or sometimes more sustained state of relation as well as the passage (and the duration of passage) of forces or intensities" (p. 2).

Every encounter changes a body's affective capacities. Relations between human and non-human bodies always generate affective responses (Duff, 2014). Musical sounds produce active affects that may cause toe-tapping, shoulders relaxing, singing, or dancing. Different bodies register and respond to affects in different ways. An affect that may arouse one might inhibit another. Take the example of walking into a room where two group members who arrived early have just been having a heated argument. You may not hear what they're saying and don't know what the fight is about, but the atmosphere in the room is highly charged (even though they stop speaking and turn towards you with polite smiles on their faces as you enter). In registering the tense affect in the room, you might move towards the conflict, ready to invite them to engage openly with one another; or you may soften your body posture in readiness to enhance gentleness and peace, depending on your relationship with these group members and your body state at the time (or you may have the urge to turn around and walk out again). Affect is the pre-reflective force that catalyses these actions (even as the actions may unfold differently in different situations and by other actors) (Swift, 2012).

Although we drew on aspects of Carolyn Pedwell's work in the last part of this book, she holds perspectives that are relevant as we explore empathising assemblages as well. According to Pedwell (2014), the process of empathy is less about feeling or recognising "the same" feeling as another person, but rather about engaging in affective qualities of relations that open us to affecting and being affected by another (human or non-human, material and conceptual, animate or inanimate) body. The notion of affective synchronisation involves tuning into affective frequencies and rhythms. This can be seen as a kind of empathising with space and time as one opens oneself up to being affected by different temporalities and spatialities (Pedwell, 2014). In this context, empathy (like other affective relations) isn't a property encapsulated within or owned by a subject.

We may ask what a body can and can't do within its relational assemblage and what it can become (Alldred & Fox, 2015). What is a body's capacity to affect and be affected? For Deleuze (1992), a body is most importantly constituted through its actions (or capacities for actions: what it can do). We consider bodies, therefore, in terms of their "capacities" rather than their "functions" (Duff, 2014). Bear in mind that a "body" refers to the composition of its relations, so we're asking: what can this collection of relations do? As we wonder what a body can do, we're wondering what relations it can compose with other human and non-human bodies (Deleuze, 1992).

As relations are always dynamic, a body's capacity to affect and be affected is not fixed, but freshly determined in each new instance. Each encounter transforms a body's relations and generates emergent affects and capacities (Deleuze, 1992). A body's power increases when it becomes more capable of entering into new relations with other bodies and affecting and being affected by other bodies. The body's capacity to be affected is understood to be synonymous with its power of action. This is how we are "becomings."

As a result, we don't (and we can't) know up front what a body can do because we don't yet know how relations will unfold, what affects might emerge, and what a body may become capable of. This is often a gradual process. Capacities are developed slowly as bodies form new connections and affective relations. A body slowly orients itself to a musical instrument, for example, relating to the scales, timbres, textures, notations, sounds, and exercises, growing in capacity to affect the instrument and be affected by it.

As Deleuze' drew on Spinoza, he understood affects as forces produced through encounters that either increase or decrease one's capacity to act (Stover, 2017). Spinoza distinguished between two kinds of encounters: those in which diverse human and non-human bodies meet such that the power of each body to act is enhanced (generating joy); and encounters that decrease or even destroy their power to act (generating sadness) because power is being employed in the efforts to reduce or eliminate the affects associated with the undesired encounter (Deleuze, 1992). Spinoza conceived of ethics as the collection of practices through which individuals seek to organise encounters in a way that maximises their experience of joyful affects (Duff, 2014). This is the type of encounter I'm referring to in this pathway, where empathy involves territorialising an assemblage as mutually affectively response-able. In an empathising assemblage, the capacity of each body-in-relation is being enhanced.

Not all assemblages function in this way. Power operates through affect, and affect is produced through power relations. Affect fuels and intertwines with cultural values. Which affective practices are highly regarded (as forms of social and cultural capital) and which are viewed with disdain? Who gets to feel and express what and when? Who is privileged, who is disadvantaged, and what does this process look like? What relations do particular affective practices produce, enact, reinforce, and disrupt? (Wetherell, 2012). As we seek to enhance the empathy of the assemblages we're part of, these are questions we need to explore.

Affect, Emotion, and Feeling

In Deleuze's (2003) view, emotion is the interpretation of an affective reaction. Affect is energy, resonance, and movement. It's an impetus for thought and interpretation. Affect motivates bodies to sense kinaesthetically the entity that is producing the affect. As soon as we've interpreted the affect as meaningful in a particular way, the affect disappears: it changes into subjective content as emotion (Massumi, 2002). Pedwell (2014) explained that emotions (like happiness or frustration) are the labels we give to certain temporarily stable (though variable) groups of feelings, thoughts, physiological responses, and affective intensities. Recognition and labelling of such states usually emerge relationally, both within interactions with other people, objects, events and forces and through how we distinguish specific assemblages of intensities/feelings/thoughts/reactions from others. Sometimes we realise in the moment that we are "having" a particular emotion, but often we figure it out retrospectively (within seconds, hours, days, or even years). As "affect" refers to emerging and dynamic intensities (instead of the recognised and named entities that are emotions), we could understand them as functioning in a less discursive and socially constructed way than emotions do. Emotions are more individualised and informed by personal biographies, while affect is often experienced collectively and before individual processes of interpretation. In this understanding, affect doesn't "belong" to a singular subject, but is constantly flowing, moving, and transmitting through multiple human and non-human bodies (Niccolini, 2016).

Affective Flows in Assemblages

Pedwell (2014) used the term "affective relations" to highlight both the relational nature of affects (they flow between and take part in constituting subjects, objects, and contexts rather than residing in people) as well as how affects form and circulate through interactions with other affects. Affective practices are tangled up in music, technology, discourses, tools, images, values, events, norms, social movements, and so on. To analyse affective practices requires consideration of how they are situated and connected (in careful, predictable, and repetitive ways or in contingent, spontaneous, and fluctuating ways). Affective practice is always dynamic and continuously holds the potential to move in many divergent directions. The interconnected patterning of affective practice can flow across a small or large number of participants, across a smaller scene, a more extensive site, throughout an institution, or even society (think about affective flows within countries as they went into lockdowns during Covid-19) (Wetherell, 2012).

Affects flow through assemblages. As discussed in the previous chapter, we can think of an assemblage as a web of forces and intensities in motion. Focusing on affects invites a way of considering energy, relational currents, flows, and flux between people that can disrupt distinctions between "interior" and "exterior" (Bondi et al., 2007). According to Gregg and Seigworth (2010), "affect arises in the midst of in-betweenness: in the capacities to act and be acted upon" (p. 2). Affective flows between elements in assemblages are rhizomatic. Affects fold, unfold, and change in predictable and unexpected ways (Massumi, 2002). Affects are becomings that produce further affective capacities in an assemblage. Subjects emerge at the intersections of affective trajectories because subjects are subjects of experience, and this intersection creates a field of conditions for the ways in which subjects emerge (Stover, 2017).

Affects play a role in territorialising, deterritorialising, and reterritorialising. For example, excitement, vulnerability, or expressiveness can be understood as specific territorialisations produced by affects in an assemblage. All assemblages are affective entities in that affective processes are always, at least in part, responsible for their formation (Duff, 2014).

Affects and Musicking

Both understandings of affect (intensities and the capacity to affect and be affected) are at play when we're musicking. "Musical bodies" are encountering one another "in affective exchanges of intensities" (Stover, 2017, p. 5) and in an emergent space that is created as the musicking unfolds. Each individual contribution needs to be understood in the context of the developing improvisation because prior contributions shape how we experience and understand the present act of expression (Swift, 2012).

Gilbert (2004) highlighted the rhizomatic character of improvisation, where musicians are free to create their own connections without being directed by a central coordinating force. This doesn't (necessarily) result in chaos or aimlessness. The improvisational assemblage has creative forces at play, but these are decentralised. Any musician can affect and be affected by any (and all) of the others. Affect is autonomous: a musician can make a sound, but they can't control how it will affect other bodies in the environment (Cummins, 2009). Another group member may respond with bright delight or dim and withdraw.

As we discussed in the previous chapter, a focus on assemblages calls for a critique of the centrality of human actors. In musical improvisations, many human and non-human actors come together to create (and be affected by) a musical process (Nesbitt, 2010). Swift (2012) also reminded us that no special place is reserved for conventional acoustic instruments in the improvisation assemblage as the shift to the digital opens up new possibilities.

When we explore the affective processes that unfold in improvisation, we are interested in repetition and change. As mentioned in Chapter 12, Deleuze and Guattari (1987) described a milieu as "a block of space–time constituted by the periodic repetition of the component" (p. 313). When the child sang to himself in the dark, the repetition of elements in the tune allowed for a sense of comfort to characterise the affective quality of the space. The process of (even very simple) musicking entails sustained, patterned activity from the assemblage of musical bodies. Musicking also involves constant change and transformation as we embellish, develop, and even destroy these patterns. As we explore what musical bodies have the capacity to do (how they can affect and be affected), we notice the affective atmospheres that give rise to shifts and transitions in the improvising group. One might explore sonic "discontinuities," such as a sudden, loud noise, the introduction of an interesting timbre, or the prompt removal of a sound. Some of these changes may be serendipitous and unintentional, while others could be deliberate attempts to change things up. These moments of difference and novelty can be catalysts for the group as a whole because they change the overall atmosphere of the improvisation (Swift, 2012). Stover (2017) argued that in musical analysis, one should begin by exploring affect and then work towards examining the human and non-human bodies performing the actions.

When we talk about bodies affecting and being affected by other bodies, this doesn't mean that a neat temporal succession of causal force relations occurs. Let's take the example of "call-and-response." Here, human and non-human musical bodies interact and impinge on one another in multiple, simultaneous ways. The "response" is also a new call that invites a range of potential responses, but not merely as part of a simple temporal flux. A call could be picked up in any number of imaginative ways by differently affectively attuned responders, enacting a multitude of new call trajectories. Any musical action here might result from an intended move towards an anticipated future event, towards something that took place a few events back, or towards something that is currently happening. In call-and-response, there's always the possibility of new affective agency. A line of flight could take us from the current ongoingness of an improvisational activity to somewhere wholly unexpected (Stover, 2017). As Matney (2021) wrote, music, the music therapist's training, the clients' experiences and the environment "have the capacity to affect and be affected by each other through events, promoting differentiation and change in future contexts" (p. 13).

Travelling and Sticky Affects

Affects in an assemblage can be free-flowing or stuck and trapped, the energy may be heating up, or cooling down and dissipating. Assemblage theory (De Landa, 2006) offers a way to map how forces come together, the directions, spaces, and speeds through which lines of intensity connect, what becomes through the assembled relations, and what is blocked from being produced. We can map and track affect intensities and explore which lines are life-affirming and which are life-destroying (Ringrose & Renold, 2014).

Kofoed and Ringrose (2012) suggested the idea of "sticky affects" (p. 10), referring to force relations that temporarily glue (or fix) certain affects to particular bodies. As an example, we might think of how the diagnostic framework of borderline personality disorder is often entangled with interpretations of a client's behaviour being manipulative (Potter, 2006). This can become a sticky "point of fixation" and can operate as a process of subjectification (the creating of a subject) that literally "puts this person in their place."

A music therapy group may proceed through what Tuckman described (and revisited with his co-author Jensen in 2010) as the "storming" phase where hostility, rivalry, negativism, aggression may coagulate temporarily before affects become more free-flowing again. Even during these times, though, participants may still be negotiating this fixity with surprising effects.

Kofoed and Ringrose (2012) contrasted sticky affects with "travelling affects," where positions, desires, and subjects are dispersed, displaced, and moved around. By noticing how affects travel and stick, we can examine the situated processes of how people and objects are created and positioned in dynamic ways. Complex relations influence who can move and be moved, affect and be affected, and who is set in place (Pedwell, 2014).

Ethically Response-able Assemblages

"Response-ability" is a term used particularly by posthuman writers, such as Haraway (2016), to refer to the ability or capacity to respond (to affect and be affected). Consider the following vignette offered by DeNora and Ansdell (2017). Eloise, a group member at SMART, initially used a drumstick to hit the cymbal. Eloise took great care not to play with too much force because when the cymbal is hit strongly with a drumstick, the loud and resonant result doesn't blend in well with the overall group sound. This constrained her ability to be physically expressive. The following week, Ansdell, therefore, offered jazz brushed instead of a drumstick and Eloise's possibilities for physical exertion expanded. With this new freedom, Eloise could engage in new gestures and movements. She was now also response-able within the group.

There are different ways in which we can be response-able. If we're exploring an empathising assemblage that is being territorialised as response-able, then does any response qualify within this process? What about violent responses? Or responses of denial or avoidance? An empathising assemblage is being territorialised as *mutually* affectively response-able. Response-ability is a relational idea. It refers not only to one's personal capacity to respond but to how one invites, welcomes, and enables the responses of others (Barad, 2010). Promoting response-ability entails being affectively open to one another, letting go of seeking controlled outcomes and answers, allowing one's assumptions of how things "should" be to fall away, being attentive to the emergent qualities of experiences, and allowing ourselves to see and hear more and differently (Cooke & Colucci-Gray, 2019).

A violent response (or a response of denial or avoidance, for example) may not lead to increasing another's capacity to respond. Ethical responses are those that facilitate an ongoing becoming of mutual response-ability. Importantly, we are looking at the flow of how this process unfolds over time (ongoing becomings). We are not just making this assessment based on one limited actualised moment. Let's think of a hypothetical example of Sandra, a music therapy client. During a session, the fire alarm goes off in the building. The music therapist asks everyone to follow the safety procedure and evacuate the building. Sandra decides to fold her arms, stand her ground, and refuse to leave the room. This is how she is expressing her ability to respond at this moment. Would this be an example of empathic response-ability? The problem here is that Sandra may, perhaps, now be standing in the way of other people trying to leave the room and the music therapist isn't willing to go without her. She is thereby hampering others' response-ability. If Sandra is harmed in the fire, her future capacities for responding may be impaired. The other group members' capacities to respond may be negatively impacted as well through the trauma they experience. In the relational view of the world that we're engaging in within this empathy pathway, our focus is on collective well-being—on the well-being of assemblages, not just individuals—and we are not looking at static moments, but at flow through time. Through that lens, it's difficult to see Sandra's behaviour as empathically enhancing the response-ability of the assemblage.

When Haraway (2016) described response-ability as the capacity to respond, she immediately tilted the kind of response she was referring to in a specific direction, relating response-ability to "living and dying well together" (p. 29). She highlighted that cultivating response-ability "requires the risk of being for some worlds rather than others and helping to compose those worlds with others" (p. 179). Noddings (2003), a philosopher of education, wrote that a relational orientation focuses on creating conditions in which people are more likely to engage in mutually supportive ways. Within these conditions, "we are led to redefining responsibility as response-ability, the ability to respond positively to others and not just to fulfil assigned duties" (p. 35).

Boyd (2016) explained how, when we say to another person, "You're not listening to me," we usually mean, "You're not responding to me and my meaning. When I tell you something, you just say whatever is on your mind without actually building on or taking into account what I just said to you." When I fully listen to another person, I respond in a way that is contingent on what they have just said. Importantly, if I expect you to listen to me, I also have a special responsibility: I must speak in a response-able way that invites your engaged reaction. If I don't do this, I'm just trying to get you to comply; I'm speaking "at" you. Response-ability involves changeability and accountability, not reactivity or engaging with the intent to change or convince the other person. In Boyd's view, when we don't speak response-ably, we shouldn't be surprised if others are not really listening to us.

Our musical responses as therapists are contingent on what the client is expressing. Our task in group work is to facilitate musicking (and other kinds of verbal and arts-based responses) that are characterised by attentively following the emerging expressions and meanings. This kind of response-ability entails co-presencing, which refers to cultivating sensitivity to others. We allow our imaginations to open up, feel, articulate, and be receptive to others' suffering. This "other" isn't separate and distanced. Response-ability is a form of becoming-with and sharing suffering as we open ourselves to encounter pain and mortality, and to learn together what living teaches us (Greenhough & Roe, 2010).

Creative arts therapists seek to enable affective responsiveness. As Sajnani (2012) highlighted, our practice encourages response-ability, which she defined as "the ability to respond amidst suffering and against oppression" (p. 189). In music therapy sessions as relational phenomena, we actively do-with, create-with, music-with, and become-with each other through intra-actions. We render each other capable as we constitute each other through our entanglement (Bozalek et al., 2018). One of the central goals of music therapy, I would argue, is to create and be in empathic assemblages where all involved can affect and be affected and where there is sensitivity to how this is taking place. As Ruud (2008) argued, the goal of music therapy is to increase possibilities for action.

References

Alldred, P., & Fox, N. J. (2015). The sexuality-assemblages of young men: A new materialist analysis. *Sexualities, 18*(8), 905–920.

Barad, K. (2010). Quantum entanglements and hauntological relations of inheritance: Dis/continuities, spacetime enfoldings, and justice-to-come. *Derrida Today, 3*(2), 240–268.

Bondi, L., Davidson, J., & Smith, M. (2007). Introduction: Geography's "emotional turn." In L. Bondi, J. Davidson, & M. Smith (Eds.), *Emotional geographies* (pp. 1–18). Ashgate.

Boyd, M. P. (2016). Calling for response-ability in our classrooms. *Language Arts, 93*(3), 226.

Bozalek, V., Bayat, A., Gachago, D., Motala, S., & Mitchell, V. (2018). A pedagogy of response-ability. In V. Bozalek, R. Braidotti, M. Zembylas, & Shefer, T. (Eds.), *Socially just pedagogies in higher education: Critical posthumanist and new feminist materialist perspectives* (pp. 97–112). Bloomsbury Academic.

Brennan, T. (2004). *The transmission of affect.* Cornell University Press.

Cooke, C., & Colucci-Gray, L. (2019). Complex knowing: Promoting response-ability within music and science teacher education. In C. Taylor & A. Bayley (Eds.), *Posthumanism and higher education* (pp. 165–185). Palgrave Macmillan.

Cummins, F. (2009). Rhythm as an affordance for the entrainment of movement. *Phonetica, 66*(1–2), 15–28.

De Landa, M. (2006). *A new philosophy of society: Assemblage theory and social complexity.* Continuum.

Deleuze, G. (1992). *Expressionism in philosophy: Spinoza* (M. Joughin, Trans.). Zone Books.

Deleuze, G. (2003). *Francis Bacon: The logic of sensation* (D. W. Smith, Trans.). Continuum.

Deleuze, G., & Guattari, F. (1987). *A thousand plateaus: Capitalism and schizophrenia.* Athlone.

DeNora, T., & Ansdell, G. (2017). Music in action: Tinkering, testing and tracing over time. *Qualitative Research, 17*(2), 231–245.

Duff, C. (2014). *Assemblages of health: Deleuze's empiricism and the ethology of life*. Springer.

Gilbert, J. (2004). Becoming-music: The rhizomatic moment of improvisation. In I. Buchanan & M. Swiboda (Eds.), *Deleuze and music* (pp. 118–139). Edinburgh University Press.

Greenhough, B., & Roe, E. (2010). From ethical principles to response-able practice. *Environment and Planning D: Society and Space, 28*(1), 43–45.

Gregg, M., & Seigworth, G. (2010). An inventory of shimmers. In M. Gregg & G. Seigworth (Eds.), *The affect theory reader* (pp. 1–28). Duke University Press.

Haraway, D. (2016). *Staying with the trouble: Making kin in the Chthulucene.* Duke University Press.

Hickey-Moody, A., & Malins, P. (2007). Introduction: Gilles Deleuze and four movements in social thought. In A. Hickey-Moody & P. Malins (Eds.), *Deleuzian encounters: Studies in contemporary social issues* (pp. 1–26). Palgrave Macmillan.

Kofoed, J., & Ringrose, J. (2012). Travelling and sticky affects: Exploring teens and sexualized cyberbullying through a Butlerian-Deleuzian-Guattarian lens. *Discourse: Studies in the Cultural Politics of Education, 33*(1), 5–20.

Massumi, B. (2002). *Parables for the virtual: Movement, affect, sensation.* Duke University Press.

Matney, W. (2021). Music therapy as multiplicity: Implications for music therapy philosophy and theory. *Nordic Journal of Music Therapy, 30*(1), 3–23.

Meelberg, V. (2009). *Sonic strokes and musical gestures: The difference between musical affect and musical emotion.* In ESCOM 2009: 7th Triennial Conference of European Society for the Cognitive Sciences of Music.

Nesbitt, N. (2010). Critique and clinique: From sounding bodies to the musical event. In N. Nesbitt & B. Hulse (Eds.), *Sounding the virtual: Gilles Deleuze and the theory and philosophy of music* (pp. 159–179). Ashgate.

Niccolini, A. (2016). Terror (ism) in the classroom: Censorship, affect and uncivil bodies. *International Journal of Qualitative Studies in Education, 29*(7), 893–910.

Noddings, N. (2003). *Happiness and education.* Cambridge University Press.

Pedwell, C. (2014). *Affective relations: The transnational politics of empathy.* Palgrave Macmillan.

Potter, N. (2006). What is manipulative behavior, anyway? *Journal of Personality Disorders, 20*(2), 139–156.

Ringrose, J., & Renold, E. (2014). "F** k rape!" Exploring affective intensities in a feminist research assemblage. *Qualitative Inquiry, 20*(6), 772–780.

Ruud, E. (2008). Music in therapy: Increasing possibilities for action. *Music and Arts in Action, 1*(1), 46–60.

Sajnani, N. (2012). Response/ability: Imagining a critical race feminist paradigm for the creative arts therapies. *The Arts in Psychotherapy, 39*(3), 186–191.

Seyfert, R. (2012). Beyond personal feelings and collective emotions: Toward a theory of social affect. *Theory, Culture & Society, 29*(6), 27–46.

Stern, D. (2010). *Forms of vitality: Exploring dynamic experience in psychology, the arts, psychotherapy, and development*. Oxford University Press.

Stover, C. (2017). Affect and improvising bodies. *Perspectives of New Music, 55*(2), 5–66.

Swift, B. (2012). Becoming-sound: Affect and assemblage in improvisational digital music making. In *Proceedings of the SIGCHI conference on human factors in computing systems* (pp. 1815–1824). https://dl.acm.org/doi/abs/10.1145/2207676.2208315?casa_token=NDBtM6R50j8AAAAA:qWIPtHWq7WjKjq8oHhSBHhGI2kY5Mls5XcNptb8vTrQvWCqVOU-hNBMarp3T045jE5khCoBl5QqA3A

Wetherell, M. (2012). *Affect and emotion: A new social science understanding*. Sage.

14

Awareness Through Mapping Relationships

As we explore empathising assemblages, we draw our attention to an assemblage, seeking to notice its affective response-ability and to wonder how we might be part of enhancing it. One way to do this is to start by mapping the assemblage. We wouldn't be seeking to draw "a map" as a static, fixed representation when doing so. We'd be taking part in an ongoing, always provisional, dynamic, unfolding mapping process. A becoming-map isn't a reproduction of what's going it; it actively produces through creating connections and multiple entryways. It's always open to continuous modification (Deleuze & Guattari, 1987). When mapping relationships we are creating knowledge rather than discovering it and posing questions instead of making definitive claims (Waterhouse, 2011).

As we map relationships, we are identifying assemblages, including elements that are human and non-human, material and abstract, animate and inanimate, as well as what may usually be thought of as "micro" (interpersonal interactions) and "macro" (systems of authority and privilege) (Fox & Alldred, 2013). We explore how these affect and are affected, examining what bodies can do and are doing, and what capacities are being produced. We also identify territorialisations and de/re-territorialisations. As we look at a range of potential connections

A. dos Santos, *Empathy Pathways*,
https://doi.org/10.1007/978-3-031-08556-7_14

and interminglings, we may notice certain moments that "glow" to use MacLure's (2013) language. We might become particularly curious about a specific connection within the situation we're exploring; a certain entanglement may suddenly feel like it wants to "take off and take over" (p. 660). Our trail of thought may shift somewhere else unpredictably. For Deleuze and Parnet (1987), this involved being open to and aware of the possibility of a jolt or disturbance as an idea peeks its head up through the crack, the in-between. This process is improvisational. There are no strict procedures here (as this would go against the spirit of what rhizomes are all about) (Masny, 2016). Rhizoanalysis can be an arts-based process (Honan & Sellers, 2006). As art offers one of the most useful sites for affect to be produced (Hickey-Moody & Malins, 2007), and as music enables the virtual (Hallward, 2006) then what better place to begin than "the power of open creativity" (Williams, 2005, p. 19) found through the arts?

Barad (2007) defined attentiveness as "the ongoing practice of being open and alive to each meeting, each intra-action so that we might use our ability to respond, our responsibility, to help awaken, to breathe life into ever new possibilities for living justly" (p. x). Attentiveness requires engaging respectfully and carefully with whatever one is exploring. It isn't possible without openness towards human and non-human others. Openness is also about creating material spaces that enable such encounters to happen. This particular cultivation of attentiveness changes us and leads to co-constitution and becoming-with (Bozalek et al., 2018).

Composer and philosopher Pauline Oliveros (2005) coined the term "deep listening," referring to listening continuously in every possible way to everything possible: relationships in music, oneself, the physical space, and to other musicians. There will always be more to hear, and the unheard is also an inherent part of listening. Pavlicevic and Impey (2013) drew on Oliveros' idea of deep listening as they attempted to reframe health and well-being practices in the context of international development. They explored "cultural listening" (including a focus on the particular ways that music is situated and experienced); "social listening" (where we respond to improvisation and shared performance as configured social spaces and embodied social patterns, and also listen to social engagements and patterns as music), and "therapeutic listening" (noticing the listening that takes place between therapists and participants, how individuals and groups listen to multiple aspects of themselves, their relationships and to being listened to). This process invites co-creation, collaboration, spontaneity, and multiplicity of meanings. As we begin exploring how we might map assemblages, we do so through attentiveness and deep listening.

What is unfolding in your assemblage (your music therapy group, organisation, multidisciplinary team, classroom, research study, community), and how response-able is it? What event do you want to map? A therapeutic process as a whole? One session? One encounter? A video excerpt? A multidisciplinary team meeting? This chapter offers a range of approaches to mapping, not as a prescriptive recipe but as open-ended, creative invitations. Some approaches may be more useful to you than others as you explore how mutually affectively response-able your particular assemblage is, depending on the context and participants, time, and tools.

Collaborative Mapping

As an individual music therapist, you could map the relationships you're noticing, but it would be limited if you did this alone. As Menkiti (2004), an influential Nigerian ethicist remarked of an Igbo African proverb: "What an old man sees sitting down, a young man cannot see standing up" (p. 325). Music therapy literature frequently features the term "collaboration." It's a defining feature of feminist orientations, culture-centred practices, and community music therapy. However, collaboration as a process has been under-articulated (Bolger et al., 2018). Rolvsjord (2010) proposed that collaboration in music therapy is a shared negotiation process between clients and music therapists, characterised by mutuality, equality, and active engagement in decision-making. Responsibility for the direction and purpose of the process is shared. Bolger et al. emphasised that we need a better understanding of collaboration in music therapy to avoid tokenism and promote empowerment, especially when working with marginalised groups. It's not the music therapist's task to force collaboration, but to offer a collaborative intent from which a shared process might emerge. Collaboration can feel chaotic, but it's necessary to sit with this, rather than fixating on a "destination" or reinforcing power imbalances.

A group could engage in a mapping process through various forms, such as a conversation, an improvisation, or an art process; it depends on how the group members explore and communicate. As we discussed in Chapter 12, assemblages are composed of human and non-human things, so as we engage in mapping an assemblage, objects can be part of the process too. We might map with lyrics, clay, paper, pens, instruments, snacks, ideas, paint, chairs, beanbags, Wi-Fi, dance moves, and open windows that let in the breeze.

In Scrine's (2018) research, we see how mapping of critical issues was accomplished through songwriting. Scrine facilitated a participatory action research project investigating gender and power with young people in schools. A pilot project was conducted as the first cycle of inquiry, seeking to explore the issues that the participants wanted to raise about gender and what might emerge through music as a medium of exploration. Through flexible, participatory, and non-confrontational music-based activities, Scrine opened up space for the participants' feelings, thoughts, and beliefs about a vast array of topics. The pilot workshops were loosely based on two main activities: songwriting and music video analysis. Songwriting offered a flexible, dynamic, and affective form of engaging. As a group, they then selected a popular music video to analyse based on themes identified during the songwriting process.

North American Indigenous scholar Lavallée (2009) used "Sharing Circles" to collect data when researching with members of the Native Canadian Centre of Toronto. The epistemology rooted in Cree culture that underlies the Sharing Circle allows for including spirituality within research. Sharing Circles are similar to focus groups but differ in that they're considered sacred and offer opportunities for transformation and personal growth. The spirits of ancestors may visit the circle to guide the process. Authors in music therapy have also called attention to the role spirituality can play in exploring music therapy situations (Moonga, 2019; Potvin & Argue, 2014).

Seehawer (2018) described how *ubuntu* methodologies could be used to generate a collaborative learning process of mutual becoming. Storytelling, for example, is central within the oral traditions of many Southern African cultures. Storytelling celebrates holistic interconnectedness, reciprocity, respect, collaboration, humility and spirituality, and, importantly, impacts practice (Kovach, 2012). Relations take place through stories in a collaborative meaning-making process. Mucina (2011) emphasised that stories in Indigenous contexts extend further than individuals and are owned by the community as a whole. Storytelling was here before individuals were. Persons are born into stories, gain from and add to stories, co-author and share stories, and stories continue after individuals are gone. Stories have neither beginning nor end. When storytelling, a person becomes active for that moment in the collective story (Datta, 2018). Storytelling unfolds in an ongoing way through all the people involved. While individual contributions are received, these are relevant only in connection with the greater whole, including community members from earlier generations and those yet unborn (Seehawer, 2018).

The land upon which a collaborative mapping process takes place can be crucial, too. In a paper entitled *Co-becoming Bawaka*, offering an "Indigenous-led understanding of relational space/place" (p. 455), the authors included indigenous elders and their family members, a group of researchers and Bawaka Country, an Indigenous Homeland in northern Australia. They explained how Bawaka Country was the heart and lead author of the paper because Bawaka enabled their meeting and learning, the stories that guided them, and the connections they discussed. Bawaka Country brought them being (as who they are now and whom they continue to become). "Country" was seen to include everything tangible and non-tangible, human and more-than-human.

Mapping "Things" (as Always-Relational Becoming-Things)

As we explore empathising assemblages, we're concerned with the flow of affects. If we want to map an assemblage then we should start with "the between" where these affects are flowing (or getting stuck). We discussed in Chapter 12 that weak relationality notices parts and then explores how those parts are related, while strong relationality sees relationships "all the way down" (Slife, 2004, p. 159). We shouldn't start with "parts" as we map out our assemblage, then. But how do we do that? It seems very abstract.

Let's remember that even if we pay attention to an individual human in an assemblage, this human is also an assemblage (of organs, bacteria, water, history, thoughts). It's always through relationships that all "things"—subjectivities, ideas, activities of musicking, experiences of trauma, experiences of ill-health, health, forms of power, meanings of objects, strategies of resistance—emerge. It's impossible to step out of relational thinking, even when we're drawing "elements" of an assemblage on a map to explore "what" is in the assemblage, how these are affecting and being affected by each other, and how affects are flowing through the assemblage.

As Stover (2017) reminded us, a turn to affect does not and cannot focus only on the middle because there are still bodies that function as nodes between which a middle is drawn. Bodies affect and are affected by one another as they participate in the encounter. If we only tried to map the between and ignored elements, it would be like exploring the encounter between or the movement from a dominant to a tonic chord without considering the dominant or the tonic, which wouldn't make sense. Deleuze's philosophy is a philosophy of action, and there is no action other than that engendered by encounters of actual bodies in actual contexts. Even when we take into account the dominant and the tonic, for example, we can only do so through considering these chords as a dominant and as a tonic in terms of their relationship with one another. Therefore, we can actually start by mapping the "things" we're noticing in the assemblage if this feels easier at the start of the process, without betraying our relational intentions.

Thisness

When mapping the "things" present in an assemblage in a way that honours strong relationality, the concept of haecceity (or "thisness") is also beneficial for helping us understand how "things" are always "relational-things." Let's begin with an example that we can draw on as we explore these concepts. Fouché and Torrance (2005) described their work in Heideveld, an area approximately 15 km from Cape Town's city centre in South Africa. This suburb is home predominantly to "Coloured" people. (This is a term used distinctly by persons constructed as mixed-race in South Africa to self-identify; it does not have the offensive connotations that it holds in other countries [Adhikari, 2005].) The older generation in Heideveld remembers being forcefully relocated to the "Cape Flats" from suburbs in the inner city in the 1960s by the Apartheid government. Residents in Heideveld experience a proliferation of social problems, including overcrowded living conditions, family fragmentation, widespread drug and alcohol dependency, unemployment, school truancy, and high levels of violence, specifically gang violence. Gangs provide young boys in particular with the emotional support that their families often cannot. Gang members find a sense of identity, acceptance, and belonging within these surrogate families, as well as power and purpose. Even though gangs terrorise the community, community members often give misinformation to the police or turn a blind eye when a gang-related incident occurs. While residents fear recrimination, there is also a mutual dependency and control through charity: gangs can offer protection, gang bosses sponsor feeding schemes, local soccer teams, and music groups.

Police in Heideveld are reticent to arrest youngsters as rehabilitation rates are exceedingly low in Southern African prisons and prison gangs recruit young adolescents. When they became aware of the music therapy work that Fouché and Torrance (2005) were facilitating in the area, they began fetching children and teenagers (from a range of different gangs), disarming them, and delivering them to music therapy. Sessions comprised both musicking (often through unstructured improvisation) and talking. Participants offered stories and explanations of life in their community and life within their gangs. While they seemed withdrawn and sceptical at first, they returned each week eagerly, perceiving music as a "cool" thing to do. In Rap and Hip-Hop culture, musicians are seen as heroes and social commentators.

As these youth were affiliated with different gangs, their contact with one another outside of sessions was usually only hostile. The therapists were also "Whiteys" from Cape Town who would typically be considered outsiders. In improvisation, however, a sense of togetherness was produced. Within the Music Therapy room, the identities and roles of gang members were shed and replaced with vulnerability and openness. Music therapy provided opportunities for exploring different ways of expressing, creating, and relating. Fouche and Torrance described how their stereotypes of gangs and adolescent gang members were challenged. They related to the participants as young persons in music therapy, which fostered a space where members could experiment with other social identities.

Adele Clarke (2005) developed a method called situational analysis. In this approach, "there is no such thing as 'context'" (Clarke, 2016b, p. 98). The (internal/external/local/global) conditions of the situation are not framing, surrounding, or contributing to it. They are seen as being in the situation; they are the situation. In situational analysis, the central question is how these conditions appear as integral components of the situation under investigation. Within situational analysis, one is invited to create three kinds of maps (Clarke, 2016a). The first map, a situational map, can be done in three phases: first, you can make a messy one, then a more ordered one, and finally, a relational one showing links. The first two phases relate to our current focus. For the messy one, you initially lay out all the significant elements in the situation of concern freely and broadly on a large piece of paper.

Clarke (2016a) recommended writing down elements such as (but by no means only) individual human actors, silent actors, collective human actors, non-human elements, discursive constructions of human actors and non-human elements, political/economic elements, temporal elements, spatial elements, significant issues and debates, sociocultural/symbolic elements, and related discourses. If we were to start a mapping process related to the Heideveld music therapy example, we might list some of the following: the suburb of Heideveld; "Coloured" people; Apartheid; unemployment; drug and alcohol dependency; family fragmentation; overcrowded living conditions; school truancy; gang members, gang-violence; identity; belonging; surrogate families; power and purpose; adolescence; protection; soccer teams; music groups; feeding schemes; police intervention; music therapists; talking; improvisation; stories; Rap and Hip-Hop culture; music therapists as outsiders; identities; stereotypes; vulnerability.

Situational maps can be created and recreated multiple times across the duration of a project; there is no one "correct" map. The process can help open up the analysis and create room for surprises. Once you have drawn a messy situational map, you can create a more ordered one, grouping the elements into categories. This method has been used in music therapy by Potvin et al. (2018), for example, in a study with informal caregivers of hospice patients.

Crucially, as we're identifying elements in the assemblage, we're not considering them to be discrete "points." The notion of "thisness" offers us a useful tool to remain in relational thinking. Thirteenth-century philosopher Duns Scotus distinguished between quiddity ("whatness") and haecceity ("thisness") (Cross, 2010). What is the object up against the wall next to my desk? It's a piano. That is its "whatness." It's not just any piano, however. What makes it *this* distinct piano? It's an upright piano that became part of my family when I was 12 years old; it is a piano that is intimately intertwined with my personal and musical development; because it is over 100 years old it currently needs a thorough overhaul. These and many other features are part of what makes this piano *this* piano. The description I've just given is relational: upright-piano-that-joined-my-family-when-I-was-12; piano-that-connects-to-my-development; piano-that-has-grown-old-over-time-and-needs-work, and so on. As I describe this piano's "thisness," I can't help but use a relational lens to do so. As I reflect on what makes it this piano, I see the assemblage lines that it's entangled in.

A haecceity (thisness) *is* an assemblage. Deleuze and Guattari (1987) illustrated the notion of haecceity by referring to a sentence by Virginia Woolf: "The thin dog is running in the road, this dog is the road" (p. 263). The dog, the road, and the running can't be separated or individuated; they compose one another. Haecceity can refer to subjects ("this is you," "this is me") and to events (this wind; this time of day; this improvisation) (St. Pierre, 2017). As a singularity ("this piano"), haecceity has "neither a beginning nor an end, origin, nor destination; it is always in the middle. It is not made of points, only of lines" (Deleuze & Guattari, 1987, p. 263).

There are some parallels to the idea of haecceity within certain Indigenous Knowledge Systems. Bird-David (1999) described how hunter-gatherers in Nayaka of the Nilgiri region of South India got to know "Nayaka-in-relatedness with fellow Nayaka" (p. 72). They got to know how each member of their social environment talked with fellows (not how each talked), how each shared with fellows (not how each shared), how each worked with fellows (not how each worked), and so on. Personhood was made as they shared relationships with human and non-human beings. They regarded themselves as nested within each other.

Let's return to the earlier list drawn from Fouche and Torrance's (2005) article and examine items such as "gang-violence" or "improvisation." We can ask: what made this gang violence, *this* gang violence in this area at this time? What was it about an improvisation that made it this improvisation in this particular session? If we reflect on the elements of "gang members" and "music therapists," we could ask: how did these gang members relate to other gang members in the group such that they could be recognised in a nested manner as gang-members-in-relation-with-other-gang-members and gang-members-in-relation-with-"Whitey"-music-therapists? It's problematic to examine any component of an assemblage in isolation because being this component within the larger whole is a core part of what it is.

Becoming

As we explore "this" element of an assemblage, we remember that it isn't static. It's a relational unfolding in terms of its flow through time and its differentiating from itself. We (and everything else) are in a constant process of becoming. As music therapists, we're familiar with this idea through Small's (1998) notion of "musicking." We know that music is not an object. Music is a verb (hence "musicking"). It's always in action. It's always becoming. We don't analyse musical objects; we analyse musical processes.

Deleuze and Guattari (1987) distinguished between the "actual" (the creature/the created) and the "virtual" (the creating). Instead of trying to explain the actual, as the expressed, the written, the sung, the played, we focus on the virtual by exploring the expressing, the writing, the singing, the playing. In other words, we are interested in becoming itself. Deleuze did pay careful attention to the mechanics of actual creation, but the reason he did so was to escape them. We may see glimpses of created creatures (the image, the sound, the work, the person), but the creature/the created is always an unfolding of the process of creating (Hallward, 2006).

A music therapist is always a becoming-music therapist; a participant is a becoming-participant; a gang member is a becoming-gang member. There are no "destinations" here (we are not becoming *a* music therapist, a person is not becoming *a* teenager). These are fluid, ongoing, transforming, interacting processes. As Zarate (2020) explained, improvisation in music therapy does not express a client's personal narrative. It is a condition through which a client comes into being (in an ongoing way). The music therapist is also becoming through improvisation, for example. Affective responsiveness comes about through and with this (and other) dynamic intertwining of lines in music therapy assemblages.

The previous chapter discussed how Deleuze (1991) understood humans as habits. When we map assemblages, we are noticing habits, happenings, utterances, desires, expressions, mannerisms, energy, patterns, events, ideas, signs, and concepts, not only as actualised products but in terms of the conditions of their emergence within this field of experience (Rolli, 2009). In Ferrando's (2019) explanation of posthumanism, she argued that we should understand the human as a verb: to humanise. The human is a process, not an essence. One is becoming human through socialisation, experience, reception, retention, or refusal. We are humanising within and through relationships all the time. All human and non-human bodies are continuously becoming as they differentiate from themselves. I'm becoming anew today in relation to who I was yesterday, or an hour ago. The world is made of verbs rather than nouns, processes rather than substances, and "happenings" rather than "things" (Shaviro, 2009). It can be enormously helpful to always see our clients as becoming, no matter how "stuck" they may seem or feel. If a person doesn't have a substantive essence, then neither health nor illness is a knowable, stable, fixed property of an individual body. Instead, we can understand these also as processes of becoming (Duff, 2014).

Agential Cuts (Why These "Things" Rather Than Other "Things"?)

Identifying an entanglement such as becoming-gang-violence-that-is-fueling-a-sense-of-belonging-within-surrogate-becoming-families is an example of an agential cut, to use language offered by Barad (2003). An agential cut is a momentary stabilisation. Through agential cuts, things come to be known as things (this creation and not that creation) (Kessler, 2019). Barad (2007) explained that agential cutting happens as part of the everyday performance of material-discursive practices within which individuals are intertwined.

Law and Urry (2004) highlighted how researchers influence the reality they study through their choice of theories and methods. We don't only make epistemological choices (how we will go about knowing things), but we make agential cuts that have ontological implications (i.e., what things are). Researchers and data are part of a research assemblage. We can think of research as "a territorialisation that shapes the knowledge it produces" (Law, 2004, p. 403). We create worlds through the theoretical-methodological approaches we adopt. There are, thus, high stakes in these endeavours (Schultze, 2017). Identifying the things we will map (and deciding that we will leave out other things) matters. The same is true for how we engage in and reflect on our music therapy sessions, selecting certain aspects to focus on and not others.

We ourselves are always entangled in the process (Shannon & Truman, 2020). In the previous chapter, we discussed how immanence refers to being "of" a situation (while transcendence refers to being "on" or "in" a situation). If we seek to map relationships, we are not doing so as an observer who is "outside" these relationships. We are *of* the relationships.

Mapping Positions

The second map used in situational analysis is the social worlds/arenas map. The goal of this map is to visualise social relations (Clarke, 2016b). A "social world" refers to a group of people who gather because of a particular shared interest, shared commitment, and willingness to act (for example, a band, people in the scholarly field of music therapy, a rugby team). An "arena" is an area of sustained concern (music, music therapy, sport) that brings multiple social worlds together over time. When making a social worlds/arenas map, you lay out the collective actors as well as the arena(s) of concern that they are engaged in (Clarke, 2016a). You make a social worlds/arenas map by putting the arena in the middle of a page (engaging in a process of group music therapy together, for example) and then writing out the different social worlds that come together around that, drawing circles with dotted lines around each one (the therapists; the participants; the care workers who sometimes attend sessions; management at the organisation; cleaning staff; and so on). You may note the main discourses and technologies of the different social worlds within the circles. As your map is taking shape, you can consider the relationships between the different social worlds. Do some overlap? (If so, visualise them in this way). Are some more central to the arena? These can be centred or enlarged (Clarke, 2016b).

In a study with a group of Aboriginal healthcare workers in an urban Aboriginal health service who described and interpreted their occupational activities, Genat et al. (2006) created a social worlds/arena map. This showed the Aboriginal healthcare workers, as well as the groups of clients, nurses, doctors, welfare workers, and managers who formed specific social worlds with particular perspectives on health worker practice. These social worlds together created the social arena of Aboriginal health worker practice. Social worlds/arenas maps help us see how there are always multiple voices involved in any situation.

The third map, a positional map, lays out the most prevalent positions taken and not taken in the situation, including axes of variation and difference. In positional maps, you don't show persons or groups, but rather seek to identify the wide range of (potentially heterogeneous and complex) discursive positions on significant issues at play in the overall situation of concern (Clarke, 2016a).

Strong et al. (2012) used situational analysis to explore how counsellors respond to the Diagnostic and Statistical Manual (DSM) (American Psychiatric Association, 2000). (They referred to the DSM-IV-TR as this was the current edition at the time.) The authors made a messy situational map after gathering data through semi-structured interviews, website discussion forums, survey responses, and research literature. The elements were organised into categories, including professions that have some influence within the situation (such as psychiatry, clinical psychology, counselling, and social work), elements relating to practices, academic/professional discourses, legal processes, and cultural/moral issues. Strong et al. then developed two relational maps: one exploring the influences of the DSM on counsellors' practices and one showing the creative responses of counsellors to expected uses of the DSM. When creating positional maps, they considered how counsellors could simultaneously hold two or more positions on the DSM (for example, the DSM enables medical services for some clients, and the DSM offers medicalised language that can be generally problematic in counselling). Strong et al. didn't aim to homogenise the voices of their participants into a coherent portrayal, but to convey positional differences on a complex situation.

Mapping Processes of Territorialising (and De/Reterritorialising)

Chapter 12 explored how a territory can be understood as a "context of lived reality" (Stover, 2017, p. 20). It is a coming-together of bodies that produces a context and is always an active expression of territorialisation. To map this process, we may ask how things and ideas, practices, and positions mark the space we're examining. What expressive acts are unfolding here? What kind(s) of territories are becoming here? What is the character of this territory (as an event)? What are the effects of the resonant space? How are we becoming within this territory? What habits are at home here? Who feels comfortable in this territory? What ruptures are happening (as deterritorialisation)? What new lines of flight are opening (as new thoughts, feelings, movements, roles, possibilities?) Are these leading to new territories? How are we being reterritorialised within these new territories?

Mapping Pathways

As we map the assemblage that we're part of, we may wonder what pathways are being travelled here. What musical journeys, conversations, connections, encounters, decisions, sensory experiences, friendships are unfolding? What are the capacities of the elements of this assemblage to travel these pathways?

Pavlicevic et al. (2015) described music therapy's "ripple effect" in a care home for persons with dementia. Music therapy resonates between a micro-level of person-to-person musicking, a meso-level where musicking occurs beyond the time of the sessions, and a macro-level within and beyond the care home. All persons within the care home (residents, cleaners, managers, visiting family members) were invited and supported to generate, sustain, and enliven social-musical pathways and networks by adding to the musical repertoire and taking part in impromptu musical interactions. Pavlicevic et al. proposed that self-identities could be reconfigured through participative therapeutic musicking from ones that restricted persons to their health condition or work status to fuller and more expansive identities.

Through their six-year ethnography of mental health and community music therapy (SMART), DeNora and Ansdell (2017) explored how musical engagement happened in real-time and how it mattered to the people involved. As mentioned earlier, SMART is a community centre alongside a psychiatric hospital. Weekly community music therapy activities were offered, such as opportunities to take part in solos with accompaniment (provided mainly by the music therapists at the keyboard), group ensemble pieces, sing-a-longs, and improvisations. A choir was then added to the mix, as well as a rock band and a music theory class. The data that DeNora and Ansdell gathered included observations of the built environment and its uses, material culture, music therapeutic crafting, repertoire, gestures, and the evolving musical space.

DeNora and Ansdell (2017) described nine individual "pathway" case studies focusing on connections made by the participants. By documenting and comparing repeated musical practices and attachments within the social network, they could map (by hand) a topology of practice over time. An individual's participation could be drawn in relation to the parts of the space they occupied (using line weight to show the depth of their musical engagement during a particular period). One of their case studies concerned Daniel, a frequent participant in SMART, whose musical contributions clustered around jazz and blues. In another case study, they described Eloise as connecting initially to the cymbal, and then how this expanded out (through her growing freedom of movement) to connections with many other things such as jazz brushes on a cymbal, friendships with other musicians, and reminiscing about travels to Turkey. These "furnishings" (such as instruments) and the connected music altered Eloise and her relation to others, drawing her into new kinds of practice and relationships through gentle processes of give-and-take. Eloise was thus afforded even more links, including more flexible use of verbal language. DeNora and Ansdell didn't include information about Daniel or Eloise in terms of other facets of their cultural, ethnic, or national becoming (for example) and how these features could either be included as part of the pathways in the maps or help us think about the maps. This may have been useful too.

Mapping Interminglings

In Clarke's (2005) situational analysis, once you've drawn your messy situational map and then a more ordered one, you can do relational analyses with it. Relations among the various elements in the situational map are key. You can systematically take each element at a time and think about it in relation to the other elements on the map, drawing lines between them (potentially with coloured markers) and specifying the nature of the relationships by describing the character of each line.

According to Clarke (2016b), this is the most important work that one can do with situational maps.

While every "thing" in an assemblage is already itself an assemblage (intermingled and entangled with all sorts of other "things"), in the mapping process, we can purposefully zoom into and play around with entanglements between the things we've identified. Deleuze and Guattari (1987) referred to "interminglings" as some of the infinitely potential connections that take place in an assemblage. What interminglings are glowing here? What is emerging as these elements responsively intra-act? We may choose to wonder what opens up and closes down in these connections. We could, for example, explore how the intermingling of different participants' stories in a music therapy group produces new ideas as they affect and are affected by each other. In a study I conducted (dos Santos, 2018) on empathy and aggression in music therapy with groups of teenagers, some of the multiple interminglings that seemed to me to "glow" in the sessions included leadership—fun (the relationships of playfulness that the teenagers had with me as the therapist was very different in nature to the ones that they had with authority figures in their lives such as parents and teachers); aggression—friendship (the participants sometimes used aggression to defend a friend); aggression—self-esteem (when they used aggressive acts to declare their self-worth in the face of insults); and physical violence—desire/wanting (when the urge to hurt another person rose within them).

Mapping Affective Flows

Imagine a group of teenagers writing a song about examination stress in music therapy. One of them offers an idea for the lyrics about how hard it is to concentrate when you're supposed to be studying but you just want to scroll through social media. The others enthusiastically agree, hopping out their chairs and gesticulating with bright eyes and broad grins. Another group member then suggests a melody for this line, and the group suddenly disintegrates, some agreeing strongly and others shaking their head as they dismiss the idea. A gravelly texture enters the affective space. A third member then offers an idea to change the song altogether so that it becomes about dropping out of school and the others sigh, slumping back in their chairs. The previous excitement seems to seep through the floor.

As we explore an assemblage, we notice affect as intensity. We think about how we are affecting and being affected, how affect moves through the assemblage, and how it travels and sticks. How do we then use these ideas in practice when attempting to be part of an assemblage that can become more mutually response-able? According to Alldred and Fox (2015), we can analyse an assemblage by exploring the ability of the assembled relations to affect or be affected rather than in terms of human agency. Clough (2004) referred to the "affect economy" of an assemblage, refers to the forces that move bodies and other relations "from one mode to another, in terms of attention, arousal, interest, receptivity, stimulation, attentiveness, action, reaction, and inaction" (p. 15). In *Beyond Good and Evil*, Friedrich Nietzsche (2009) wrote, "thanks to music, the passions enjoy themselves" (p. 75). How do we trace these passions? How is "bursting," "disintegrating," and "seeping" shifting, unfolding, affecting, and being affected?

We can develop mindfulness to affect within an event. When we are exploring how empathic an assemblage may be, we could do so in the moment, through giving "living attention" (Brennan, 2004, p. 32) to the ongoing unfolding of the assemblage. We can also do so in retrospect, as we look back or stand back (and think about how we wish to move forward). Willink and Shukri (2018) reflected on how one could analyse a research interview situation within the interview itself through affective attunement, and afterwards through reflective affective analysis. I would argue that their approach is helpful in exploring music therapy, too.

The term "affect attunement" is referred to widely in music therapy literature. For Willink and Shukri (2018), affective attunement refers to how individuals orient themselves relationally within encounters, people, events, spaces, histories, and futures through somatic interplay, which is the capacity to feel and notice (often through sound) and to tune into the non-representational. We are sensing "the intimate unfolding of becoming together" (p. 201) as it is happening. Affect attunement occurs in the "ecology of relational happenings" (Alifuoco, 2017, p. 37) that occur within a music therapy session. Affects can sometimes be felt as shifting tonalities in the relational dance. The affective lens calls attention to embodied experiences and highlights vitality affects (Stern, 2010). We sense the choreography of the affective. Affective attunement is a practice, and we become better at it with practice.

Foregrounding affective dimensions is a specific discipline of presence. It requires sustained attention, humility, letting go of control and authority, and purposefully denaturalising our emotional responses (to try to get free of our "hard-wired settings" and to stay alert, attentive, and curious instead of just seeing and hearing what we're always used to seeing and hearing, and noticing what we've already decided should matter). We become mindful of compelling moments, even if they are unexpected or unrecognisable. We are inviting response-ability.

Willink and Shukri (2018) employed reflective affective analysis when reflecting back on affect within a situation. This involves retrospectively reflecting on how affects, intensities, and rhythms animated the encounter. As a method, reflective affective attunement requires an aesthesic sensibility. Aesthesics refers to the study of perception with all the senses.

Mapping Affective Lines of Flight

In relation to SMART, Ansdell explained (in DeNora & Ansdell, 2017) how he once walked across to the adjacent mental health unit to fetch songbooks from the music therapy room and, along the way, met a woman who was shuffling along the corridor towards the locked doors. She appeared anxious and tense. Pushing the buzzer, she spoke to the security guard who opened the door, telling her that she had two hours' leave but may not go far. As Ansdell walked along, he realised that he and the woman were going in the same direction. As she walked away from the mental health unit, he noticed how her gait gained a greater sense of freedom. Her pace quickened, her shoulders relaxed, and she looked up for the first time. She seemed to have a stronger sense of purpose. As she walked into the SMART cafe, someone called out "Bridget!", inviting her to come in, sit down and have a cup of coffee, and commenting on how great it was that she came. Bridget smiled and sat down next to this long-term member of SMART, Lucy. Seated at the piano, Andsell noticed how Bridget continued to relax further. She shared a songbook with Lucy and began to sing. During the tea break, Bridget told Gary that she had been admitted to the hospital due to overwhelming anxiety and agoraphobia. Music therapy was one thing that helped her feel human again, but she had been nervous about coming to SMART. It relieved her to find how informal and friendly the group was. Bridget found a line of flight, both in terms of the physical space from the mental health facility to SMART space and from patient to musician.

Mapping Response-Ability

As we consider empathy, we map an assemblage with interest in how mutually affectively response-able it is. Comparisons between assemblages that are empathising to greater and lesser degrees do not function as binaries but as continuums. A less mutually affectively response-able assemblage is shut off, rigid and inflexible, and blocks entry. It's authoritarian and focuses on singularities and unitary identities. Within it, the emphasis lies on elements as static and molar. It's more fixed and is limited in its capacity to affect and be affected. Affects tend to stick. It can be inward-looking, lethargic, and much remains the same. It may prefer using solutions that worked in the past. People and things may seem to become diminished in some way. Certain people may wish to change other people. The assemblage may prompt retreat, dismissiveness, avoidance, or silencing.

A more mutually affectively response-able assemblage is open, flexible, accessible, and invites in people, ideas, events. It's connected and open to expanding. It is also collaborative and acknowledges multiplicities and becoming. This assemblage is fluid and can affect and be affected. Affects tend to move more fluidly. A mutually affectively response-able assemblage produces vitality, learning, growing, and changing. It is one where unexpected things can happen, and these tend to be welcome. All are transformed within such an assemblage.

References

Adhikari, M. (2005). Contending approaches to coloured identity and the history of the coloured people of South Africa. *History Compass, 3*(1), 1–16.
Alifuoco, A. (2017). 'Alive' performance: Toward an immersive activist philosophy. *Performance Philosophy, 3*(1), 126–145.
Alldred, P., & Fox, N. J. (2015). The sexuality-assemblages of young men: A new materialist analysis. *Sexualities, 18*(8), 905–920.
American Psychiatric Association. (2000). *Diagnostic and statistical manual of mental disorders* (4th ed., text rev.). American Psychiatric Association.
Barad, K. (2003). Posthumanist performativity: Toward an understanding of how matter comes to matter. *Signs: Journal of Women in Culture, 28*(3), 801–831.
Barad, K. (2007). *Meeting the universe halfway: Quantum physics and the entanglement of matter and meaning.* Duke University Press.
Bird-David, N. (1999). "Animism" revisited: Personhood, environment, and relational epistemology. *Current Anthropology, 40*(S1), S67–S91.
Bolger, L., McFerran, K. S., & Stige, B. (2018). Hanging out and buying in: Rethinking relationship building to avoid tokenism when striving for collaboration in music therapy. *Music Therapy Perspectives, 36*(2), 257–266.
Bozalek, V., Bayat, A., Gachago, D., Motala, S., & Mitchell, V. (2018). A pedagogy of response-ability. In V. Bozalek, R. Braidotti, M. Zembylas, & T. Shefer (Eds.), *Socially just pedagogies in higher education: Critical posthumanist and new feminist materialist perspectives* (pp. 97–112). Bloomsbury Academic.
Brennan, T. (2004). *The transmission of affect.* Cornell University Press.
Clarke, A. (2005). *Situational analysis.* Sage.
Clarke, A. E. (2016a). Introducing situational analysis. In A. E. Clarke, C. Friese, & R. Washburn (Eds.), *Situational analysis in practice* (pp. 11–76). Routledge.
Clarke, A. E. (2016b). From grounded theory to situational analysis: What's new? Why? How? In A. E. Clarke, C. Friese, & R. Washburn (Eds.), *Situational analysis in practice* (pp. 84–118). Routledge.
Clough, P. T. (2004). Future matters: Technoscience, global politics, and cultural criticism. *Social Text, 22*(3), 1–23.
Cross, R. (2010). Recent work on the philosophy of Duns Scotus. *Philosophy Compass, 5*(8), 667–675.

Datta, R. (2018). Traditional storytelling: An effective Indigenous research methodology and its implications for environmental research. *AlterNative: An International Journal of Indigenous Peoples, 14*(1), 35–44.

Deleuze, G. (1991). *Empiricism and subjectivity: An essay on Hume's theory of human nature* (C. Boundas, Trans.). Columbia University Press.

Deleuze, G., & Guattari, F. (1987). *A thousand plateaus: Capitalism and schizophrenia.* Athlone.

Deleuze, G., & Parnet, C. (1987). *Dialogues.* Athlone.

DeNora, T., & Ansdell, G. (2017). Music in action: Tinkering, testing and tracing over time. *Qualitative Research, 17*(2), 231–245.

dos Santos, A. (2018). *Empathy and aggression in group music therapy with adolescents: Comparing the affordances of two paradigms* [Doctoral dissertation, University of Pretoria].

Duff, C. (2014). *Assemblages of health: Deleuze's empiricism and the ethology of life.* Springer.

Ferrando, F. (2019). *Philosophical posthumanism.* Bloomsbury Academic.

Fouché, S., & Torrance, K. (2005). Lose yourself in the music, the moment, Yo! Music therapy with an adolescent group involved in gangsterism. *Voices: A world forum for music therapy, 5*(3).

Fox, N. J., & Alldred, P. (2013). The sexuality-assemblage: Desire, affect, anti-humanism. *The Sociological Review, 61*(4), 769–789.

Genat, B. with Bushby, S., McGuire, M., Taylor, E., Walley, Y., & Weston, T. (2006). *Aboriginal healthworkers: Primary health care at the margins.* University of Western Australia Press.

Hallward, P. (2006). *Out of the world: Deleuze and the philosophy of creation.* Verso.

Hickey-Moody, A., & Malins, P. (2007). Introduction: Gilles Deleuze and four movements in social thought. In A. Hickey-Moody & P. Malins (Eds.), *Deleuzian encounters: Studies in contemporary social issues* (pp. 1–26). Palgrave MacMillan.

Honan, E., & Sellers, M. (2006). *So how does it work?—Rhizomatic methodologies.* Paper presented at the AARE Annual Conference, Adelaide. Retrieved from http://www.aare.edu.au/publications-database.php/5086/so-how-does-it-work-rhizomatic-methodologies

Kessler, N. (2019). *Ontology and closeness in human-nature relationships.* Springer.

Kovach, M. (2012). *Indigenous methods: Characteristics, conversation, and contexts.* University of Toronto Press.

Lavallée, L. F. (2009). Practical application of an indigenous research framework and two qualitative indigenous research methods: Sharing Circles and Anishnaabe symbol-based reflection. *International Journal of Qualitative Methods, 8*(1), 21–40.

Law J. (2004). *After method: Mess in social science research.* Routledge.

Law, J., & Urry, J. (2004). Enacting the social. *Economy and Society, 33*(3), 390–410.

MacLure, M. (2013). Researching without representation? Language and materiality in post-qualitative methodology. *International Journal of Qualitative Studies in Education, 26*(6), 658–667.

Masny, D. (2016). Problematizing qualitative research: Reading a data assemblage with rhizoanalysis. *Qualitative Inquiry, 22*(8), 666–675.

Menkiti, I. (2004). On the normative conception of person. In K. Wiredu (Ed.), *A companion to African philosophy* (pp. 324–331). Blackwell Publishing.

Moonga, N. U. (2019). *Exploring music therapy in the life of the batonga of Mazabuka Southern Zambia* [Master's dissertation, University of Pretoria].

Mucina, D. (2011). Story as research methodology. *AlterNative: An International Journal of Indigenous Peoples, 7*(1), 1–14.

Nietzsche, F. (2009). *Beyond good and evil* (I. Johnston, Trans.). Richer Resources Publications.

Oliveros, P. (2005). *Deep listening: A composer's sound practice.* IUniverse.

Pavlicevic, M., & Impey, A. (2013). Deep listening: Towards an imaginative reframing of health and well-being practices in international development. *Arts & Health, 5*(3), 238–252.

Pavlicevic, M., Tsiris, G., Wood, S., Powell, H., Graham, J., Sanderson, R., Millman, R., & Gibson, J. (2015). The 'ripple effect': Towards researching improvisational music therapy in dementia care homes. *Dementia, 14*(5), 659–679.

Potvin, N., & Argue, J. (2014). Theoretical considerations of spirit and spirituality in music therapy. *Music Therapy Perspectives, 32*(2), 118–128.

Potvin, N., Bradt, J., & Ghetti, C. (2018). A theoretical model of resource-oriented music therapy with informal hospice caregivers during pre-bereavement. *Journal of Music Therapy, 55*(1), 27–61.

Rolli, M. (2009). Deleuze on intensity differentials and the being of the sensible. *Deleuze Studies, 3*(1), 26–53.

Rolvsjord, R. (2010). *Resource-oriented music therapy in mental health care.* Barcelona Publishers.

Schultze, U. (2017). What kind of world do we want to help make with our theories? *Information and Organization, 27*(1), 60–66.

Scrine, E. (2018). *Music therapy as an anti-oppressive practice: Critically exploring gender and power with young people in school* [Doctoral dissertation, University of Melbourne].

Seehawer, M. K. (2018). Decolonising research in a Sub-Saharan African context: Exploring Ubuntu as a foundation for research methodology, ethics and agenda. *International Journal of Social Research Methodology, 21*(4), 453–466.

Shannon, D. B., & Truman, S. E. (2020). Problematizing sound methods through music research-creation: Oblique Curiosities. *International Journal of Qualitative Methods, 19*. https://doi.org/10.1177/1609406920903224

Shaviro, S. (2009). *Without criteria: Kant, Whitehead, Deleuze and aesthetics*. MIT Press.

Slife, B. D. (2004). Taking practice seriously: Toward a relational ontology. *Journal of Theoretical and Philosophical Psychology, 24*(2), 157.

Small, C. (1998). *Musicking: The Meanings of Performing and Listening*. Wesleyan University Press.

St. Pierre, E. A. (2017). Haecceity: Laying out a plane for post qualitative inquiry. *Qualitative Inquiry, 23*(9), 686–698.

Stern, D. (2010). The issue of vitality. *Nordic Journal of Music Therapy, 19*(2), 88–102, 570.

Stover, C. (2017). Affect and improvising bodies. *Perspectives of New Music, 55*(2), 5–66.

Strong, T., Gaete, J., Sametband, I. N., French, J., & Eeson, J. (2012). Counsellors respond to the DSM-Iv-tR. *Canadian Journal of Counseling and Psychotherapy, 46*(2), 85–106.

Waterhouse, M. (2011). *Experiences of multiple literacies and peace: A Rhizoanalysis of becoming in immigrant language classrooms* [Doctoral dissertation, University of Ottawa, ON].

Williams, J. (2005). *Understanding poststructuralism*. Acumen.

Willink, K. G., & Shukri, S. T. (2018). Performative interviewing: Affective attunement and reflective affective analysis in interviewing. *Text and Performance Quarterly, 38*(4), 187–207.

Zarate, R. (2020). *Critical social aesthetics and clinical listening ←→ Cultural listening as a method for arts-based research* [Paper presentation]. World Congress of Music Therapy, Pretoria, South Africa.

15

Enhancing Response-ability

In one sense, response-ability is at play all the time. Every bit of matter is in a response-able relationship with other bits. Matter is always touching and being in touch (Barad, 2015). In another sense, we can acknowledge that there are relational forces that enhance or decrease the capacity for action and that practices can become more or less response-able (Ringrose et al., 2019). When we intra-act more response-ably as part of the world, we are responsive to the possibilities that could help us and it flourish (Bozalek et al., 2018).

In this chapter, we'll look at a range of ways for enhancing empathising within our music therapy assemblages. Once we've mapped the assemblage we are part of (understanding the "things," relationships, positions, affects, and entanglements at play, as well as how territorialising is taking place), we can ask specific questions about how we can be part of enhancing empathising in a number of practical ways.

A. dos Santos, *Empathy Pathways*,
https://doi.org/10.1007/978-3-031-08556-7_15

Response-able Access and Inclusion

We can't claim to have response-able music therapy practices if access is exclusive. How do we respond with someone who hasn't gained entry in the first place? Involvement in musicking in general can offer resources for gaining a better quality of life (Ruud, 1997). Still, many are kept away from aspects of musicking by social, economic, and/or structural constraints (Small, 1998). Poverty may limit individuals' access to musical instruments, concert attendance or paid streaming services, and communities' opportunities to participate in choirs, orchestras, and music education opportunities. Traditions of elitism, gender expectations, social values, and patterns of social capital distribution may also constrain access to musicking. In addition, access depends on the varying abilities and possibilities people and groups have for using what musicking affords. Some may experience a lack of musical proficiency as a barrier to participation in musicking. Many other factors, such as disability, social anxiety, performance anxiety, low self-esteem, or depression, can also make participation more difficult. According to Rolvsjord (2006b), the reduction or elimination of these constraints that can stand in the way of people's use of music as a health resource must be a chief concern for music therapists. As far back as 1998, Ruud called for music therapy "to align with those forces in society that work towards creating a space for human empowerment, self-insight, personal growth, solidarity, and social networking and with those that work toward alleviating structural forces blocking possibilities of action" (p. 5).

While music therapy is a recognised and regulated health care profession in my country, South Africa, the public health services do not include music therapy (yet). It is an option that is mostly only available to people who can afford to access private health care (approximately 17.2% of the population) (Statssa, 2019). Through a community music therapy project, Fouché and Stevens (2018) took part in the process of co-creating musical spaces where young people and those around them could access resourced relationships. MusicWorks facilitated this process as a non-profit organisation based in Cape Town, South Africa. MusicWorks seeks to enable access to music therapy services for children who wouldn't be able to take part otherwise. They aimed to participate in enhancing the resilience of children and their communities through offering opportunities for active engagement in musicking (spanning individual and group music therapy sessions to community performances). At MusicWorks, music therapists and community musicians work together in mutually response-able ways.

In the United States, Thomas (2020) described how adolescents from under-resourced communities struggle to define and empower themselves within environments that neither value nor support them. School-based music ensembles are still over-represented by White, upper-middle-class students. Thomas collaboratively explored the feasibility of a community-based music-making intervention with Black/African American teenagers from these communities that could create culturally responsive opportunities for engaging with peers in a safe, supportive, and meaningful way as part of a holistic health continuum. Through processes such as actively creating and sharing a music video, they enhanced their capacity to act by uplifting their voices, connecting with peers and others in their community, and accessing resources in sustainable ways.

Community participation is crucial for recovery from mental illness. Yet, individuals living with mental illness face a greater risk of being excluded, and ongoing initiatives are necessary to mitigate this (although government strategies and policies in most countries still neglect this). Social inclusion supports recovery through affording the expansion or reterritorialisation of health assemblages. As Duff (2014) explained, when persons living with mental illness experience social inclusion, this affords ongoing enhancements in the range of affects, bodies, events, processes, and objects they may affect and be affected by.

The Covid-19 pandemic required many music therapists to think even more creatively about ensuring safe access to their services. Telehealth has been one solution (albeit with limitations). Music therapists such as Knott and Block (2020) wrote about their virtual music therapy services model that includes appropriate, accessible formats, which can best meet clients' needs and capacities. Molyneux et al. (2020) also described how they continued offering music therapy sessions virtually to provide continuity, support, and connection for people in the United Kingdom living with dementia and their companions when they were isolated at home during the national lockdown. However, as Agres et al. (2021) highlighted, not all clients find online sessions convenient, accessible, or user-friendly, and technology can serve as an obstacle. Online sessions can be draining, frustrating, and limiting. As we open up access to develop more response-able practice, we need to be continuously thinking flexibly about how technology may be enhancing and/or restricting the health assemblages we are part of.

Response-able Services

Healthcare services are not always collaborative, communal, and mutually response-able; they are frequently the opposite (Kinman et al., 2004), generating stigma and even trauma or re-traumatisation (Adams et al., 2010; Hamilton et al., 2016; Henderson et al., 2014; Knaak et al., 2017; Thornicroft et al., 2007). For example, epistemic injustice as a specific form of psychiatric harm occurs when the experiential knowledge of service users is not taken seriously. Staff may also experience harm within healthcare systems. Mckeown and Spandler (2017) called for a truth and reconciliation process in mental health systems and services where service users, survivors and refuses of services, and staff who work/ed with them can begin the work of healing in relation to hurtful effects of the healthcare system.

Spandler and Stickley (2011) asked how to create compassionate healthcare systems. Compassion operates in and through relationships, which are, in turn, constituted within the particular social milieus that may be more or less conducive to compassion. Unfortunately, a compassion deficit is often particularly acute in mental healthcare systems. Spandler and Stickley highlighted that health assemblages need to be ones in which hope can flourish. They suggested we move towards this via radical acceptance of people's lived experiences of distress, sadness, and madness—no matter how challenging or painful—within a hopeful, relational environment.

Examples of response-able services in music therapy include Moonga's (2019) process of consulting community members of Mazabuka in Zambia on how a culturally sensitive music therapy project could be developed and Abad and Williams' (2005) process of engaging community elders in planning to adapt a music therapy intervention with indigenous families in Australia. Other examples include Tuastad and Stige's (2018) rock band with three individuals who had been released from prison and a music therapist, where the role of the therapist was to facilitate equality, participation, and mutuality, Hasler's (2017) community engagement when planning music therapy work with young people in foster care, Thompson's (2012) collaborative approach to work with families of children with Autism, Rickson's (2010) collaborative planning and evaluation with caregivers of students with disabilities, and Hunt's (2005) school-based collaborative project with teenagers from refugee backgrounds. Music therapists have emphasised the importance of working responsively with teenagers in particular, using flexible boundaries (Austin, 2010), appropriate techniques and technologies (Derrington, 2012) and genres (Alvarez, 2012), and creating opportunities for them to direct the trajectory of the process (McFerran & Teggelove, 2011).

Response-able Material Places and Spaces

All assemblages are, at least in part, material achievements (Deleuze & Guattari, 1987). How can we be part of enhancing the response-ability of the physical space where we host music therapy (or our classrooms, the places we have team meetings, or the spaces where we conduct research)? How are we responding to our spaces, and how are our spaces responding to us? How can our spaces be more play-able? Can we enhance mutual flexibility in these relations, so that affects flow and take shape freely (rather than clogging, sticking, and getting trapped)?

Place is constituted relationally through an intimate web of processes, associations, and transactions. Places and people are enmeshed (Duff, 2012). Attempts at social inclusion and participation for recovery often involve the generation of what Pinfold (2000) called "landscapes of care" (p. 201). These include a mixture of formal and informal community resources, "place attachments," healthcare initiatives, and individual support that generate social inclusion. The term "therapeutic landscapes"

(Williams, 2017) has also been used. Restorative environments can afford a range of mental health benefits, including alleviating anxiety and stress (Korpela et al., 2008), reducing the incidence and severity of mental illness symptoms (Milligan & Bingley, 2007), greater personal meaning and belonging (Boyd et al., 2008), enhanced personal security and safety (Parr & Philo, 2003), and increased self-efficacy (Cattell et al., 2008). Aspects like natural lighting, secure, homelike, barrier-free environments, and technology enable skilled use of one's surroundings to contribute to healing experiences (DuBose et al., 2018).

Music scaffolds spaces so they can afford different kinds of sharing and affiliation (Van der Schyff & Krueger, 2019). Increased attention is being given to the sound environment of health care (Wood, 2016), including musical and non-musical sounds (Kittay, 2008). When Pavlicevic et al. (2015) explored music therapy in a care home for persons with dementia, she showed how the music therapists in their study remained cognisant of the care home's architecture. The authors discussed how arranging a space differently encouraged safe moving, dancing, and embodied musical engagement. Music-making often extended beyond the music therapy room, impacting the general atmosphere and mood of the care setting.

Fouché and Stevens (2018) explained how their music therapy work often occurs in unusual places and spaces. Conventional approaches to music therapy take place in private, protected areas to ensure confidentiality and focus on the therapeutic relationship (Aigen, 2018). However, community music therapy takes place wherever it's needed, perhaps a private room but sometimes in a more public space like a corridor, a park, a shared recreational area. The therapeutic frame is more permeable and fluid.

In Duff's (2014) study, participants journeying through mental illness recovery highlighted their experiences of place in affective tones. Some local places provoked negative affective responses, while others generated more positive ones (inspiring enjoyable feelings, offering a sense of motivation or empowerment, and enhancing their capacity to act and manage their stress and problems). These places included gardens, parks, bookshops, a local church, homes, and kitchens.

Response-able Objects

Everyday objects play an important role in shaping our trajectories (Duff, 2014). As discussed in Chapter 12, agency isn't a function that is unique to human bodies; it's distributed in and among assemblages of human and non-human bodies. As McPhie (2018) wrote, "freed from limiting notions of agency, things behave" (p. 306). We can think about the therapeutic properties of matter itself.

Sunelle Fouché (personal communication, 2020) discussed the role of a cupboard at a school where she and her team from MusicWorks conducted a community music therapy project. After weeks of carrying many djembes to and from the room where the sessions were held, she was finally offered a storage cupboard to use. This freed up energy, space, and time with many knock-on effects. As a line of becoming within the assemblage, the cupboard intermingled with her levels of energy, planning options, time available, and quality of session beginnings. New qualities of interaction emerged within the assemblage. A fresh openness for responsiveness was produced between Sunelle and the children in the group. This is not because Sunelle "became more empathic." Empathising was a property of the assemblage that became more flexible. Affects could flow more freely through the intertwining of humans and non-humans.

Stuart Wood (2016) described how he met a woman, PJ, who had an undiagnosed tremor and wanted to learn about herself, see what she could still do and what she could still play. After beginning to explore in music therapy, PJ commented how she experienced becoming disabled as being placed on life's scrap heap, but music made her beautiful again. PJ then brought a washboard and some old thimbles to her next group session, which were incorporated into the musical improvisation. The rest of the group was inspired to bring their bits of "scrap" (saucepans, foot massagers, chains, plant pots) to use within the creative musical process as acts of reclaiming. As their process evolved, they spent their Friday mornings visiting the local scrap yard, collecting objects they could dismantle, manipulate, and play. Wood noted how the spirit of the project was confident and generous. The group decided to share publicly their earthy, scruffy, handmade "junk band." The concert included a mix of prepared sections and free improvisations. As they played, the group members transformed as much as the salvaged scrap metal objects had. Patients became jazz musicians with strength and dexterity. As their music was witnessed, their place in the community was changing and complexifying.

Music therapists adapt instruments when clients struggle to use them in conventional forms (Crowe & Ratner, 2012). This can increase clients' affective capacities. Electronic instruments, technological applications, assistive music technology, and accessible Digital Musical Instruments have featured prominently in recent efforts to enhance the playability of musical instruments and equipment within music therapy practice (Knight, 2013; Viega, 2019). Stensæth (2018) reminded us to consider the agency of technology in music therapy. Through technology, persons with diverse abilities can experience enhanced capacity to engage in a wide range of musicking experiences (Frid, 2019). While such objects in assemblages are valued, this does not mean that humans are disregarded or undervalued. Humans and objects co-create.

Response-able People

Response-able people can affect and be affected. This capacity is generated within relations. Rietveld et al. (2018) argued that a specific ability might be necessary to act on a particular affordance. Without being able to read English, I couldn't respond to the possibility this sentence is offering of being read.

Ansdell (2014) discussed an example of musicking with a woman called Mary, who had no speech and significant difficulties moving her body. As Ansdell held out a Greek bell for her to play, Mary slowly looked at the bell, then reached her hand out stiffly and with great effort, nudging the bell, which clanged loudly. In doing so, her ears and mind reached out to gain skilful access to the affordances of the musicking situation. However, she did need some help to gain full access to this situation, and this is the role that Ansdell played. He held the bell out for Mary to strike rhythmically within the simple song he had written and was singing, making his own musical experience available for Mary to co-opt as her own as she lived through the experience of the melody, sharing musical space and time with him.

The improvisational way of working that Paul Nordoff and Clive Robbins pioneered showed how music isn't just something to play; it can be inhabited as an enormous world to live, play, and work in. Nordoff and Robbins explored how best to help establish such musical worlds for children with particular physical, developmental, emotional, and intellectual needs. By carefully listening and attuning to each child, they realised that they could draw them into a shared musical world where the child felt comfortable and increasingly engaged with other people and the broader world of musical culture. "Musical things" in a musical world afford "para-musical" things such as expression, contact, relationship, collaboration, and performance. Through their music therapy work with hundreds of children, Nordoff and Robbins showed that many important affordances were available and could be made use of "in music" that were not accessible to them "outside music." This makes therapeutic change possible (Ansdell, 2016).

It is still too frequently the case that the mechanisms of music therapy are attributed to the music (as an autonomous object) as well as to the expert-therapist. Rolvsjord (2006a) contended that the client should be highlighted as one who is using music in particular ways as a health resource as they engage in the relational and dynamic process of musicking with the therapist and potentially other group members. Rolvsjord (2014) emphasised the "competent client" as a way of attempting to deconstruct the grand narrative of "expert therapist and weak client." A competent client can affect and be affected. In music therapy, clients show musical competence (through communicative musicality, musicianship, and musical experience skills). Clients also hold their own theories of change that are culturally contextualised, based on the wisdom of daily life experience, illness, and disability, and navigating a healthcare system. Clients are experts of experience. Rolvsjord also explained how clients show competence through reflexive moment-to-moment interactions with the therapist, actively directing how the session unfolds (by taking the initiative, controlling goals and activities, nurturing relationship with the therapist). She cautioned therapists, reminding us that we can play a disabling role in sessions, even with good intentions, for example by trying to "fix" the client, holding stigmatising attitudes, and having an individualistic focus (trying to change the client instead of the system that is harming them).

Response-able Groups, Communities, and Social Networks

Community participation and social inclusion facilitate recovery by promoting access to a range of affective, social, and material resources that can be used in the everyday recovery journey. Social capital entails the collections of bonds of trust, cooperation, and reciprocity that make up social life. These resources fuel an array of coordinated actions (Duff, 2014).

Liesel Ebersöhn (2019), an educational psychologist studying resilience, noticed two schools in an informal, urban settlement community struggling with high levels of poverty. Both schools had few classrooms, limited learning materials, and inadequate sanitation. Yet one was struggling, and the other was thriving. Ebersöhn heard from the deputy principal of the first school that many of their learners were experiencing teenage pregnancies. Bullying was taking place, and many learners were dropping out of school or engaging in criminal activity. However, teachers at the second school explained to Ebersöhn how they had partnered with social workers who visited learners at their homes when they were absent from school. They'd also forged connections with social development officers who helped the learners' parents apply for social grants. In collaboration with clinic nurses, they ensured that children who needed healthcare received treatment. A network of non-governmental organisation partnerships had been established to provide in-service teacher training and after-school programs. Through connections with small and large businesses, the library was supplied with books and computers and a counselling room could be furnished. Teachers at this school engaged in collective information-seeking, collective problem-solving, and collaborative negotiation strategies as they moved towards collective resilience.

This research led Ebersöhn (2019) to her theory of flocking, which she placed within a framework of "relationship-resourced resilience." Flocking entails linking to build on inherent strengths. She explained how marginalised people's interdependent cultural–relational beliefs can guide collective decision-making as groups select flocking as a pathway to resilience (rather than responses of fight, flight, freeze, faint, or swarm), bolstering health and well-being outcomes amid inequality. The power of flocking doesn't lie in its ability to eliminate poverty. Flocking "clothes those made vulnerable systematically over ages of systemic exclusion with the benefits of cultural beliefs of communal agency to buffer the collective" (p. x). Relationships and resources can change an environment's capacity to enable resilience: connected people using available resources to respond to persistent risk. As an indigenous theory, relationship-resourced resilience brings pluralism to the resilience discourses of the dominant Global North. We can think of flocking as a form of territorialising (through a collective appraisal of need and through solidarity). The first school did not territorialise through flocking, while the second school did.

Response-able Musicking

Improvisational music therapy has always sought to be response-able through techniques such as matching (where the therapist creates music that fits with the client's way of playing), reflecting (matching the client's feelings, moods, or attitudes) (Bruscia, 1987), grounding (creating containing, stable music that serves as an anchor for the client), and dialoguing (where client and therapist communicate through musical play) (Wigram, 2004). In Kantor's (2020) reflection on his work with persons with profound intellectual disability, he noted how verbal communication could frequently be one-sided, and carers may tend to speak on behalf of the persons they interact with as a compensatory mechanism. However, in musical interaction, communication is reciprocal and responsive, with various interaction patterns. Silence may be a valuable part of the music, rather than uncomfortable "emptiness." In receptive music therapy approaches, the listening experience focuses on the client's presentation, needs, and responses (Grocke & Wigram, 2006). In Wood's (2006) matrix model, he argued that any form of music-making can be used therapeutically, from individual or group sessions to workshops, concert trips, performances, or even tuition, as part of the interconnected matrix of therapeutic possibilities that musicking offers. This approach is logical because people and music are interconnected webs of relationships.

In a paper by Sajnani et al. (2017), Zarate also offered an approach to response-able musicking. She termed her approach "critical social aesthetics" (CSA), integrating musical aesthetics with cultural meaning. Musical themes and images are voiced within the connection between the client and therapist through improvisation, which is understood as a cultural-relational dynamic. The therapist and client are both seen as sociocultural beings represented in musical themes, and the production of cultural metaphors is welcomed. If a client expresses their "fear of the worst happening," for example, Zarate listens to their social story, their acts of intention, and how the sounds emerge from the relationship to how they are perceiving the symptom. The client becomes actively engaged in improvising their perceptions of their experience not just as a clinical symptom but also as a social narrative. Client and therapist both participate in a dialogue through singing about, playing as or to the anxiety symptom. They have the opportunity to reflect on the story as it connects to experiences of oppression, and insights into the social narrative of the symptom can emerge. The therapist's own socialised biases may arise, as well as feelings about experiences of difference and oppression, requiring reflexivity on their behalf.

Response-able Concepts and Theories

I hope this book offers an example of how we can move towards thinking with more response-able concepts in music therapy. How are our concepts and theories taking shape in dynamic ways, and how can we be part of enhancing their flexibility so that they do socially just work in diverse contexts? When we hold onto rigid theories too tightly (sometimes through passionate investment or early-career nerves, or as we become more set in our ways over the years), we can prioritise theory over the client's experience. Kenny (1999) reminded us that theory needs to be response-able in living, breathing, relevant ways.

There are debates in music therapy about how we should (or shouldn't) construct, use, borrow, develop, narrow, and diversify theory (Daveson et al., 2008). Often, theory, research and practice develop in response-able relation. The term "indigenous music therapy theory" has been used to describe theory that grows out of practice directly instead of being imported from outside the field. Ghetti (2012) offered an example of a theory that grew responsively out of practice and functions in a response-able way. Within music therapy for procedural support during invasive medical procedures, Ghetti noticed that rationales and assumptions regarding how music therapy was functioning were conflicting. She, therefore, developed a working model that could guide a music therapist as they reflexively engage in the process of continually assessing a client's responses during the procedure within the unfolding context. How a medical procedure is perceived depends on the interrelationships of person—procedures—context—music—therapist—other medical staff—physical space (and so on). By being sensitive to these interrelations, the music therapist can refocus their intervention lens in an ongoing, dynamic way (changing aspects of the music, their focus of attention, and the interaction between client and therapist). Importantly, we see how this theory of how music therapy "works" in procedural support does not become a rigid frame imposed on the work. The theory is designed to be inherently response-able.

Music therapy theories and concepts can become (and in some ways are becoming) more culturally response-able. As an example, Low et al. (2020) explored how music therapy comes into conversation with the pluralistic healthcare culture in Malaysia. While this allows for Eurocentric music therapy approaches and Indigenous-based music practices to coexist and even thrive alongside each other, Low et al. emphasised that music therapists need to continuously and critically examine their education, knowledge, and practices so that the field can develop in responsive ways.

The process of creating meaning together is undeniably complex. When moving from the United Kingdom to South Africa, Pavlicevic

(2004), for example, asked how music therapy would find itself in this new space. Concepts of healing, music, living, the world, and "being a person" mean differently than they do in other areas. As a result, theories of "psychopathology," "wellness," and "healing" need to be formed contextually (Nwoye, 2015) (and are continuously forming). Ratele (2017) argued that nothing prohibits us from drawing on and travelling between theories and concepts from different schools of thought and worldviews; the boundaries between them can be permeable; and some will be more or less relevant at other times, in different contexts and with different clients. Theory is heterogeneous terrain.

Response-able Recovery-Tracking

How are therapeutic needs identified and represented? How is the client articulating distress, sadness, madness, or illness? What is "pathological," "unhealthy," "dysfunctional," and what isn't? Do we decide what the primary therapeutic needs seem to be on behalf of clients, or is this process a mutually response-able one? Is there room for a client to respond to diagnostic decisions that a multidisciplinary team member has reached and communicated to them? How? Is the "problem" deemed to be situated within the client, their social/political/historical/economic context, or the assemblage as a whole? Who reaches that conclusion? This section uses the term "recovery-tracking" to replace the more conventional ideas of "assessment" and "progress monitoring and evaluation" as these terms have more hierarchical baggage and less response-able undertones (and overtones).

Exploring Needs

In community music therapy, the idea of "needs" has been problematised and reconsidered (Wood, 2016). CoMT embraces an account of needs that isn't centred on individual clinical symptoms but is more contextually response-able, including notions of artistic, social, or political needs. The deficit in a situation could be systemic rather than (only) physiological or psychological, and therefore, it may necessitate a systemic response. The idea of "needs" may also be rejected altogether in CoMT, as music therapists respond instead to individuals' and groups' abilities, resources, and potentials. While the expert-client relationship is still the dominant view within a medical model, a social model assumes an empowerment perspective and emphasises mutual relationships (Rolvsjord, 2010).

Psychiatric diagnoses can assist and impede therapeutic work (Spandler, 2014). They may offer insight, and they may be oppressive. Diagnoses shape our assumptions about the problem and how we might address it. However, this may be accomplished by plucking suffering out of its context, losing the situated person along the way, and limiting possibilities for imagining alternative ways of exploring experience. In other words, we need a response-able approach.

Whitehead-Pleaux et al. (2017) emphasised how culture has to be taken into consideration when conducting assessment in music therapy, by asking clients questions (through an interpreter if necessary) about their culture(s) of heritage, religion and spirituality, relationships between culture and socio-economic status, generational cultures, cultures surrounding gender expression, gender identity and sexual orientation, cultures of disability, cultures associated with trauma survivorship, and any other culture that the client identifies with for a particular reason. Enquiring about these layers of cultural affiliation offer mere starting points, but they are crucial to beginning (and developing) a therapeutic journey that is response-able. This explorative process will inform the music used in sessions and how the client's experiences are worked with.

Tracking Recovery

In music therapy, we are not tied to assessing a client's progress; we can explore progression (Wood, 2016), for example from more isolated forms of musicking to more socially connected ones. Many music therapists assume an improvisational stance, and improvisation doesn't have clear stepping stones or a clear destination. Assemblages generate emergent properties (Anderson et al., 2012; De Landa, 2006; Price-Robertson & Duff, 2015). We may balance goals related to rehabilitation or recovery while also valuing artistic experience, lines of flight, and quality of life. The notion of "functional recovery" describes how everyday improvements in quality of life may come about despite the ongoing lingering experiences of symptoms of mental illness. This more dynamic idea pushes back against the definition of recovery as "cure" in the form of complete symptom remission. It also recognises that people may lead healthy, fulfilling, and productive lives regardless of ongoing symptoms of mental illness. Recovery looks more like a process than an endpoint.

Bateson (1991) contended that "great teachers and therapists avoid all direct attempts to influence the action of others and, instead, try to provide the settings or contexts in which some (usually imperfectly specified) change may occur" (p. 254). Although many clients do come to therapy with the explicit wish to experience a reduction in distressing symptoms, it may be necessary to support this through a broader change in the relationship between the individual and their contexts so that we avoid the risk of treating "the symptom to make the world safe for the pathology" (p. 296). Individual well-being is inseparable from quality of relationships, and enhancing connectedness is, therefore, an essential task of therapy (Tramonti, 2019).

A more ecological approach sees health as relational (emerging between us) and performed (as we actively take part, pursue, and cultivate it). Being "healthy," "ill," or "disabled" is a condition or state that at least partly depends on one's environment (Ansdell, 2016). Changes in the ecology of an environment can relate to changes in objective and/or subjective health states. Health and illness have key physical dimensions, but this is only a very limited start of any adequate account. We need to attend to the full ecology of health and illness and how they are continuously constructed, undermined, or maintained by social and cultural forces. While traditional music therapy thinking places "the client" at the centre of a system, an ecological approach centres nothing but recognises that we are all interdependent (Wood, 2016). According to Price-Robertson et al. (2017), relational recovery is rooted in the idea that humans are always interdependent. Relationships are not vital because they shape the separate individuals "within" them. Relationships themselves are recovering.

"Assessment" then can't simply be a process of determining the capacities of an individual; we are asking what the assemblage can do. A strength of this approach lies in how our attention can be drawn to the wide variety of human and non-human entities, affects, forces, and relations that are active and hold potential within any process of recovery (McPhie, 2019). According to Bruscia (2014), health in music therapy relates to one's "ecological wholeness" (p. 107).

When we understand recovery in this way, we can see how it advances or retreats in the many events of everyday life: an interaction with a barista at a coffee shop; selecting a playlist to listen to during family dinner; in the silence between passengers on a train; in the texture of prickly grass pressing into one's palms just before launching into a handstand. Each of these events reveals the forces through which transformation in an assemblage expresses the recovering body. These are the real conditions of mental health and recovery: creating meaning in the experience of recovery while cultivating an art of becoming-well (Duff, 2014).

Response-able Teaching and Researching in Music Therapy

We began this chapter by reflecting on enhancing access and inclusivity, primarily related to clients accessing music therapy services. We also need to discuss who is gaining access to music therapy education spaces, who isn't, and why that may be the case. As Fansler et al. (2019) documented, people of marginalised identities are over-represented as service users yet underrepresented in those who are music therapists. While they wrote from the context of the United States, the same is true in my country of South Africa.

As bell hooks (2014) wrote in *Teaching to transgress*, "any radical pedagogy must insist that everyone's presence is acknowledged…There must be an ongoing recognition that everyone influences the classroom, that everyone contributes" (p. 23). Inside the classroom, how are we (or aren't we) rendering each other capable (as Bozalek [2020] asked)? Guyotte and Kuntz (2018) promoted pedagogical practices that invite "slippage." As we become unbalanced (and teeter on the edge of slipping), we might let go of certainty and open up possibilities for becoming differently. In such classrooms, opportunities are created for students and teachers to (re)consider assumptions, critique, interrogate and/or embrace preconceived ideas within an ongoing journey of repositioning. This ethical endeavour is situated within socio-historical claims regarding what can be known and how we can access this knowledge.

In the teaching environment, how response-able are our learners and teachers? Does the physical classroom promote rigidity, discomfort, and limited movement, or does it invite flexible flow? Do students feel they can speak their minds freely? When teaching online, do the platforms often freeze and stutter, and do students struggle with having insufficient mobile data to join classes, which constrains their response-ability? Do teachers have the response-ability to flexibly adapt their syllabi?

Fansler et al. (2019) described the approach to music therapy pedagogy within their programme at Slippery Rock University in Pennsylvania, USA. Through queering music therapy pedagogy, they intentionally destabilised rigid structures and fixed categories (such as teacher/learner), seeking fluidity and reciprocity instead. As such, they urged educational stakeholders to move into more generative liminal spaces where participants could experience radically inclusive relationships. They pointed out that education is political (drawing on Freire, 1970/2000), involving interpretation and change, reflection and action. The approach to queering music therapy pedagogy that Fansler et al. held included discarding certainty; questioning what is unremarkable and unremarked; problematising what is normalised, included, and excluded; critically examining how one responds when reading, listening, observing, and interacting; wondering about what is difficult to accept and bear and why that may be the case; and disorienting perceptions. They asked how these processes translated into practices, such as auditions (what is valued and accepted and what isn't?); how response-able the curriculum is (does it question the hegemony of the Western canon? How does it regulate what is thinkable, knowable, recognisable? How are concepts of health, illness, pathology, therapy, music challenged?

Are people defined within rigid "population groups"?). Fansler et al. described the programme as being in a constant state of becoming.

Let's turn our attention to research. Some areas of study within music therapy are particularly response-able. Stige (2002) described participatory action research (PAR) as taking place through democratic decision-making and proactive attempts on behalf of the researchers to engage participants as active role-players in every aspect of the study. PAR unfolds through a cyclic, collaborative approach of assessing, planning, acting, evaluating, and reflecting. This allows participants to develop through the process, reconsider the problem, and act on new strategies. As such, PAR holds the potential to improve the lives of marginalised groups (Elefant, 2017).

PAR has a close relationship with Community Music Therapy (Stige & Aarø, 2011). Within a CoMT framework, Bolger (2015) sought to work with three groups of marginalised young people and their supporting communities in Melbourne, Australia, to explore the process of collaboration collectively. She wanted to know why elements of a collaborative project were most meaningful to the young people involved and to understand the process of collaboration in a participatory music project in a way that acknowledged contextual variation and honoured complexity. In their three groups, the participants, whom Bolger referred to as "players," elected respectively to form a band; undertake a song-writing project; and participate in a series of individual and small-group projects, including song sharing and instrument playing, singing, song-writing, and community performance. As the players reflected on the collaboration process, the aspects that emerged as most meaningful for them were choice, having a tangible purpose, pathways of moving towards independence, and the music therapist's collaborative support style. Changed self-concept and the development of skills and mastery were the most meaningful outcomes for the participants.

Drawing on a disability studies perspective, Metell (2014) emphasised how music therapists need to ask who benefits from our research. According to Stone and Priestley (1996), several principles guide emancipatory disability research. These include surrendering claims to objectivity and overt political commitment to the struggles of disabled people for self-emancipation. Willingness to undertake research with disabled people only where it produces a practical benefit to their self-empowerment and/or the removal of disabling barriers is essential, ensuring full accountability to disabled people and their organisations, and adopting plurality of methods in response to changing needs. In all aspects of music therapy—clinical and community work, teaching and learning, and research—we are ethically compelled to reflect on response-ability and to participate in territorialising our assemblages as more mutually response-able.

References

Adams, E., Lee, A., Pritchard, C., & White, R. (2010). What stops us from healing the healers: A survey of help-seeking behavior, stigmatisation and depression within the medical profession. *International Journal of Social Psychiatry, 56*(4), 359–370.

Agres, K. R., Foubert, K., & Sridhar, S. (2021). Music therapy during COVID-19: Changes to the practice, use of technology, and what to carry forward in the future. *Frontiers in Psychology, 12*, 1317.

Aigen, K. (2018). Community music therapy. In G. McPherson & G. Welch (Eds.), *Special needs, community music, and adult learning* (pp. 138–154). Oxford University Press.

Alvarez, T. T. (2012). Beats, rhymes and life: Rap therapy in an urban setting. In S. Hadley & G. Yancy (Eds.), *Therapeutic uses of rap and hip-hop* (pp. 117–128). Routledge.

Anderson, B., Kearnes, M., McFarlane, C., & Swanton, D. (2012). On assemblages and geography. *Dialogues in Human Geography, 2*(2), 171–189.

Ansdell, G. (2014). *How music helps in music therapy and everyday life*. Ashgate.

Ansdell, G. (2016). *How music helps in music therapy and everyday life*. Routledge.

Austin, D. (2010). When the bough breaks: Vocal psychotherapy and traumatized adolescents. In K. Stewaer (Ed.), *Music therapy and trauma: Bridging theory and clinical practice* (pp. 176–187). Satchnote Press.

Barad, K. (2015). Transmaterialities: Trans*/matter/realities and queer political imaginings. *GLQ: A Journal of Lesbian and Gay Studies, 21*(2–3), 387–422.

Bateson G. (1991). *A sacred unity: Further steps to an ecology of mind.* HarperCollins.

Bolger, L. (2015). Being a player: Understanding collaboration in participatory music projects with communities supporting marginalised young people. *Qualitative Inquiries in Music Therapy, 10*(3), 77–126.
Boyd, C., Hayes, L., Wilson, R., & Bearsley-Smith, C. (2008). Harnessing the social capital of rural communities for youth mental health: An asset-based community development framework. *Australian Journal of Rural Health, 16*(4), 189–193.
Bozalek, V. (2020). Rendering each other capable: Doing response-able research responsibly. In K. Murris (Ed.), *Navigating the postqualitative, new materialist and critical posthumanist terrain across disciplines* (pp. 135–149). Routledge.
Bozalek, V., Bayat, A., Gachago, D., Motala, S., & Mitchell, V. (2018). A pedagogy of response-ability. In V. Bozalek, R. Braidotti, M. Zembylas, & T. Shefer (Eds.), *Socially just pedagogies in higher education: Critical posthumanist and new feminist materialist perspectives* (pp. 97–112). Bloomsbury Academic.
Bruscia, K. (1987). *Improvisational models of music therapy*. Thomas.
Bruscia, K. (2014). *Defining music therapy* (3rd ed.). Barcelona.
Cattell, V., Dines, N., Gesler, W., & Curtis, S. (2008). Mingling, observing, and lingering: Everyday public spaces and their implications for wellbeing and social relations. *Health and Place, 14*(3), 544–561.
Crowe, B. J., & Ratner, E. (2012). The sound design project: An interdisciplinary collaboration of music therapy and industrial design. *Music Therapy Perspectives, 30*(2), 101–108.
Daveson, B., O'Callaghan, C., & Grocke, D. (2008). Indigenous music therapy theory building through grounded theory research: The developing indigenous theory framework. *The Arts in Psychotherapy, 35*(4), 280–286.
De Landa, M. (2006). *A new philosophy of society: Assemblage theory and social complexity.* Continuum.
Deleuze, G., & Guattari, F. (1987). *A thousand plateaus: Capitalism and schizophrenia.* Athlone.
Derrington, P. (2012). "Yeah, I'll do music!" Working with secondary-aged students who have complex emotional and behavioural difficulties. In J. Tomlinson, P. Derrington, & A. Oldfield (Eds.), *Music therapy in schools* (pp. 195–211). Jessica Kingsley.
DuBose, J., MacAllister, L., Hadi, K., & Sakallaris, B. (2018). Exploring the concept of healing spaces. *HERD: Health Environments Research & Design Journal, 11*(1), 43–56.

Duff, C. (2012). Exploring the role of 'enabling places' in promoting recovery from mental illness: A qualitative test of a relational model. *Health & Place, 18*(6), 1388–1395.
Duff, C. (2014). *Assemblages of health: Deleuze's empiricism and the ethology of life*. Springer.
Ebersöhn, L. (2019). *Flocking together: An indigenous psychology theory of resilience in Southern Africa.* Springer.
Elefant, C. (2017). Reflections: Giving voice: Participatory action research with a marginalized group. In B. Stige, G. Ansdell, C. Elefant, & M. Pavlicvic (Eds.), *Where music helps: Community music therapy in action and reflection* (pp. 199–216). Routledge.
Fansler, V., Reed, R., Bautista, E., Arnett, A., Perkins, F., & Hadley, S. (2019). Playing in the borderlands: The transformative possibilities of queering music therapy pedagogy. *Voices: A World Forum of Music Therapy, 19*(3).
Fouché, S., & Stevens, M. (2018). Co-creating spaces for resilience to Flourish. *Voices: A World Forum for Music Therapy, 18*(4).
Freire, P. (1970/2000). *Pedagogy of the oppressed* (30th Anniversary ed.). Bloomsbury Press.
Frid, E. (2019). Accessible digital musical instruments: A review of musical interfaces in inclusive music practice. *Multimodal Technologies and Interaction, 3*(3), 57.
Ghetti, C. (2012). Music therapy as procedural support for invasive medical procedures: Toward the development of music therapy theory. *Nordic Journal of Music Therapy, 21*(1), 3–35.
Grocke, D., & Wigram, T. (2006). *Receptive methods in music therapy.* Jessica Kingsley.
Guyotte, K., & Kuntz, A. (2018). Becoming openly faithful: Qualitative pedagogy and paradigmatic slippage. *International Review of Qualitative Research, 11*(3), 256–270.
Hamilton, S., Pinfold, V., Cotney, J., Couperthwaite, L., Matthews, J., Barret, K., Warren, S., Corker, E., Rose, D., Thornicroft, G., & Henderson, C. (2016). Qualitative analysis of mental health service users' reported experiences of discrimination. *Acta Psychiatrica Scandinavica, 134*, 14–22.
Hasler, J. (2017). Healing rhythms: Music therapy for attachment. In A. Hendry & J. Hasler (Eds.), *Creative therapies for complex trauma* (pp. 135–153). Jessica Kingsley.
Henderson, C., Noblett, J., Parke, H., Clement, S., Caffrey, A., Gale-Grant, O., Schulze, B., Druss, B., & Thornicroft, G. (2014). Mental health-related

stigma in health care and mental health-care settings. *The Lancet Psychiatry, 1*(6), 467–482.

hooks, b. (2014). *Teaching to transgress*. Routledge.

Hunt, M. (2005). Action research and music therapy: Group music therapy with young refugees in a school community. *Voices: A World Forum for Music Therapy, 5*(2).

Kantor, J. (2020, June). "How well do I know you?": Intersubjective perspectives in music therapy when working with persons with profound intellectual and multiple disability. *Voices: A World Forum for Music Therapy, 20*(2).

Kenny, C. (1999). Beyond this point there be dragons: Developing general theory in music therapy. *Nordic Journal of Music Therapy, 8*(2), 127–136.

Kinman, C. J., Finck, P., & Hoffman, L. (2004). Response-able practice. In T. Strong & D. Pare (Eds.), *Furthering talk* (pp. 233–251). Springer.

Kittay, J. (2008). The sound surround. *Nordic Journal of Music Therapy, 17*(1), 41–54.

Knaak, S., Mantler, E., & Szeto, A. (2017). Mental illness-related stigma in healthcare: Barriers to access and care and evidence-based solutions. *Healthcare Management Forum, 30*(2), 111–116.

Knight, A. (2013). Uses of iPad® applications in music therapy. *Music Therapy Perspectives, 31*(2), 189–196.

Knott, D., & Block, S. (2020). Virtual music therapy: Developing new approaches to service delivery. *Music Therapy Perspectives, 38*(2), 151–156.

Korpela, K. M., Ylén, M., Tyrväinen, L., & Silvennoinen, H. (2008). Determinants of restorative experiences in everyday favorite places. *Health & Place, 14*(4), 636–652.

Low, M. Y., Kuek Ser, S. T., & Kalsi, G. K. (2020). Rojak: An ethnographic exploration of pluralism and music therapy in post-British-colonial Malaysia. *Music Therapy Perspectives, 38*(2), 119–125.

McFerran, K., & Teggelove, K. (2011). Music therapy with young people in schools: After the Black Saturday fires. *Voices: A World Forum for Music Therapy, 11*(1).

Mckeown, M., & Spandler, H. (2017). Exploring the case for truth and reconciliation in mental health services. *Mental Health Review Journal, 22*(2), 83–94.

Mcphie, J. (2018). I knock at the stone's front door: Performative pedagogies beyond the human story. *Parallax, 24*(3), 306–323.

McPhie, J. (2019). *Mental health and wellbeing in the Anthropocene: A posthuman inquiry*. Palgrave Macmillan.

Metell, M. (2014). Dis/Abling musicking: Reflections on a disability studies perspective in music therapy. *Voices: A World Forum for Music Therapy, 14*(3).
Milligan, C., & Bingley, A. (2007). Restorative places or scary places? The impact of woodland on the mental well-being of young adults. *Health and Place, 13*, 799–811.
Molyneux, C., Hardy, T., Lin, Y., McKinnon, K., & Odell-Miller, H. (2020). Together in sound: Music therapy groups for people with dementia and their companions—Moving online in response to a pandemic. *Approaches: An Interdisciplinary Journal of Music Therapy*, Advance Online Publication.
Moonga, N. U. (2019). *Exploring music therapy in the life of the batonga of Mazabuka Southern Zambia* [Master's dissertation, University of Pretoria].
Nwoye, A. (2015). African psychology and the Africentric paradigm to clinical diagnosis and treatment. *South African Journal of Psychology, 45*(3), 305–317.
Parr, H., &, Philo, C. (2003). Rural mental health and social geographies of caring. *Social & Cultural Geography, 4*(3), 471–488.
Pavlicevic, M. (2004). Music therapy and the polyphony of near and far... *Voices: A World Forum for Music Therapy, 4*(1).
Pavlicevic, M., Tsiris, G., Wood, S., Powell, H., Graham, J., Sanderson, R., Millman, R., & Gibson, J. (2015). The 'ripple effect': Towards researching improvisational music therapy in dementia care homes. *Dementia, 14*(5), 659–679.
Pinfold, V. (2000). Building up safe havens…all around the world': Users' experiences of living in the community with mental health problems. *Health and Place, 6*(2), 201–212.
Price-Robertson, R., & Duff, C. (2015). Realism, materialism, and the assemblage: Thinking psychologically with Manuel DeLanda. *Theory & Psychology, 26*(1), 1–19.
Price-Robertson, R., Obradovic, A., & Morgan, B. (2017). Relational recovery: Beyond individualism in the recovery approach. *Advances in Mental Health, 15*(2), 108–120.
Ratele, K. (2017). Four (African) psychologies. *Theory & Psychology, 27*(3), 313–327.
Rickson, D. (2010). *The development of music therapy schools consultation protocol for students with high or very high special education needs* [Doctoral dissertation, University of Wellington].
Rietveld, E., Denys, D., & Van Westen, M. (2018). Ecological-enactive cognition as engaging with a field of relevant affordances. In A. Newen, L.

De Bruin, & S. Gallagher (Eds.), *The Oxford handbook of 4E cognition* (pp. 41–70). Oxford University Press.

Ringrose, J., Osgood, J., Renold, E., & Strom, K. (2019). PhEmaterialism: Response-able Research & Pedagogy. *Reconceptualizing Educational Research Methodology, 3*(2).

Rolvsjord, R. (2006a). Whose power of music? *British Journal of Music Therapy, 20*(1), 5–12.

Rolvsjord, R. (2006b). Therapy as empowerment: Clinical and political implications of empowerment philosophy in mental health practises of music therapy. *Voices: A World Forum for Music Therapy, 6*(3).

Rolvsjord, R. (2010). *Resource-oriented music therapy in mental health care.* Barcelona Publishers.

Rolvsjord, R. (2014). The competent client and the complexity of dis-ability. *Voices: A World Forum for Music Therapy, 14*(3).

Ruud, E. (1997). Music and quality of life. *Nordic Journal of Music Therapy, 6*(2), 86–91.

Sajnani, N., Marxen, E., & Zarate, R. (2017). Critical perspectives in the arts therapies: Response/ability across a continuum of practice. *The Arts in Psychotherapy, 54*, 28–37.

Small, C. (1998). *Musicking: The Meanings of Performing and Listening.* Wesleyan University Press.

Spandler, H. (2014). Letting madness breathe?: Critical challenges facing mental health social work today. In J. Weinstein (Ed.), *Mental Health: Critical and radical debates in social work* (pp. 29–38). Policy Press.

Spandler, H., & Stickley, T. (2011). No hope without compassion: The importance of compassion in recovery-focused mental health services. *Journal of Mental Health, 20*(6), 555–566.

Statssa. (2019). *Statistical release. General household survey.* Retrieved from: http://www.statssa.gov.za/publications/P0318/P03182019.pdf

Stensæth, K. (2018). Music therapy and interactive musical media in the future: Reflections on the subject-object interaction. *Nordic Journal of Music Therapy, 27*(4), 312–327.

Stige, B. (2002). *Culture-centered music therapy.*

Stige, B., & Aarø, L. E. (2011). *Invitation to community music therapy.* Routledge.

Stone, E., & Priestley, M. (1996). Parasites, pawns and partners: Disability research and the role of non-disabled researchers. *British Journal of Sociology, 47*(4), 699–716.

Thomas, N. (2020). Community-based referential music making with limited-resource adolescents: A pilot study. *Music Therapy Perspectives, 38*(2), 112–118.

Thompson, G. (2012). Family-centered music therapy in the home environment: Promoting interpersonal engagement between children with autism spectrum disorder and their parents. *Music Therapy Perspectives, 30*(2), 109–116.

Thornicroft, G., Rose, D., & Kassam, A. (2007). Discrimination in health care against people with mental illness. *International Review of Psychiatry, 19*(2), 113–122.

Tramonti, F. (2019). Steps to an ecology of psychotherapy: The legacy of Gregory Bateson. *Systems Research and Behavioral Science, 36*(1), 128–139.

Tuastad, L., & Stige, B. (2018). Music as a way out: How musicking helped a collaborative rock band of ex-inmates. *British Journal of Music Therapy, 32*(1), 27–37.

Van der Schyff, D., & Krueger, J. (2019). *Musical empathy, from simulation to 4E interaction. Music, sound, and mind.* Brazilian Association of Music Cognition.

Viega, M. (2019). Globalizing adolescence: Digital music cultures and music therapy. In K. McFerran, P. Derrington, & S. Saarikallio (Eds.), *Handbook of music, adolescents, and wellbeing* (pp. 217–224). Oxford University Press.

Whitehead-Pleaux, A., Brink, S., & Tan, X. (2017). Culturally competent music therapy assessments. In A. Whitehead-Pleaux & X. Tan (Eds.), *Cultural intersections in music therapy: Music, health, and the person* (pp. 271–283). Barcelona Publishers.

Wigram, T. (2004). *Improvisation: Methods and techniques for music therapy clinicians, educators and students.* Jessica Kingsley.

Williams, A. (2017). *Therapeutic landscapes.* Routledge.

Williams, J. (2005). *Understanding poststructuralism.* Acumen.

Wood, J. (2016). *Interpersonal communication: Everyday encounters* (8th ed.). Cengage Learning.

Wood, S. (2006, November). "The matrix": A model of community music therapy processes. *Voices: A World Forum for Music Therapy, 6*(3).

Wood, S. (2016). *A matrix for community music therapy practice.* Barcelona Publishers.

Part IV

Relational Empathy

Situated awareness of emotion co-storying.

16

A Stance of Recognising That We Are Selfing Through Relationships

Our first two empathy pathways—insightful and translational—grew from an ontology that considers people first and then sees relationships as developing when those people interact with one another. When we moved along to empathising assemblages, we took a different view: relationships come first, and everything stays relational. One could argue that we've now set up a problematic binary: individuals versus relationships. In fact, we could accuse an approach that only considers relationships of *not* being fully relational: it's only when entities and relationships come into relation with each other that we've reached a fully relational ontology. This is where we find a pathway of relational empathy.

The theories that inform this final part of the book come from diverse perspectives. My argument is not that they neatly align with one another. I'm laying them down as stepping stones to create a path we can walk down. I have purposefully assumed a generous, inclusive, and flexible theoretical approach here, as many varied orientations offer valuable resources for building a notion of relational empathy as situated awareness of emotion co-storying.

A. dos Santos, *Empathy Pathways*,
https://doi.org/10.1007/978-3-031-08556-7_16

Valuing Both Relationships and Selves

Some authors contend that both entities and relationships can be ontologically co-primal (Wildman, 2010). Selves and the relational webs they are inextricably embedded in are both of equal value and are ontologically irreducible (Kessler, 2019). We emerge as a particular Self through our relationships, and we also influence those relationships. As Hycner (1993) wrote, "We are as much a-part-of, as well as apart from, other human beings" (p. 8). Holdstock (2000) echoed this: "The self is part of the social world, and the social world is part of the self" (p. 100). We could similarly think of music. A piece of music is indeed made up of musical elements, but we hear those elements in particular ways because of their relationships within the piece (Levitin, 2006). George Herbert Mead (1934) wrote that selves only exist in relationship to other selves. In McCormack and McCance's (2016) formulation of person-centred practice they explain person-centredness as involving the acknowledgement of four modes of being: being in relation; being in a social world; being in place; and being with self.

To maintain that subjectivity consists solely of a person's consciousness or mind is reductionist, as no subject is an isolated unit. Social actors emerge through social interaction (Crossley, 2018). A mind is an emergent property of a person-environment relationship (Stolorow et al., 2002). Also, it is equally reductionist to claim that consciousness exists only as it is formed through relations. Consciousness and relationality are necessary for each without one having ontological priority (Archer & Donati, 2015). We are always connected to the world because consciousness is the consciousness of something (as argued from a phenomenological perspective by authors such as Husserl [1977] and Heidegger [1962]). Consciousness is, therefore, not a substance but a relation (Crossley, 2018). We are selves-in-the-world. The English word "consciousness" (derived from the Latin conscious, combining con- "together" and scio, "to know") means to think with or to know with others (Lewis, 1990).

Other authors, like Gergen (2018), don't argue for a co-primal ontology explicitly but for a relational one, although he acknowledges that beings emerge from a relational substrate. According to Gergen (2009, p. 133), "the word 'I' does not index an origin of action, but a relational achievement." From the perspective of relational sociology, Archer and Donati (2015) emphasised how a human is always a subject-in-relation. In Donati's (2011) words, "in the beginning there is the relation" out of which "subjects and objects are defined relationally" (p. 17). A person is their relationships (Gibson, 2017), and a human being can't exist without relations with others: "Were the relation with the other to be suspended, so too would be the relation with the self" (p. 13). As Mitchell (1991), a relational psychoanalyst, wrote, "When we feel most private, most deeply 'into' ourselves, we are in some other sense most deeply connected with others through whom we have learned to become a self" (p. 130). If no one is watching a child, they may feel as if nothing is really happening. If no one sees them jump into a pool, there isn't much point in doing it. When a child cries, "Watch me! Watch me!" they are not begging for attention; they are calling out for existence itself (Montgomery, 1989).

Veissière et al. (2020) described the process of "thinking through other minds (TTOM)" (p. 1). In Chapter 5 (when we were travelling along a pathway of insightful empathy), we explored the idea of theory of mind (TOM) (human's ability to ascribe mental states, feelings and intentions to themselves and others), within which we found theory theory (TT), the idea that humans acquire literal theories about how to ascribe mental states to others. Veissière et al. argued that the construct of TT is derived within individualistic Western contexts and fails to work well cross-culturally. Many non-Western cultures don't emphasise individual intentions and mental states. As a dualistic cognitivist construct, TT also doesn't adequately take into account embodied cognition or the cooperative nature of cognition and behaviour. According to proponents of TTOM, humans gain expectations (rich repertoires of beliefs reflecting action readiness), habits, and norms through immersive engagement with cultural practices, including epistemic resources that indicate to them what is salient, relevant and reliable.

From the perspective of community music therapy, Stige (2006) commented, "The point is not that collaboration and community are better or more important than the individual, but that the individual must be understood in relational and non-atomistic terms" (p. 133). Ansdell and Stige (2015) explained community music therapy as "'joined-up music therapy' that works across the individual-communal continuum" (p. 597). In music therapy in general (not only community music therapy), we become participants through relationships. We can't analyse "the client," "the music therapist," or "the music" in isolation. Post-Cartesian approaches to musicking that resist dichotomies between internal/external and subjective/objective have been presented by authors such as Schiavio (2015) and Funk and Coeckelbergh (2013). It makes little sense to consider singular cause-and-effect outcomes between music, therapist, and client. There is relational reciprocity between all three "points" of these triangles. We can only address questions about how music therapy "works" by examining interrelationships (Garred, 2001, 2006). Everything that happens in therapy is a self-with-other-in-action story (DeYoung, 2014).

Selfing

When meeting a client for the first time, part of our assessment process (and part of our ongoing work) entails wondering: who is this person before me? The term "Self" can be used with a capitalised 'S' to refer to an individual-in-relation. This Self is a subject with their own telos, individuality, and agency within relational exchanges (Kessler, 2019). Alternatively, Gergen (2009) used the term "being" to express how we are always in motion, carrying a past as we move through the present into a future. From a Gestalt perspective, we can also draw on the idea of "selfing" as an ongoing process. We are always selfing as we creatively adjust within our environments (Mann, 2020). Through this lens, there wouldn't be "a narcissist" or "a mother," "a musician," or "an addict". We would consider "a person behaving narcissistically within their relational situation"; "a person engaged in mothering"; "an individual who is musicking"; or "a person taking part in addictive behaviours." Our question then, would then shift from "Who is this person before me?" to "How is this person selfing?" and "What is their history of selfing through relationships with others in their lives?" We, as music therapists, are also required to enter each therapeutic relationship reflecting on how we are selfing.

Contact is our very first reality (Yontef, 1993). We are with others before we have a notion of being apart from them (Benjamin, 1995). As Winnicott (1965) so famously articulated, "There is no such thing as an infant" (p. 39). However, through our relationships, a sense of subjective singularity begins to form (Donati, 2011). A child learns to experience their Self as an object in others' experience and then as an object within the flow of their own experience (Crossley, 2010). As caregivers provide and receive responsiveness and recognition, patterns of co-created action develop. Through a relationship of shared awareness, each party is recognised as a Self by the other through affectively meaningful experiences. These acts of recognition confirm that each is seen, known, and understood, that each can move the other and be moved by them and that they both matter to one another (Benjamin, 2004). Socialisation doesn't just happen "to us," though; we play an active role in it. We are actors in the processes forming us, achieving autonomy within our social relations, but never autonomy from them (Crossley, 2010).

I emerge as a music therapist through relationships with those who seek my services. I become a writer through relation with you as a reader; a mother through the relations with my children; a wife through the relationship I have with my husband; a lecturer through relations with my students. My client may become as musically perseverative in relation to me if I don't adequately entrain and offer alternatives, clearly communicate an ending, or as I frame their playing as perseverating instead of as a potentially culturally appropriate repeating rhythm. A person's healthy, creative, and vibrant "music child" (to use Nordoff and Robbin's terminology) may emerge in a session as we create musical-relational opportunities that enable that Self to come into being. As music therapists, we find ourselves playing in a specific way with a client that we wouldn't have done if we were on our own (Pavlicevic, 1999). Learners in classrooms emerge as accomplished or lagging in relation to their teacher's expectations (Braun, 1976). Through the lens of the social model of disability, we've awoken to how bodies become differently abled within various physical and social spaces. If a social environment values verbal communication, for example, an individual who communicates in only non-verbal ways would become more disabled (Ansdell, 2016). Even nations become through connections with other nations (Said, 1978). Identities are always co-produced (Fine, 1997).

Even as we emerge with a sense of being a subjective Self, this "Self" is relational in terms of how we distinguish our "I," "Me," and future Self. Today's "I" and "Me" are not the same as they were yesterday, last year, or in childhood. As "I" and "Me" shift, so too does my idea of who I will be in the future. A client's sense of "Me" after they have just experienced the death of a family member may differ from their "Me" before this experience; bereavement has become part of their "I"; and their future self may seem now to have an irrevocably different trajectory (Archer, 2015).

Through musicking, we become as certain "Selves." As Loaiza (2016) stated, "A sonic-musical world is brought forth in correspondence and within a participatory process of individuation" (p. 418). Musical dynamics can be facilitators of change by co-creating new meaning as therapist and client continually coordinate and regulate themselves in relation to each other's emotions and intentions (Johns, 2018).

Participation Is Ontological

The relational substrate out of which we emerge as a Self can be thought about in a number of ways, as participation, intersubjectivity, dialogue, confluence, and narrative. Let's begin by considering participation. According to Orange (2002), empathy is "a mode of understanding by participation" (p. 699). As we move towards exploring how this takes shape through participation in co-storying emotions, we'll first dwell a bit on how participation can be ontological, in other words, how we are constantly emerging (selfing) through participation in a range of practices (ways of doing things), within which our emotional lives take shape.

Pavlicevic (1997) reminded us that in music therapy improvisations, we aren't just focusing our attention on how a client spontaneously plays music but on how they are "in the world" (p. 122). Merleau-Ponty (1962) used the term "involved subjectivity." A person is always "in the middle of things." Our previous involvement in practices provides us with a starting point. Merleau-Ponty explained that life is lived before it is thought. We don't act on situations; we act in them. Our perceptions, emotions, thoughts, and behaviours are parts of practical activities that dovetail with each other, other people's behaviours, and events as part of organised activity. What we think and believe is relative to our situation. Heidegger (1962) phrased this as being "thrown" into the world. Our knowledge is always based on how we're already involved in the particular situation we're exploring. (This includes how we arrive at knowledge about ourselves.)

Even solitary activities are social practices. "Inner" experiences are part of shared worlds of practice (Westerman, 2013). Take the example of playing a piece of music alone in your practice room. You purchased this instrument from another person; you may practise to perform in front of others or to work on a piece that you'll use with a client in a session; you may be renting the space you're practising in; you're playing in response to musical traditions. Any impulse to do something or not to do something that we sense in our inner dialogue is a "voice" from a relationship (Gergen, 2009). Thought is inherently social and dialogical (Crossley, 2010).

Musicking is always situated activity (Stige, 2012). We engage in musicking as we take part in various practices, including listening, playing, composing, improvising, performing, interpreting, and reflecting. Stige (2006) was interested in how individuals and communities reciprocally form each other in social and cultural contexts. He highlighted the importance of participation as collaborative activity, as located in a relational understanding of humanity and resonating with understandings of health as the capacity for participation and mutual care. Participation in music therapy involves more than "being there" and "joining in." Stige contended for a stronger relational approach to understanding participation, explaining how some definitions of music "stress that music invites—or maybe even is—participation" (p. 128). Small's (1998) contention that music makes more sense as a verb than a noun also feeds the view that music is participation: "To music is to take part" (p. 9).

Social understanding is an ongoing and dynamic process of participatory sense-making. We participate in making sense together. Our interactions have a life of their own (think about the last time you and your conversation partner felt reluctant to end a phone call even though you both felt that the conversation had come to an end). Processes of interaction influence, modify and create (at least in part) the intentions of those involved. Through participatory sense-making in social interactions, people can access ways of making meaning that would be inaccessible if they were on their own. Participatory sense-making not only transforms individual meaning-making processes but gives shape to the individuals who are taking part (Loaiza, 2016).

Each therapeutic relationship has music that is true for the individuals involved in it. Thematic musical material emerges through reciprocal interaction between the client(s) and therapist as an organic process that can't be predetermined or created deliberately. It grows out of and feeds into the therapeutic relationship (Aigen, 2005). The music develops through shared active involvement as an open structure shaped by and shaping the participants (Schiavio & De Jaegher, 2017). According to Ansdell (1995, p. 26), "we communicate with words to convey our meaning, whereas we improvise music to find something meaningful between us." In other words, musical interactivity is a form of participatory sense-making. Musicking offers us meaning and world-making resources (DeNora, 2000, 2011).

Intersubjectivity (The Version That Considers Intersubjectivity to Be the Precondition for Subjectivity)

Intersubjectivity has already appeared a few times in this book. Here, we draw on an understanding of the concept that refers to how a shared experiential world between us generates our subjectivity. Much of the flavour of this form of intersubjectivity has come from relational psychoanalytic thought. Orange (2002) explained intersubjectivity as a relational context within which all experience gains its form. An intersubjective field is not a mode of experiencing or sharing experiences; it is the precondition for any experience to happen. Intersubjectivity within a more fully relational stance has been described by authors such as Scott (2011). He focused on what is between, rather than just on what is inside (first-person) or outside (third person). Orange (2009) described intersubjective systems theory as the view that "personal experience always emerges, maintains itself, and transforms in relational contexts" (p. 237).

We gain a sense of our subjectivity within an intersubjective field (Froese & Di Paolo, 2011; Orange et al., 2015). Individuation is always intersubjective because realising our independence is dependent on another person recognising it (Benjamin, 1988). Intersubjectivity is a process of mutual recognition. I experience my subjectivity in your presence because you recognise me as a subject, as I recognise you (Daly, 2016). We co-experience dimensions of ourselves through one another. The other's expressions invite, animate, sound, and resound our own (Rosan, 2012). We make each other through our expectations, responses, listening, witnessing, inviting, dismissing, positioning, and so on.

From a relational psychoanalytic perspective, Benjamin (2004) described the idea of the "third." For some authors, the third refers to the context within which we emerge, and for others, it is a property that emerges from dyadic interaction (Barsness, 2017). To address this debate, Benjamin distinguished between the "one-in-the-third" and the "third-in-the-one." The one-in-the-third captures the idea of the harmonic or rhythmic sense of oneness or mutual accommodation. Here, two people share a dance (as seen, for example, in the reciprocal patterns of eye gaze, gestures, communicative vocalisations, movements, and mirroring). This resonates with Malloch and Trevarthen's (2009) notion of communicative musicality. Both partners are simultaneously creating this musical pattern and surrender to being shaped by it. In a musical improvisation, we experience this one-in-the-third as we accommodate each other and accommodate to the co-created rhythm between us. We influence each other, and we are influenced by what has been established between us.

In terms of the third-in-the-one, Benjamin (2004) acknowledged that there is also space for differentiation in a relationship. Caregivers don't always perfectly attune to their infants. They offer their version—their understanding—of the infant's expression. The response has the caregiver's signature. Through this differentiation, the infant becomes able to decouple their own and the other's self-state. If a parent exclaims as a child falls over, hurts themselves, and begins to wail, their exclamation tends not to hold the same degree of disorganisation that the child's expression does. The child will experience this response as "nearly like me, but not quite." In music therapy, we accomplish this through matching our clients, reflecting back key elements, but not in precisely the same manner (Wigram, 2004). Differentiation of the self takes place within connectedness. The one-in-the-third and the third-in-the-one are interconnected and depend on each other.

Within the content of psychoanalysis, Stern (2005) explained how, from an intersubjective approach, neither the analyst nor the patient can ever take a third-person perspective (discovering the psychic reality within the other's mind). There is no objective reality that stands outside of the intersubjective realm. The patient's psychic reality is not "discovered" but is determined by the active relational processes taking place at the moment. (Transference and countertransference would then be subsumed here under the category of intersubjectivity.) The subject matter of a session emerges from the participants' creative interplay. We are always in the middle of what is unfolding, and we don't know what will happen next. The emphasis lies in creating, adjusting, and exploring the intersubjective field. According to Stern, three intersubjective moves are important in the therapeutic process. We sound each other out to see where each of us is in the intersubjective field (exploring the relationship moment-by-moment to see where it is, where it's going, and how we are positioned within it); we explore how we are sharing and avoiding sharing experiences to become known (or not known) as the intersubjective field enlarges and offers fresh exploration pathways and new ways to become understandable to one another, and we define and redefine ourselves by using each other's reflections of ourselves (forming ourselves through each other's eyes).

Stolorow and Atwood (1996) argued for understanding all pathology within the unique intersubjective contexts within which it originates (which calls into question the very idea of psychodiagnosis). These authors contended that what should be diagnosed is the function of an entire system, not a patient's psychological organisation. The intersubjective field of the therapeutic relationship becomes a developmental second chance.

Dialogue Is Ontological

Philosopher Buber (1923/1970) believed that humans have an innate and primary "longing for relationship" (p. 77). He explained that "[i]n the beginning is the relation" (p. 78), and we "take our stand in relation" (Buber, 1958/2000, p. 20). Buber (1923/1970, 1947/2002) understood dialogue as an ontological state of "between-ness," out of which dialogical selves develop. Similarly, Bakhtin (1984) wrote that two voices are the minimum for life and existence.

The dialogical relationship is oriented towards what Buber (1958/2000) called the I-Thou relationship, which differs from an I-It relationship. In an I-It relationship, the object of our experience is something we can collect, analyse, classify, use, and manipulate. There is some distance between I and It. The experiencing I is an objective observer. However, in an encounter between I and Thou, we enter a mutual and reciprocal relationship in which we are both transformed. When I encounter you as Thou, it is as if an entire universe exists within you. "We" is the common ground between I and Thou.

Dialogue can be understood as the intersubjective realm between persons (Buber, 1965). It can't be reduced to what takes place within an individual psyche or in the dynamics of a group. Everything that participants play, say, and do in a music therapy session, for example, exists in response to things that have been played, said, and done before and in anticipation of things that will come about in response. We "fall into" the flow of dialogue rather than create dialogue (Gadamer, 1976). Dialogue has a nature and dynamic of its own that's irreducible to the participants involved. The dialogue calls forth the words that are used within it (Merleau-Ponty, 1962).

Buber (1965) wrote that humans seek to have a presence in the being of another through dialogue. As we are recognised, accepted, and affirmed by another, this allows us to be. Within dialogue, genuine listening entails encouraging another to form and express their own meanings. These may be vastly different from one's own. Through listening, one creates space for each other as whole people (Gordon, 2011). In dialogic listening, we accept the other person as a person of worth simply because they are human being (Floyd, 2010). Through engaging in dialogue, we afford respect and dignity to each other, such that we are each confirmed as persons. The African Zulu greeting "Sawubona" literally means "I see you," through which one acknowledges and values the presence of another (Caldwell & Atwijuka, 2018). We don't grow as a self through relationship to ourselves, but through becoming present through relationships with others. We call each other into being. We are doing just this when we meet and match musically in improvisations within music therapy sessions. We confirm the presence and being of the person we are musicking with and express that we see and hear them. The other not only becomes a self for me; the other becomes a self with me. Gergen (2009) wrote, "the removal of affirmation is the end of identity" (p. 168). We are not passive in this process, though. As therapists and clients respond to one another, they are recurrently saying, "Yes, you can make that of me, but no, you can't make this of me" (Muran, 2007).

According to Buber (1947/2002), inclusion is an extension and fulfilment of confirmation. It entails us both apprehending the other and apprehending ourselves. Buber (1969) contended that psychological suffering results from being alienated from such dialogic relations. He wrote, "sicknesses of the soul are sicknesses of relationship" (p. 150). For Buber (1990), therapeutic dialogue is grounded in an ontological understanding of interrelatedness. The problems identified in the therapeutic relationship are not the client's problems or mine as the therapist. They are ours because they are mutually influenced. Therapists are always involved in complex dialogical influence and response processes with clients. In therapy, we get to know one another as we participate in the situation. This is about knowing *with* another rather than knowing *about* another (Seeley & Reason, 2008). We could think of dialogue as a shared rhythm, where each party mutually tunes into the other and feels when to move and what the other is about to do.

Robarts (2006) referred to meeting in the music therapy relationship as the "Creative Now." New emotional landscapes can be created as we hear ourselves in fresh ways. Interactive musical improvisation is a kind of shared play and every therapeutic relationship is unique. Through playing instruments and/or vocalising within this musical interplay, we experience ourselves coming into being in new ways.

Co-action Within Confluences

Any meaningful action is a form of co-action (Gergen, 2009). In themselves, the actions of an individual hold no meaning. Your response makes my action meaningful in a particular way (and your response is also an action inviting further response). If I smile at you and you smile back, I have been friendly, but if you retreat and avoid eye contact, I have been presumptuous. While there is a tendency to attribute characteristics to individuals (Mary is aggressive; Anthea is kind), one's behaviour is not aggressive if others perceive it as playful, or kind if others see it as manipulative. The attribute is located in the relationship between the participants.

Gergen (2009) argued that "causes" and "effects" are mutually defining. He proposed the idea of a confluence, which can be thought of as a situation that offers particular kinds of sense-making. We eat at a dinner party, while we dance or offer applause at a concert not because someone told us to, but because these actions just seem to make sense to us in those spaces. There would be no joy if there were no conditions that we defined as worthy of joy. When conditions that justify anger multiply, then anger proliferates. As a client in a therapy session, I share my vulnerabilities. I do so as a participant in a confluence of relationships within which this behaviour becomes meaningful.

In the study I conducted a study with teenagers at a high school in Eersterust, South Africa, who were referred to music therapy for aggression towards peers, teachers, and family members, I was interested in exploring the characteristics of the music therapy space as a confluence. In contrast to the participants' classroom confluences, the music therapy sessions became a confluence where willing participation, gentle negotiation, playfulness, a sense of group ownership, support, and acceptance made sense (dos Santos, 2018).

Narrative Practices Are Ontological

Stories are everywhere. We remember in stories, dream and daydream in stories, anticipate, hope, doubt, despair, believe, plan, review, criticise, envy, learn, communicate, hate, and love in stories. As Gergen (2009) so clearly articulated, "every relationship is a story in the making" (p. 133). Illness narratives are polyphonic (made up of multiple voices) and heteroglossic (speaking in multiple codes). They entail a merging of the many voices of medical professionals, loved ones, fellow ill persons, support groups, perspectives encountered online, and so on (Frank, 2015). Narratives are not interpretive frameworks placed over a person's life; narratives are what that life is (Goldie, 2000). We render the world and the self intelligible and interdependent through stories.

Narratives are representations of a series of events that are connected in a coherent and culturally informed way (Le Poidevin, 2005). Narratives are constantly forming, being negotiated, reshaped, and dissolving (Gergen & Gergen, 2010). For the most part, we don't create our stories; they create us. We emerge as Selves through stories. Human consciousness, and the narrative of selfhood, is not their source but their product (Schafer, 1992). In therapy, we are continuously telling and retelling, shaping, and being shaped by stories.

Narratives in Musicking

Narrative is frequently understood as a verbal medium, while music is primarily nonverbal. When we think of musicking as a narrative process, we may feel a bit stuck as we wonder: What's the plot? Who are the characters? What's the unfolding action? Some have argued that we can understand music construction similarly to plot construction (Maus, 1997). According to Klein (2004), instead of focusing on a sequence of actions by actors, musical narratives entail an unfolding plot of expressive states.

In songs, we find a narrative unfolding that blends the verbal and verbal. Programmatic music also makes narrative more explicit by directly imitating familiar musical content (a bird call, a hunting horn), through approximate imitation (of rustling leaves), and through musically symbolising visual material (the qualities of clouds or lightning). In non-programmatic music, no explicit storyline may accompany the music. However, if we draw on Klein's (2004) understanding of musical narrative, we can still understand how the story of the music unfolds through connections between feeling states. Stories are not in the "object" of a piece of music. We make stories through musicking (and musical stories make us).

Musicking can play a vital role in the narrative emergence of Selves because it offers building blocks of subjectivity. Musical materials afford "terms and templates for elaborating self-identity—for identity's identification" (DeNora, 2006, p. 145). Eyre (2007) offered the example of her work with a man who had a diagnosis of schizophrenia. The content of his speech was poor, and the expressed narrative of his life story could be characterised as having many gaps. Through music improvisation (and creating visual and verbal narratives), he could re-narrate himself.

When individuals face degenerative conditions, a sense of having a "spoiled identity" may emerge as they face increasing loss of their capacities. Through greater isolation that brings fewer opportunities for negative constructions of self to be challenged, an identity of a "salvaged self" may form. In the context of complex disability, music therapy can provide a relational context within which a person can generate and sustain an alternative Self. Within the relational story between the therapist and/or other group members, an individual's defining and redefining of Self is validated. Burland and Magee (2014) showed how music technology could be used in therapeutic settings to facilitate identity formation and connect with experiences of Self that were present before a person's current struggles. Electronic music technology can increase access to music for those with highly restricted movement, for example. Persons adapting to disability or experiencing degenerative conditions can remain active participants in musicking. Access to a broad palette of musical sounds is available through electronic music technology, which offers material that can facilitate a sense of connection to and subjective building blocks for culturally relevant and age-related forms of narrative expression.

We Become as Selves Who Are Polyphonic Multi-beings

If we are becoming a Self through emerging from many different relationships, practices, dialogues, confluences, and stories, then we have opportunities to self in diverse ways. In every relationship, we gain potentials for action as other people's actions offer us models for what is possible; and we draw from the many forms of co-action that we participate in (as we learn to play the piano, we gain another way of being). We emerge, therefore, with an enormous range of possibilities for being (Gergen, 2009).

One might then wonder how individuals keep some sense of continuity of being if they constantly emerge as participants within different practices, and these practices are changing and multiple. Despite this fluidity, we do often experience being a somewhat consistent Self in different contexts and relationships. Westerman (2013) suggested that we look for trends in how a person engages in practices. The question of concern would not be, "What is this person like?" or "What is this person's disposition that they bring to situations?," but "In what ways does this person engage in situations?" and "How do they tend to relate their thoughts, feelings, and behaviours to what is taking place?" (These trends would still have emerged out of their relational history, as opposed to being an inherent quality of their character.)

Some therapists work towards helping clients gain a more coherent, stable, or anchored sense of being. Think of the term mental "disorder." Researchers of cognitive dissonance assume an underlying need to reduce inconsistency in people's thoughts. Alternatively, others understand clients as multi-beings involved in many social relationships, through which they are emerging as different Selves (Gergen & Gergen, 2010). Gergen (2009) argued that we find a world of complexity and conflict behind the facade of coherence, unity, and wholeness. "Do I contradict myself?" asked Walt Whitman (1855). "Very well, then I contradict myself. I am large, I contain multitudes" (verse 51).

Civilisation is held together primarily by the everyday demands to remain intelligible within our relationships. ("This is the way we do it here, so we expect you to present yourself in a relatively consistent way.") When integration and coherence of identity are valued (through co-active agreement), we may tend to engage as a more stable and predictable Self. That is not because this is all we are. It's because this is a shape we take on within this particular relational context. Gergen (2009) claimed that, in such a situation, we would then be a partial person (a fragment of our multi-being). (This differs from the idea of wearing a mask, underneath which our "real self" resides.) We gain a sense of authenticity and comfort from feeling known and accepted. We, therefore, often choose to remain "in character" rather than break expectations. When the bubble is burst and we "step out of character"—bringing in other fragments of our multi-being—difficult dynamics and even anguish may result, but this is also an invitation into a new conversation and a new adventure. Music therapy offers rich opportunities to welcome clients as multi-beings.

Plotting the Path of Relational Empathy

In the realm of insightful empathy, you and I understood ourselves as inherently separate individuals, but we believed we could gain access to each other's inner emotional worlds. Translational empathy rested on the same view that we are inherently separate, yet we did not presume to be able to see into one another's inner lives. We honoured otherness. When we explored empathic assemblages, we looked solely at interminglings and entanglements. Empathising related to the responsive capacities of the assemblage, and we didn't take much account of individual selves. Here, as we explore relational empathy, we direct our concern towards a relational foundation *and* we are interested in the selves that emerge from it (as well as how they influence the relationship). Our stance is one of acknowledging this dynamic, fully relational process. From this vantage point, we don't consider emotions as simply dwelling inside a person. Emotion is a story that plays out between people. We will delve into this further in the following chapter.

Relational empathy is situated awareness of how we are co-storying. Fishman (1999) explained that "there is no possibility of restoring to the concept of empathy the illusory idea of privileged, objective, and uncontaminated access into the inner world of another person" (p. 378). We achieve understanding *with* one another. I am co-storying with you. Within relational–cultural theory, relationships are considered the primary means of change. As therapists interact with clients, they recognise they are active participants in their clients' contexts. Authors such as Judith Jordan (2000) have highlighted the importance of mutual empathy in therapy. Mutual empathy is not only about knowing another person's subjective experience but it is about the experience of connectedness. Both parties affect each other, and knowledge of this is valuable for each of them. When we empathise, we are both changed. Clients need to witness therapists being moved by them, demonstrating that they matter. This can occur through the reciprocal attunement and connection facilitated in music therapy (Kobin & Tyson, 2006).

A critique of this approach of mutual empathy could be that therapy should centre the client. If we are focusing on reciprocal connection and mutual empathy, then the client is also being expected to empathise with the therapist, which could be inappropriate, unethical, and even unsafe. This is not what mutual empathy means in therapy, however. Mutual empathy can be ethical and therapeutically safe (Rolvsjord, 2004). Jordan (2000) explained that in mutual empathy (as we acknowledge we are co-storying together), the therapist is not there to be healed by the client (although through the relationship they will grow too); the client's role is not to take care of the therapist; and the therapist is not totally open, completely disclosing, or entirely spontaneous with no limits. The roles of the client and therapist are different. Careful, reflexive, and ethical considerations are still at play. Mutual empathy is about the authentic, engaging quality of the relationship.

References

Aigen, K. (2005). *Being in music: Foundations of Nordoff-Robbins music therapy.* Barcelona Publishers.

Ansdell, G. (1995). *Music for life.* Jessica Kingsley.

Ansdell, G. (2016). *How music helps in music therapy and everyday life.* Routledge.

Ansdell, G., & Stige, B. (2015). Community music therapy. In J. Edwards (Ed.), *Oxford handbook of Music Therapy* (pp. 595–621). Oxford University Press.

Archer, M. (2015). The relational subject and the person: Self, agent, and actor. In P. Donati & M. Archer (Eds.), *The relational subject* (pp. 85–122). Cambridge University Press.

Archer, M., & Donati, M. (2015). The plural subject versus the relational subject. In P. Donati & M. Archer (Eds.), *The relational subject* (pp. 33–78). Cambridge University Press.

Bakhtin, M. (1984). *Problems of Dostoevsky's poetics* (C. Emerson, Trans.). University of Minnesota Press.

Barsness, R. E. (2017). *Core competencies of relational psychoanalysis: A guide to practice, study and research.* Routledge.

Benjamin, J. (1988). *The bonds of love.* Pantheon Books.

Benjamin, J. (1995). *Like subjects, love objects.* Yale University Press.

Benjamin, J. (2004). Beyond doer and done to: An intersubjective view of thirdness. *The Psychoanalytic Quarterly, 73*(1), 5–46.

Braun, C. (1976). Teacher expectation: Sociopsychological dynamics. *Review of Educational Research, 46*(2), 185–213.

Buber, M. (1923/1970). *I and Thou* (W. Kaufmann, Trans.). Touchstone.

Buber, M. (1947/2002). *Between man and man* (R. Smith, Trans.). Routledge.

Buber, M. (1958/2000). *I and Thou* (R. Smith, Trans.). Scribner.

Buber, M. (1965). *The knowledge of man.* Harper and Row.

Buber, M. (1969). *A believing humanism: Gleanings.* Simon and Schuster.

Buber, M. (1990). *A believing humanism: My testament* (M. S. Friedman, Trans.). Humanities Press.

Burland, K., & Magee, W. (2014). Developing identities using music technology in therapeutic settings. *Psychology of Music, 42*(2), 177–189.

Caldwell, C., & Atwijuka, S. (2018). "I see you!"—The Zulu insight to caring leadership. *The Journal of Values-Based Leadership, 11*(1), 13.

Crossley, N. (2010). *Towards relational sociology.* Routledge.

Crossley, N. (2018). Networks, interactions and relations. In F. Depelteau (Ed.), *The Palgrave handbook of relational sociology* (pp. 481–498). Palgrave Macmillan.

Daly, A. (2016). *Merleau-Ponty and the ethics of intersubjectivity*. Palgrave Macmillan.

DeNora, T. (2000). *Music in everyday life*. Cambridge University Press.

DeNora, T. (2006). Music and self-identity. In A. Bennett, D. Shank, & J. Toynbee (Eds.), *The popular music studies reader* (pp. 141–147). Abingdon.

DeNora, T. (2011). *Music in action: Selected essays in sonic ecology.* Ashgate.

DeYoung, P. A. (2014). *Relational psychotherapy: A primer.* Routledge.

Donati, P. (2011). *Relational sociology: A new paradigm for the social sciences.* Routledge.

dos Santos, A. (2018). *Empathy and aggression in group music therapy with adolescents: Comparing the affordances of two paradigms* [Doctoral dissertation, University of Pretoria].

Eyre, L. (2007). The marriage of music and narrative: Explorations in art, therapy, and research. *Voices: A World Forum for Music Therapy, 7*(3).

Fine, M. (1997). Witnessing whiteness. In M. Fine, L. Weis, L. Powell, & L. Wong (Eds.), *Off White: Readings on race, power, and society* (pp. 57–66). Routledge.

Fishman, G. (1999). Knowing another from a dynamic systems point of view: The need for a multimodal concept of empathy. *The Psychoanalytic Quarterly, 68*(3), 376–400.

Floyd, J. J. (2010). Listening: A dialogic perspective. In A. D. Wolvin (Ed.), *Listening and human communication in the 21st century* (pp. 127–140). Wiley-Blackwell.

Frank, A. (2015). Practicing dialogical narrative analysis. In J. A. Holstein & J. F. Gubrium (Eds.), *Varieties of narrative analysis* (pp. 33–52). Sage.

Froese, T., & Di Paolo, E. (2011). The enactive approach: Theoretical sketches from cell to society. *Pragmatics & Cognition, 19*(1), 1–36.

Funk, M., & Coeckelbergh, M. (2013). Is gesture knowledge? A philosophical approach to epistemology of musical gestures. In H. De Preester (Ed.), *Moving imagination: Explorations of gesture and inner movement* (pp. 113–132). John Benjamins.

Gadamer, H. (1976). *Philosophical hermeneutics* (D. E. Linge, Trans.). University of California.

Garred, R. (2001, November). The ontology of music in music therapy. *Voices: A World Forum for Music Therapy, 1*(3). https://voices.no/index.php/voices/article/view/1604/1363

Garred, R. (2006). *Music as therapy: A dialogical perspective.* Barcelona Publishers.
Gergen, K. (2009). *Relational being: Beyond self and community*. Oxford University Press.
Gergen, K. (2018). Human essence: Toward a relational reconstruction. In M. van Zomerpen & J. Dovidio (Eds.), *The Oxford handbook of Human essence* (pp. 247–260). Oxford University Press.
Gergen, K., & Gergen, M. (2010). Scanning the landscape of narrative inquiry. *Social and Personality Psychology Compass, 4*(9), 728–735.
Gibson, J. (2017). A relational approach to suffering: A reappraisal of suffering in the helping relationship. *Journal of Humanistic Psychology, 57*(3), 281–300.
Goldie, P. (2000). *The emotions: A philosophical exploration.* Oxford University Press.
Gordon, M. (2011). Listening as embracing the other. *Educational Theory, 61*(2), 207–219.
Heidegger, M. (1962). *Being and time*. Harper & Row.
Holdstock, T. L. (2000). *Re-examining psychology: Critical perspectives and African Insights*. Routledge.
Husserl, E. (1977). *Cartesian meditations* (D. Cairns, Trans.). Kluwer Academic.
Hycner, R. (1993). *Between person and person: Toward a dialogical psychotherapy.* Gestalt Journal Press.
Johns, U. T. (2018). Exploring musical dynamics in therapeutic interplay with children: A multilayered method of microanalysis. *Nordic Journal of Music Therapy, 27*(3), 197–217.
Jordan, J. (2000). The role of mutual empathy in relation/cultural therapy. *Psychotherapy in Practice, 56*(5), 1005–1016.
Kessler, N. (2019). *Ontology and closeness in human-nature relationships.* Springer.
Klein, M. (2004). Chopin's Fourth Ballade as musical narrative. *Music Theory Spectrum—The Journal of the Society for Music Theory, 26*(1), 23–55.
Kobin, C., & Tyson, E. (2006). Thematic analysis of hip-hop music: Can hip-hop in therapy facilitate empathic connections when working with clients in urban settings? *The Arts in Psychotherapy, 33*(4), 343–356.
Le Poidevin, R. (2005). Narrative. In T. Honderich (Ed.), *The Oxford companion to philosophy* (pp. 638–639). Oxford University Press.
Levitin, D. (2006). *This is your brain on music.* Dutton.
Lewis, C. S. (1990). *Studies in words.* Cambridge University Press.

Loaiza, J. (2016). Musicking, embodiment and participatory enaction of music: Outline and key points. *Connection Science, 28*(4), 410–422.
Malloch, S., & Trevarthen, C. (2009). Musicality: Communicating the vitality and interests of life. In S. Malloch & C. Trevarthen (Eds.), *Communicative musicality* (pp. 1–10). Oxford University Press.
Mann, D. (2020). *Gestalt therapy: 100 key points and techniques.* Routledge.
Maus, F. E. (1997). Narrative, drama, and emotion in instrumental music. *The Journal of Aesthetics and Art Criticism, 55*(3), 293–303.
McCormack, B., & McCance, T. (2016). *Person-centred practice in nursing and health care: Theory and practice.* Wiley.
Mead, G. H. (1934). *Mind, self and society* (Vol. 111). University of Chicago Press.
Merleau-Ponty, M. (1962). *Phenomenology of perception* (C. Smith, Trans.). Routledge & Kegan Paul.
Mitchell, S. (1991). Contemporary perspectives on self: Toward an integration. *Psychoanalytic Dialogues, 1*(2), 121–147.
Montgomery, M. (1989). *Saying goodbye: A memoir for two fathers.* Alfred A. Knopf.
Muran, J. (2007). A relational turn on thick description. In J. C. Muran (Ed.), *Dialogues on difference: Studies of diversity in the therapeutic relationship* (pp. 257–274). American Psychological Association.
Orange, D. (2002). There is no outside: Empathy and authenticity in psychoanalytic process. *Psychoanalytic Psychology, 19*(4), 686.
Orange, D. (2009). Intersubjective systems theory: A fallibilist's journey. *Annals of the New York Academy of Sciences, 1159*(1), 237–248.
Orange, D., Atwood, G., & Stolorow, R. (2015). *Working intersubjectively: Contextualism in psychoanalytic practice.* Routledge.
Pavlicevic, M. (1997). *Music therapy in context.* Jessica Kingsley.
Pavlicevic, M. (1999). *Music therapy: Intimate notes.* Jessica Kingsley.
Robarts, J. (2006). Music therapy with sexually abused children. *Clinical Child Psychology and Psychiatry, 11*(2), 249–269.
Rolvsjord, R. (2004). Therapy as empowerment. *Nordic Journal of Music Therapy, 13*(2), 99–111.
Rosan, P. (2012). The poetics of intersubjective life: Empathy and the other. *The Humanistic Psychologist, 40*, 115–135.
Said, E. (1978). *Orientalism: Western concepts of the Orient.* Pantheon.
Schafer, R. (1992). *Retelling a life.* Basic books.
Schiavio, A. (2015). Action, enaction, inter (en) action. *Empirical Musicology Review, 9*(3–4), 254–262.

Schiavio, A., & De Jaegher, H. (2017). Participatory sense-making in joint musical practices. In M. Lesaffre, M. Leman, & P. Maes (Eds.), *The Routledge companion to embodied music interaction* (pp. 31–39). Routledge.

Scott, C. (2011). *Becoming dialogue: Martin Buber's concept of turning to the other as educational praxis* [Doctoral Thesis, University of British Colombia].

Seeley, C., & Reason, R. (2008). Expressions of energy: An epistemology of presentational knowing. In P. Liamputtong & J. Rumbold (Eds.), *Knowing differently: Arts-based and collaborative research methods* (pp. 25–46). Nova Science Publishers.

Small, C. (1998). *Musicking: The Meanings of Performing and Listening.* Wesleyan University Press.

Stern, D. (2005). Intersubjectivity. In E. Person, A. Cooper, & G. Gabbard (Eds.), *Textbook of psychoanalysis* (pp. 77–92). American Psychiatric Publishing.

Stige, B. (2006). On a notion of participation in music therapy. *Nordic Journal of Music Therapy, 15*(2), 121–138.

Stige, B. (2012). Health musicking: A perspective on music and health as action and performance. In R. MacDonald, G. Kreutz, & L. Mitchell (Eds.), *Music, health, and wellbeing* (pp. 183–195). Oxford University Press.

Stolorow, R., & Atwood, G. (1996). The intersubjective perspective. *Psychoanalytic Review, 83*(2).

Stolorow, B., Atwood, G., & Orange, D. (2002). *Worlds of experience: Interweaving philosophical and clinical dimensions in psychoanalysis.* Basic.

Veissière, S. P., Constant, A., Ramstead, M. J., Friston, K. J., & Kirmayer, L. J. (2020). Thinking through other minds: A variational approach to cognition and culture. *Behavioral and Brain Sciences, 43.*

Westerman, M. (2013). Making sense of relational processes and other psychological phenomena: The participatory perspective as a post-Cartesian alternative to Gergen's relational approach. *Review of General Psychology, 17*(4), 358–373.

Wigram, T. (2004). *Improvisation: Methods and techniques for music therapy clinicians, educators and students.* Jessica Kingsley.

Wildman, W. J. (2010). An introduction to relational ontology. In J. Polkinghorne (Ed.), *The Trinity and an entangled world: Relationality in physical science and theology* (pp. 55–73). Wm. B. Eerdmans Publishing.

Winnicott, D. (1965). *The maturational processes and the facilitating environment.* International Universities Press.

Yontef, G. M. (1993). *Awareness, dialogue & process: Essays on Gestalt therapy.* The Gestalt Journal Press.

17

Awareness of Emotions as Co-Storying

Emotions derive their meaning from the relationships in which they occur, and they shape these relationships in return (Boiger & Mesquita, 2012). As Fuchs (2013) wrote, emotions "are not inner states that we experience only individually or that we have to decode in others, but primarily *shared states*" (p. 223). Interaffectivity refers to this notion that our emotions are not excessively ours. They are formed and continue to be shared interpersonally, and we are participants in a shared space of emotional attunement.

Relational Development of Emotions

Researchers within a dynamic systems approach have shown how an infant's emotion development emerges through the accrual of interactional sequences of attuning, sharing, offering, and responding, not because of discrete, prewired emotion "programmes" (Camras & Witherington, 2005; Fogel et al., 1992). An infant's distress signal doesn't communicate an emotionally communicative message in itself. It's rendered emotionally meaningful in the context of a caregiver's

A. dos Santos, *Empathy Pathways*,
https://doi.org/10.1007/978-3-031-08556-7_17

responses. Ongoing interactions with others influence adults' emotions just as significantly.

Barrett (2017), a psychologist and neuroscientist specialising in the study of emotion, explained how we learn concepts early in life that we use to make sense of our experiences. Our brains are constantly trying to guess what's happening in the world around us to prepare for any necessary action. Using past experiences, our brains construct hypotheses, or simulations, by imposing meaning on what is arriving through our senses. A primary tool that brains use to guess the meaning of these bits of sensory input is concepts. We experience a bee as a bee because we've learned the concept of what a bee is. We may listen to and understand Western music by learning the concept of an octave that is divided into twelve equally spaced pitches. Our brain uses the same process to interpret interoceptive sensations within the body (a stomach-ache, a gnawing feeling, a tight chest, tingling behind the eyes). According to Barrett, we experience inner sensations or feelings that vary in valence (pleasant or unpleasant) and arousal (calmness or agitation). Using the concepts we've learned, our brain interprets what that sensation means. When a child yells and pushes their sibling who is trying to take their toy, a caregiver may exclaim, "I see you're very angry!" offering the child an emotional concept—anger—as an interpretation of their sensations. Instead of understanding emotions as reactions to the world, in her theory of constructed emotion, Barrett wrote that "an emotion is your brain's creation of what your bodily sensations mean, in relation to what is going on around you in the world" (p. 30). Emotions are "a product of human agreement" (p. xiii) and we are active constructors of our emotions within the context of our relationships.

Emotions are constructed within relational themes. Anger and guilt, for example, may feel similarly unpleasant but are linked to profoundly different themes (for example, a theme of goals being blocked or a theme of failure to repay certain obligations, respectively). Different relational themes are activated in different situations (sometimes multiple, even contradictory, themes are activated within one situation). Different emotion themes are available across cultures (Kitayama et al., 2006).

Musical concepts also enable us to form and identify emotional experiences. We don't use music simply to express an emotional state. Through music, we also constitute emotional states as music is "a resource for the identification work of 'knowing how one feels'—a building material of 'subjectivity'" (DeNora, 2000, p. 57).

Relational Regulation

Emotion regulation is a process that helps us achieve functional (culturally appropriate) emotional experiences, enabling successful navigation of the social world and the maintenance of relationships (De Leersnyder et al., 2013). Humans pool their emotional resources and regulate emotions together. In dyads, optimally each member openly communicates with the other, who responds in turn, maintaining connection even when things become difficult. The sensitive responses of another help us regulate intense or potentially overwhelming affective sensations, and through this co-regulation, we eventually learn how to regulate ourselves.

Dyadic models of emotion regulation show how partners (initially infants and caregivers, but then all partners, such as therapist and client) mutually coordinate their strategies for coping with emotions (Fosha, 2001). Social baseline theory explained how people reply on close others to distribute emotion processing and for emotional load-sharing (Kappas, 2013; Lougheed et al., 2016). It appears that interpersonal regulation is primary from infancy to old age. Individual regulation grows out of it and is dependent on it.

Both parties influence each other within a field of mutual influence. For example, while studies of maternal postpartum depression have shown how a mother's limited capacity to offer reciprocal parenting can disrupt her infant's emotion regulation (Blandon et al., 2008), research has also shown how premature birth results in low regulation and high negative emotionality of the infant which disrupts a mother's ability to read her infant's expressive signals and to engage in reciprocal dialogue (Poehlmann et al., 2011).

Being in Emotions

The experienced space around us is constantly charged with emotional qualities. We may sense a threatening atmosphere, that "something is in the air," that we could "cut the tension in the room with a knife." We may walk into a space that feels serene, therefore breathing out and relaxing our shoulders. These feelings are not just background states. They relate to how we find ourselves in the world with others. Fuchs (2013) argued that we essentially feel sensations through the medium of our body as we come into context with the feelings and atmospheres around and between us and as we resonate with each another.

Orange (2002) discussed how pain is commonly referred to in Western contexts as being "inside" a person as a private state. On the contrary, as embodied subjects, we do not "have" pain; we are "in" pain. Pain is a component of a relational and communicative intersubjective system. We show and tell others where our pain is, how it feels, and their response influences our experience of pain. Even when we are alone, we may say "Ouch!" to an implicit hearer. Orange wrote of a sob that "was not an outer sign of an inner pain state, but an experiencing and embodied subject's participation" (p. 693). Empathy then entails living in the experiential world that the other person and I create together. As we gain a situated awareness of how we live and participate in this world together, new meanings and ways of participating can open up.

Butler (2017) wrote about models of temporal interpersonal emotion systems (TIES), of which there are two categories. The first is "between person" models, where the emotional processes in one person interact with those in another. In the second, interpersonal state models, emotions are understood as dynamic latent states that generate interdependent emotional experiences over time for two or more individuals. As an example of this second model, one wouldn't think of two people's anger as being separate observable states that interact. There would only be one anger state that both partners are sharing in, which explains the ebb and flow of both of their experiences, expressions, and embodied reactions. Within this process, there is dynamic flexibility of changes moment-to-moment arising from feedback loops between partners and within partners, with developmental flexibility unfolding over the long term.

This plays out not only between dyads. As an example with triads, Hollenstein et al. (2016) studied emotion patterns in families with and without adolescents experiencing depression in Australia. Families with a depressed adolescent showed a wider range of affect; more unpredictable triadic affective sequences; they spent longer in discrepant affective states and returned to these faster; and spent less time in matched affective states and were slower to return to these, particularly when engaged in problem-solving interactions. Relationships shape emotions, and emotions shape relationships.

Group-shared emotions have also been researched (terms such as "collective mood" or "affective climate" are also used here), particularly as they emerge through inclination (as people who tend to participate in emotion stories similarly engage in joint action), interaction, institutionalisation, identification, and participatory sense-making. Group-shared emotions can be acute or chronic, converging or diverging. We can explore whether and how a group recognises an emotion story, how they regulate it and potentially repeat patterns over time (Menges & Kilduff, 2015).

Emotions as Social Performances

Gergen (2009) presented a case for emotions as social performances. Emotions are not correspondences between the external world and our inner reality or "inner" experiences about events in our surroundings. They are social practices and parts of practical activities, organised by the stories we tell and enact. Emotions are "parts of ways of doing things in the world" (Westerman & Steen, 2007, p. 328). We "do" emotions within relational scenarios rather than "feel" them.

We don't have terms like "joy," "love," or "anger" because they represent events in the world. We have relationships through which we create these realities and through which they become important to us. Emotions are intelligible only as they occur within particular relational traditions and depend on our relational histories. The performance of an emotion depends on cultural education. It is only when an emotion is well-performed that we accept it as authentic. Emotion performance is a crafted achievement (Gergen, 2009). We use gestures, words, facial expressions, and posture (music, symbols, and so on) to perform an emotion in a way that another can recognise. When a person hasn't yet mastered the craft of emotion performance, we may feel mystified by what they are attempting to express and by the very existence of the emotion itself. To recognise a particular emotion, the performance also needs to occur within socially specified circumstances. If we shouted out with ecstasy in the middle of the cereal aisle at the grocery store, those around us might step away nervously. Emotion display rules vary widely across cultures (Ip et al., 2021).

Instead of wondering whether someone is truly in love, genuinely angry, or sincerely happy, we could wonder whether they are fully engaged in performing that emotion. What this allows us to do is to shift our focus from the emotions "within" isolated individuals (a belief that is lodged within the Western tradition) to a concern with relational processes (through which individuals come into being). We shift our attention from the dancers to the dance (Gergen, 2018).

Emotions Within Societies and Cultures

Every emotion is a diverse social phenomenon. Five major approaches have emerged for studying emotions within sociology (Bericat, 2016). First, cultural theories view emotions as social feelings. Societies have emotional cultures, vocabularies, feeling rules, and display rules (Illouz et al., 2014; Robinson, 2014), although individuals still strategically negotiate their emotions within these (Lively & Weed, 2014). Second, symbolic interactionists focus on feelings as social: feelings are formed and sustained by group processes and can't be reduced to the particular individuals who feel them (Fields et al., 2006). The self is a product of ongoing social interactions, and identity work is an inherently emotional process. Emotional arousal is considered to be informed by self-identity. We experience "positive" emotions when others confirm our self-concept and "negative" emotions (such as shame, guilt, distress, anxiety, and anger) when others negate our self-concept. Third, ritual theories maintain that group emotions are generated through rituals, understood as social gatherings within which people hold a shared focus of attention, shared values, and shared emotions. Such gatherings generate collective effervescent and group consciousness (Collins, 2004; Summers-Effler, 2006). Fourth, structural theories of emotion explain how the relational dimensions of structure and power generate specific emotions in the course of social interaction, with actors who hold more power and status experiencing "positive" emotions, while those with lower power and status experiencing "negative" emotions (Thamm, 2004). Finally, exchange theories explain the world of emotions by focusing on social interaction as a process in which actors exchange resources. People feel good when rewards exceed costs, and they have unpleasant feelings when costs exceed investments. The type and intensity of emotion one feels are also influenced by the kind of exchange, the characteristics of the social network, the power of the actors, the norms of justice that are at play, and whether expectations are met (Turner & Stets, 2006).

Cultures do emotions differently. Norms and ideals regarding what emotions people should experience and would like to experience motivate the generation of those emotions (and the practices linked to expressing those emotions). Cultures also promote situations that invite norm-adherent emotions or discourage or avoid situations that invite norm-inconsistent emotions (for example, greater incidents of compliment-giving occur in cultures that value feeling pride and enhancing self-esteem). People select situations that lead to ideal emotions. Cultures also do emotions differently through interpersonal regulation (supporting and rewarding certain emotional expressions and ignoring or even punishing others). This occurs through socialisation practices and co-regulation between adults (Mesquita et al., 2016).

In a study on the prevalence of affective disorders, De Vaus et al. (2018) found that clinical depression and anxiety rates were four to ten times higher in Western countries than in Asia. They made the case that this is linked to different ways of thinking about emotion in different cultures. Eastern holistic principles of change (acknowledging that the world exists in continuous flux), contradiction (every experience is associated with its opposite), and context (all things interconnect) profoundly shape emotional experiences.

Relational Understandings of Emotion in Interdependent Cultures

Some authors within Western contexts have written about relational approaches to understanding emotions as if this is a new idea. It certainly isn't. While current Western thought still tends to locate emotions inside the brains of individuals, this is a recent and far from universal view. In more interdependent cultures, emotions, thoughts, and actions are considered as arising from the relationship between individuals. Emotions are an interdependent project (Mesquita et al., 2016). In a study with participants from West Sumatra by Levenson et al. (1992), emotions were described as necessarily involving other people and were not recognised as occurring when a participant was alone. Uchida et al. (2009) studied how Japanese and American participants described emotions. Japanese participants focused on emotion as the interaction between people, while the American participants saw emotions as inner experiences. In studies by Masuda et al. (2008, 2012), Westerners rated the target's facial expressions without taking into account the facial expressions of people surrounding the target because they viewed emotions as being within people. Japanese participants used the facial expressions of the surrounding people to judge the target's emotions because they interpreted emotions to be between people. For example, if people around the target appeared angry or sad, then a smiling target was deemed to be less happy.

In another study by Mesquita et al. (2010), North American and Japanese participants reflected on situations where they felt offended. North American participants framed offence situations as threats to their self-worth and autonomy and as rejection. Their goal was to then address this situation by reaffirming the sense of self and retaliating. The Japanese participants interpreted episodes of offence as relationship threats that called for a greater understanding of the other's motives. They sought to interpret the situation from the other's perspective. This required remaining calm, doing nothing, or moving away.

The DSM-5 (American Psychiatric Association, 2013) classifies depression as a mood disorder characterised by persistent feelings of sadness and hopelessness and/or loss of interest in activities that used to be pleasurable, as well as potential feelings of worthlessness or guilt, slowing down of thought, diminished ability to concentrate, changes in appetite and weight, loss of energy, and suicidal ideation. This approach places depression within an individual. Transcultural studies (for example, a large study conducted by the World Health Organisation [Gureje et al., 1997]) highlighted how somatic complaints (for example, feelings of pain, sleep disturbances, burning, numbness, and heaviness) lie at the core of depression much more frequently than psychological concerns. People in many cultures experience depression most prominently as a bodily disturbance, commonly understanding these disturbances as being inextricably intertwined within interpersonal states and conflicts. Western culture splits mental and somatic, inner body and external world. Different approaches are found in cultures with members who don't experience themselves as discrete individuals, but as parts of communities. Disturbances of mood and well-being are then understood as bodily, interpersonal, and potentially atmospheric (Fuchs, 2013). Kitanaka (2012) explained that in traditional Japanese psychopathology, for example, social atmospheres and surrounding climate are regarded as carriers of mental illness, constituting the "in-between" that mental disorders originate from.

Emotions as Stories

Let's focus now on our primary concern: emotions are stories. As Rosaldo (1984) wrote, "Feelings are not substances to be discovered in our blood but social practices organised by stories that we both enact and tell" (p. 143). Stories don't only show or represent emotions, but emotion itself "is the acceptance of, the assent to live according to, a certain kind of story" (Nussbaum, 1988, p. 226). We link sensations in our body, expressions, character, and the situation as a structured episode and as one that relates to a larger unfolding narrative. It is this unfolding gestalt of feelings, thoughts, actors, events, and conditions that constitutes an emotion (Kleres, 2011). As we grow up, we learn about emotions through participating in stories within different contexts. To analyse emotions, we need to ask who is acting, how, to whom, and what is happening (Kleres, 2011).

What is taking place in a music therapy session is that we are active participants in emotion stories. Stolorow (2014) referred to this as "emotion dwelling." We dwell in the emotion story as it unfolds between us. As we become aware of this—as situated participants, not as a distant analytic exercise—we engage in relational empathy. Rather than holding a position of authority, the therapist collaborates with the client, as meaning is explored in the confluence of the therapeutic process (Gergen, 2000).

Let's consider an example of a parent in a neonatal ICU engaging in "kangaroo care" and musical holding with their infant. They are nurturing their relationship (through which the infant is coming into being as an infant and the parent is coming into being as a parent), co-regulating, and taking part in emotion stories of stress, hope, grief, and security. Premature infants are already social and musical beings (Haslbeck, 2016). They require regulatory input from others to sustain their basic physiological and homeostatic processes (Ham & Tronick, 2009). They can already share subtle forms of vitality within live improvised music (Haslbeck, 2016). Creative music therapy with parents and premature infants can offer opportunities for emotional co-storying without being overwhelming. The therapist's role in facilitating this is to listen responsively, reflect, and adapt to lubricate meaningful interactions with the infant and between the parents and infant by supporting the participants' sensitivity to one another. In terms of the parents, Haslbeck described how this often entails nurturing capacities that are already there but have been challenged by anxiety and uncertainty.

If we are in a difficult emotional encounter, one where attempts at insightful empathy may leave us thinking: "I just can't understand what this person is feeling or why they are feeling like this" or perhaps even, "This person has hurt me, and I can't find a way to feel empathy or behave compassionately towards them right now," relational empathy offers us another way. We can ask: "What story are we in together right now?," "How might we change this story?," "How are we co-storying?," as well as "What part may I be playing in this story?" We are all participants.

Participants in emotion stories may both perceive a situation similarly. Maybe there is some conflict between group members, and an anger story is quite prominent in the group. Sometimes multiple emotion stories may be at play, and the scenario as a whole is one of complexity, fragmentation, or friction. This is where music excels as an emotion co-storying medium, as it can coherently hold together diverse musical–emotional co-action in a way that retains a sense of overall connected relationality. In a musical improvisation, we can include a stable rhythmical grounding, layers of cross-rhythms, a tonal centre with textured dissonance, and a wide range of timbres while still experiencing the musicking as being coherent.

Perhaps a client is taking part in an emotion story of overwhelm in a session and a music therapist is taking part in an emotion story of calmness. We can understand overwhelm as co-action in the sense that the client may be experiencing an embodied sensation (unpleasant and agitated), and this is interpreted (by themselves, by the therapist, or by both) through concepts that were learned through sociocultural processes and performed in the current context of the interaction with the therapist who is responding by recognising and affirming the client's emotion story. The therapist is experiencing an embodied sensation (calm and pleasant) and has also interpreted this through the concepts at her disposal. She performs this in socioculturally recognisable ways, and the client may respond with recognition by saying, "Why are you sitting there calmly? This really is just completely overwhelming!" We have both an overwhelm story and a calm story between the therapist and the client. While co-regulation may occur as these stories merge and become more cohesive, the stories may also interact by amplifying each other, becoming more distinctly defined, or clashing even more intensely within the overarching relational story. All these dynamics can be taken into account as we reflect empathically on how we are co-storying.

There is another matter that we need to address as well. If relational empathy entails situated awareness of emotion co-storying, does that mean empathy can only occur in face-to-face encounters between people? Would we be able to cultivate relational empathy through song lyric analysis in a session when the song's writer is not in the room with us, for example? In his book *Relational being*, Gergen (2009) put a spotlight on the assumption that when reading a book, you may think that there are three separate entities: the writer, the book, and the reader. If we reconsider this, however, we realise that writers are drawing on words that are not their own (they borrow from countless sources and shape them in this new text). The place where the author "begins" and "ends" is not clear. As a reader encounters words on a page and reflects on what they mean in their own lives, do the words still "belong" to the author or have they become the reader's words? There's no clear separation in the moment of reading between writer, book, and reader (we are also connected to traditions of words and ideas that have come before us and will carry into the future). Reading a book, exploring the lyrics of a song, watching a film, or working with an image are all relational experiences and, therefore, potential sites for relational empathy. As we analyse lyrics together in a session, we are part of an emotion story playing out between the writer and ourselves and between each other as participants in the session.

In our daily emotion encounters, we are taking part in emotion co-storying all the time. What distinguishes relational empathy from these everyday acts is when we become *aware* of this process, not just as a cognitive or analytic exercise, but through situated awareness as a participant in the process. We wonder: What is going on with us? What is happening emotionally in the "us"? Through relational reflexivity, participants can orient themselves to how they are emerging from the relationship and how the encounter between their emerging selves generates a particular reality (Donati, 2011). Gergen (2009) described synchronic sensitivity as fine-tuned responsiveness to each other's actions. We affirm what has passed between us and invite what is to follow within the intertwining of our actions. Relational emotion reflexivity entails orienting ourselves to how we are engaging in emotion co-storying. We may gain awareness of this through multiple means, for example, through verbal reflection, or noticing embodied experiences (as I feel how my body retreats when you come closer to me and how you feel your own shape shrink slightly in return, for example). We can become aware through noticing what is unfolding in our musical experiences (as our musical momentum surges with intensity) or through imagery (as we draw the emotion story playing out between us).

When Co-Storying is Difficult

Authors such as Fuchs (2013) and Ağören (2021) have offered perspectives on mental illness as involving an alteration to a person's being-with-others, lived body, and lived space. As Fuchs explained, depression is not an "inner" psychological disorder but a "detunement" (numbing) of the resonant body that normally mediates attunement and participation in shared social spaces. There is a partial decoupling between the person and their environment. A depressive person has limited expressive capacities and offers fewer clues for others to respond to. Their own mimetic and resonance abilities are lacking. As depression deepens and the alluring qualities of the environment fade, the person can no longer be moved by situations and other people, leading to an inability to feel emotional atmospheres at all, and there is a loss of participation in the shared space of emotion attunement. Where feelings of guilt and shame persist, these serve to sever connection even further. Because we gain our sense of self through our relationships, loss of resonance with others also impacts our sense of self. This depersonalisation tends to be a core feature in severe depressive episodes. Fuchs argued that depressed human beings have lost their ability to engage actively with the world (to co-story).

Ağören (2021) acknowledged Fuch's (2013) perspectives, but then attempted to further push a radical agenda by locating depression with the individual-environmental system. According to Ağören, depression is a particular affective style generated by an impaired social world. Depression isn't produced by an individual brain but is brought forth by the relationships between an individual and their environment. It's the integrated system that feels (Colombetti & Roberts, 2015). Atmospheres of exclusion, for example, can disturb and disorient, closing possibilities for action and connection, and shutting down potentials for co-storying.

As individuals, we are not just passively affected by our environment. We play an active role in shaping emotion stories. When we find ourselves in situations where we feel unable to do so, then depression may arise as an emergent. For Ağören (2021), disconnection is not a state of non-feeling. Disconnectedness could itself be a feeling (which emerges in a world that does not afford possibilities for co-storying experiences of connectedness and belongingness).

Children with early and significant trauma histories may show ongoing attachment and other interpersonal difficulties, including the reduced capacity to reciprocate in social relationships (Hussey et al., 2008). Complex trauma links early, multiple, and prolonged interpersonal childhood victimisation experiences and attachment disturbances. Resultant trauma-related dysregulation can include problems with impulse control, aggression towards self and others, anxiety, hypervigilance, coercive cycles of interaction, and social withdrawal. These deficits in regulatory and relational processes impact a child's ability to engage in genuine give-and-take relationships. Music therapy can be particularly effective at facilitating the development of skills for social reciprocity, and musicking can function as a catalyst for developing healthy relationships and attachments, in other words, rebuilding capacities for co-storying.

References

Ağören, G. C. (2021). Understanding depressive feelings as situated affections. *Emotion Review, 14*, 55–65.

American Psychiatric Association. (2013). *Diagnostic and statistical manual of mental disorders* (5th ed.). American Psychiatric Association.

Barrett, L. F. (2017). *How emotions are made: The secret life of the brain.* Houghton Mifflin Harcourt.

Bericat, E. (2016). The sociology of emotions: Four decades of progress. *Current Sociology, 64*(3), 491–513.

Blandon, A. Y., Calkins, S. D., Keane, S. P., & O'Brien, M. (2008). Individual differences in trajectories of emotion regulation processes: The effects of maternal depressive symptomatology and children's physiological regulation. *Developmental Psychology, 44*, 1110–1123.

Boiger, M., & Mesquita, B. (2012). The construction of emotion in interactions, relationships, and cultures. *Emotion Review, 4*(3), 221–229.

Butler, E. A. (2017). Emotions are temporal interpersonal systems. *Current Opinion in Psychology, 17*, 129–134.

Camras, L. A., & Witherington, D. C. (2005). Dynamical systems approaches to emotional development. *Developmental Review, 25*, 328–350.

Collins, R. (2004). *Interaction ritual chains*. Princeton University Press.
Colombetti, G., & Roberts, T. (2015). Extending the extended mind: The case for extended affectivity. *Philosophical Studies, 172*(5), 1243–1263.
De Leersnyder, J., Boiger, M., & Mesquita, B. (2013). Cultural regulation of emotion: Individual, relational, and structural sources. *Frontiers in Psychology, 4*, 55.
DeNora, T. (2000). *Music in everyday life*. Cambridge University Press.
De Vaus, J., Hornsey, M. J., Kuppens, P., & Bastian, B. (2018). Exploring the East-West divide in prevalence of affective disorder: A case for cultural differences in coping with negative emotion. *Personality and Social Psychology Review, 22*(3), 285–304.
Donati, P. (2011). *Relational sociology: A new paradigm for the social sciences*. Routledge.
Fields, J., Copp, M., & Kleinman, S. (2006). Symbolic interactionism, inequality, and emotions. In J. Stets & J. Turner (Eds.), *Handbook of the sociology of emotions* (pp. 155–178). Springer.
Fogel, A., Nwokah, E., Dedo, J. Y., Messinger, D., Dickson, K. L., Matusov, E., & Holt, S. A. (1992). Social process theory of emotion: A dynamic systems approach. *Social Development, 1*, 122–142.
Fosha, D. (2001). The dyadic regulation of affect. *Journal of Clinical Psychology, 57*(2), 227–242.
Fuchs, T. (2013). Depression, intercorporeality, and interaffectivity. *Journal of Consciousness Studies, 20*(7–8), 219–238.
Gergen, K. (2000). The coming of creative confluence in therapeutic practice. *Psychotherapy: Theory, Research, Practice, Training, 37*(4), 364.
Gergen, K. (2009). *Relational being: Beyond self and community*. Oxford University Press.
Gergen, K. (2018). Human essence: Toward a relational reconstruction. In M. van Zomerpen & J. Dovidio (Eds.), *The Oxford handbook of Human essence* (pp. 247–260). Oxford University Press.
Gureje, O., Simon, G., & Üstün, T. (1997). Somatization in cross-cultural perspective: A world health organization study in primary care. *American Journal of Psychiatry, 154*, 989–995.
Ham, J., & Tronick, E. (2009). Relational psychophysiology: Lessons from mother–infant physiology research on dyadically expanded states of consciousness. *Psychotherapy Research, 19*(6), 619–632.
Haslbeck, F. (2016). Three little wonders: Music therapy with families in neonatal care. In S. Lindahl Jacobsen & G. Thompson (Eds.), *Music therapy*

with families: Therapeutic approaches and theoretical perspectives (pp. 19–44). Jessica Kingsley.

Hollenstein, T., Allen, N. B., & Sheeber, L. (2016). Affective patterns in triadic family interactions: Associations with adolescent depression. *Development and Psychopathology, 28*(1), 85–96.

Hussey, D. L., Reed, A. M., Layman, D. L., & Pasiali, V. (2008). Music therapy and complex trauma: A protocol for developing social reciprocity. *Residential Treatment for Children & Youth, 24*(1–2), 111–129.

Illouz, E., Gilon, D., & Shachak, M. (2014). Emotions and cultural theory. In J. Stets & J. Turner (Eds.), *Handbook of the sociology of emotions* (Vol. II, pp. 221–244). Springer.

Ip, K. I., Miller, A. L., Karasawa, M., Hirabayashi, H., Kazama, M., Wang, L., Olson S. L., Kessler D., & Tardif, T. (2021). Emotion expression and regulation in three cultures: Chinese, Japanese, and American preschoolers' reactions to disappointment. *Journal of Experimental Child Psychology, 201*, 104972.

Kappas, A. (2013). Social regulation of emotion: Messy layers. *Frontiers in Psychology, 4*, 51.

Kitanaka, J. (2012) *Depression in Japan: Psychiatric cures for a society in distress.* Princeton University Press.

Kitayama, S., Mesquita, B., & Karasawa, M. (2006). The emotional basis of independent and interdependent selves: Socially disengaging and engaging emotions in the U.S. and Japan. *Journal of Personality and Social Psychology, 91*(5), 890–903.

Kleres, J. (2011). Emotions and narrative analysis: A methodological approach. *Journal for the Theory of Social Behaviour, 41*(2), 182–202.

Lively, K. J., & Weed, E. A. (2014). Emotions management: Sociological insight into what, how, why, and to what end? *Emotion Review, 6*(3), 202–207.

Levenson, R. W., Ekman, P., Heider, K., & Friesen, W. V. (1992). Emotion and autonomic nervous system activity in the Minangkabau of West Sumatra. *Journal of Personality and Social Psychology, 62*(6), 972–988.

Lougheed, J. P., Koval, P., & Hollenstein, T. (2016). Sharing the burden: The interpersonal regulation of emotional arousal in mother—daughter dyads. *Emotion, 16*(1), 83.

Masuda, T., Ellsworth, P., Mesquita, B., Leu, J., Tanida, S., & Van de Veerdonk, E. (2008). Placing the face in context: Cultural differences in the perception of facial emotion. *Journal of Personality and Social Psychology, 94*(3), 365–381.

Masuda, T., Wang, H., Ishii, K., & Ito, K. (2012). Do surrounding figures' emotions affect judgment of the target figure's emotion? Comparing the eye-movement patterns of European Canadians, Asian Canadians, Asian international students, and Japanese. *Frontiers in Integrative Neuroscience, 6*, 72.
Menges, J. I., & Kilduff, M. (2015). Group emotions: Cutting the Gordian knots concerning terms, levels of analysis, and processes. *Academy of Management Annals, 9*(1), 845–928.
Mesquita, B., de leersnyder, J., & Boiger, M. (2016). The cultural psychology of emotions. In L. Feldman Barrett, M. Lewis, J. Haviland-Jones (Eds.), *Handbook of emotions* (pp. 393–411). The Guilford Press.
Mesquita, B., Karasawa, M., Banjeri, I., Haire, A., & Kashiwagi, K. (2010). *Emotion as relationship acts: A study of cultural differences* (Unpublished manuscript). University of Leuven.
Nussbaum, M. (1988). Narrative emotions: Beckett's genealogy of love. *Ethics, 98*(2), 225–254.
Orange, D. (2002). There is no outside: Empathy and authenticity in psychoanalytic process. *Psychoanalytic Psychology, 19*(4), 686.
Poehlmann, J., Schwichtenberg, A. J., Bolt, D. M., Hane, A., Burnson, C., & Winters, J. (2011). Infant physiological regulation and maternal risks as predictors of dyadic interaction trajectories in families with a preterm infant. *Developmental Psychology, 47*, 91–105.
Robinson, D. (2014). The role of cultural meanings and situated interaction in shaping emotions. *Emotions Review, 6*(3), 189–195.
Rosaldo, M. (1984). Toward an anthropology of self and feeling. In R. Sweder & R. LeVine (Eds.), *Culture theory: Essays on mind, self, and emotion* (pp. 137–157). Cambridge University Press.
Stolorow, R. (2014). Undergoing the situation: Emotional dwelling is more than empathic understanding. *International Journal of Psychoanalytic Self Psychology, 9*(1), 80–83.
Summers-Effler, E. (2006). Ritual theory. In J. Stets & J. Turner (Eds.), *Handbook of the sociology of emotions* (pp. 135–154). Springer.
Thamm, R. (2004). Towards a universal power and status theory of emotion. *Advances in Group Processes, 21*, 189–222.
Turner, J., & Stets, J. (2006). Sociological theories of human emotions. *Annual Review of Sociology, 32*, 25–52.
Uchida, Y., Townsend, S. S., Rose Markus, H., & Bergsieker, H. B. (2009). Emotions as within or between people? Cultural variation in lay theories of

emotion expression and inference. *Personality and Social Psychology Bulletin, 35*(11), 1427–1439.
Westerman, M. A., & Steen, E. M. (2007). Going beyond the internal-external dichotomy in clinical psychology: The theory of interpersonal defense as an example of a participatory model. *Theory & Psychology, 17*(2), 323–351.

18

Emotions and Meanings in Layers of Co-Storying

Our tasks in relational empathy are to explore what emotion stories are taking place and to examine our co-storying processes. I suggest that some steps from narrative analysis could help us do this. We'll draw from these approaches to guide us in investigating emotion stories.

Making and Reflecting on Emotion Stories

We can analyse stories in different ways. Labov (1972) offered ideas for examining a story's structure and how it's told, while Riessman (2008) provided guidance for looking at a story's content and themes and how the interaction between storytellers and listeners collaboratively creates meaning. These models are not mutually exclusive and can be combined in practice (Esin, 2011).

A. dos Santos, *Empathy Pathways*,
https://doi.org/10.1007/978-3-031-08556-7_18

What's This Emotion Story About?

When using Labov's (1972) process of analysing a narrative, we identify a range of story elements. First, we reflect on the abstract (What's this story about? How could we summarise it?). For our situated awareness of emotion co-storying, we would wonder what emotion is present in the interaction we're interested in. As discussed in the previous chapter, we use concepts we've learned through our social experiences to make sense of embodied sensations (Barrett, 2017) and perform our emotions within co-action (Gergen, 2009). To describe an emotion (or multiple emotions) that are playing out between us then requires reflection on what embodied sensations we are experiencing, what concepts we are using to make sense of these as emotions (and whether we share these concepts or not), and how we are performing emotions as actions (in ways that each other may or may not recognise).

Books such as *The dictionary of obscure sorrows* by John Koenig (2021) and *The book of human emotions* by Tiffany Watt Smith (2016) offer a wide range of emotion concepts from cultures worldwide. *Atlas of the heart* by Brene Brown (2021) also provides rich descriptions of emotion concepts within Western cultural frames. Drawing on a broader pallet of emotion concepts enables what Barrett (2017) termed emotional granularity. While her focus was on being able to identify inner experiences with more nuance to facilitate emotion regulation and enhanced interaction with others, when understanding emotions as stories between us, we can also learn to explore and label these stories as they unfold between us with increased richness, subtlety, and transcultural flexibility. In music therapy, we work in both verbal and non-verbal ways. Some clients employ spoken language to express themselves, and others predominantly use other forms of expression, such as non-verbal vocalisations, body movements and gestures, eye movements, and images. We facilitate a creative expressive space where our identification of what emotion is at play can be explored and articulated in multiple ways.

What's the Story Scenario?

Often there are multiple emotion stories at play. Each participant may be engaging in a different emotion story or even multiple emotion stories (we are multi-beings, after all). Emotion stories are also constantly responding to one another and being shaped by each other. When this is the case, it's helpful to reflect not only on each individual emotion story but on the scenario as a whole. How could we describe the scenario in which these different stories impact each other? Is it unfolding as a negotiation? Conflict? Misattunement? An invitation to richer, more multifaceted co-storying? Dialogue that expands participation or debate that seeks a winner? A power-struggle? A process of silencing? Generous openness to multiple stories? What is happening as potentially contradictory stories interact? Friction? Amplification? Is emotional co-regulation occurring as emotion stories pull each other into greater alignment?

What's the Setting of This Emotion Story?

In Labov's (1972) second step, we explore orientation. When and where did this emotion story happen? Who are the participants? In terms of emotion co-storying within music therapy sessions specifically, we ask who the participants are as emergent relational beings who are Selfing through this therapeutic relationship. We reflect on the music therapy confluence. What is this meaning-making space about? What does it enable? How do emotions generated here make sense to the participants involved? How are the emotion stories unfolding across time within this process? How is this therapeutic journey situated within broader socio-political contexts? For example, if a client were to express feeling worthless, we may wonder: within what emotion story is this true? What does the world of this story feel like? What are the rules of this world? (Orange, 2008).

What Are the Events in This Emotion Story?

The third step in Labov's (1972) narrative analysis entails examining the steps in the story and what complicating actions occurred along the way ("and then what happened?"). In exploring emotion stories in music therapy sessions, we would notice the unfolding process of musicking, as well as gestures, eye contact, facial expressions, movements, touch, spoken words (including the tone of the words), images that are created, and the qualities of each of these aspects (momentum, flow, pauses and stuckness, rhythms, textures, vitality affects, and flow). We would be aware of invitations and responses, synchrony, and dyssynchrony. We would reflect on the sequences, turn-taking, patterns, and links between these events and consider them as intersubjective expressions of co-action. As we tune into how interactional coordination is unfolding rhythmically, we are gaining an awareness of the process of participatory sense-making as emotion co-storying.

What Happened at the End?

Labov's (1972) fourth step invites exploration of how the story resolved (or didn't). The arc of an emotion story may reach a natural ending (for example, as the music crescendos to a bursting close and the client and therapist release a deep sigh of catharsis). While we facilitate conclusions of sessions so that a client doesn't leave amid active distress, there may still be a sense that an emotion story is incomplete, especially when a client is facing a complex, ongoing struggle in their life.

How Are We Telling This Emotion Story?

Riessman's (2008) interactional-performative model highlights co-storying processes that take place as the narrator and listener are creating meaning together. We can use a range of questions to guide our exploration here. Through what mediums are we telling this story? How is the story co-constructed? How is the interaction between us creating the story? Why was the story formed in this way? A story calls us as participants into positions, and we remember that positions are always relational (Seu & Cameron, 2013). How, then, are we positioning ourselves in the story and being positioned by the story? Once in these positions (for this moment), how are we seeing the world from this subject position? How does this positioning influence our co-storying? Are we telling this emotion story flexibly or rigidly? How much do we cohere or diverge in our descriptions of this emotion story? Is one of us dominating the storytelling process? While, as therapists, we recognise that we're co-storying, we also critically reflect on this process to make sure we're not taking over the story. We're wondering with the client rather than defining the narrative or forcing alternative stories.

Were We Reflecting on the Emotion Story as We Were Co-Creating It?

For emotion storying to include relational empathy, there needs to be some situated awareness of an emotion story and how it is unfolding between us. Labov's (1972) fifth step explores whether the story folded back to include some awareness and reflection. Our task as relationally empathic music therapists is to notice emotion stories reflexively, how co-storying is occurring, and what trajectories we're moving along as these unfold. When working with clients to enhance their relational empathic capacities, we can intentionally create opportunities for them to grow in awareness of their participation in emotion co-storying.

Layers of Stories

Emotion stories unfold through three interconnected layers. Firstly, when we enter an encounter, we've already been involved in a range of other co-storying processes within relationships we're part of outside this space. We can think of these as story horizons. Secondly, we can tune into the moments of emotion co-storying as micro-stories in the present. Thirdly, story arcs develop as many micro-stories connect.

Story Horizons

Clients and music therapists are already participants in multiple other relationships and the emotion stories generated within them. They are engaged in story horizons that form part of and are informed by their broader family contexts, sociocultural and political contexts, musicking practices, health and illness practices, and religious and spiritual lives, to name only a few. As we examine emotion stories in therapy, we also hold in mind that these are parts of all sorts of other potentially complex stories created between us in the session (musicking stories, health stories, illness stories, and so on). In the words of Kappas (2013), "dealing with nested layers is messy" (p. 7). Notably, the new ways of co-storying that we create in music therapy sessions (perhaps as reparative experiences) can become part of our clients' story horizons within their other relationships as well.

Participating in Horizons of Emotion Co-Storying in Families

Emotion stories within families are one of the most studied facets of emotion in society. There are many emotion stories in every family. Kinship systems, authority relations, perceptions of parenthood, notions of developmental phases, and gender roles can all play a role in constructing the emotional tone in families (Lutz & White, 1986). Close interpersonal relationships are the space where people experience the most frequent and intense emotions, both pleasant and unpleasant. Distress in intimate relationships, characterised by significant difficult emotions, is a common presenting problem for people coming to therapy (Berscheid & Ammazzalorso, 2001). If our client lives with or interacts with their family, we may explore with them what stories seem to be most prominent in their family. While we continuously acknowledge that clients are Selfing as multi-beings (in families with others who are multi-beings), we recognise that they may experience some story trends and patterns of positioning within these. Some stories may be particularly familiar to them ("Oh, I know *this* story well. We get angry with each other all the time.").

Close relationships are defined as those in which participants frequently influence each other's behaviours, they affect many types of each other's behaviours, the magnitude of their influence is substantial, and their interaction patterns have taken place over some time (Berscheid & Ammazzalorso, 2001). When partners are highly interdependent, they get to know each other increasingly well and can quite accurately predict how each other will respond in various situations. These predictions guide behaviours. When expectations are violated, emotions are generated, we may be unsure how to react, and our well-being can feel threatened (although this could also enhance well-being). Some individuals regulate relational closeness so as to regulate emotion stories proactively (Schoebi & Randall, 2015).

We bring experiences from previous relationships into current relationships. Adults with a secure adult attachment style tend to have different expectations within their relationships than adults with an insecure attachment style (Shaver et al., 1996). We may have difficulty articulating this, though, as it often lies outside our conscious awareness. If we can become aware of emotion stories, in terms of story horizons and how these influence emotion micro-stories at the moment, we can gain knowledge into how we are selfing and a more profound understanding of our relationships (Berscheid & Ammazzalorso, 2001).

In family emotion stories, there could also be processes of ambivalence splitting. As humans, we can experience being in multiple, competing emotion stories simultaneously, requiring us to tolerate ambiguity and paradox, which tend to be difficult (Brown, 2021). We may unconsciously polarise these in close relationships, so they belong to the stories of two separate people instead of reconciling them: "You're the angry one, and I'm the resigned one"; "You're the adventurous one, and I just want security and comfort." Once both parties step back into the fullness of the paradox of the whole story, they can experience Selfing as more integrated multi-beings. The pattern of polarising can start to shift even when one party in the relationship chooses to embrace the whole story (Perel, 2018).

Perhaps a client does not live with their family. Rather than considering what emotion stories play out in their home, we may focus on emotion stories within the institution where they reside or within the homeless shelter they are currently staying at, for example. It can also be valuable to gain some understanding of the emotion stories that are prominent in a client's school or work environment. If we are facilitating a music therapy process with a team, we may choose to explore what emotion stories they are part of within their organisation as a whole.

Participating in Horizons of Emotion Co-Storying Within Cultural and Gendered Practices and Contexts

Our sociocultural contexts provide differences in normative and habitual ways of being and interacting. As we discussed in the previous chapter, emotion stories and the elements that comprise them vary widely across cultures, and the stories that shape emotion interpretation also differ significantly. A smile, for example, may signal warmth and friendliness in one context and approval in another. A client may be part of story horizons that value autonomy and view anger as a suitable tool within separation processes. Another client may take part in story horizons that value harmony and relational interdependence, within which anger is discouraged (Boiger & Mesquita, 2012).

Emotion stories are often gender-coded (linked to beliefs about what is and isn't appropriate, typical, and natural for people who identify as a particular gender). Emotion stories are one way that individuals obtain a gendered sense of Self as emotion standards are part of defining the core of "masculinity" and "femininity" (Shields, 2000). This can take place in complex ways within specific contexts. For example, in a study with marginalised men in an informal settlement in Durban, South Africa, Meth (2009) described how interconnections of politics, place, and performances of masculinity shaped emotions. She referred to this as the relational nature of emotional geographies. Lived experiences of housing, gender, ethnicity, lack of education, employment policies, and power relations shaped emotions in the lives of the men in the study. Stories of inferiority played out in relations of otherness, poor conditions, and politics of abandonment. Fear stories were intertwined with experiences of violence and crime, the supernatural, and material vulnerability. Despair, hopelessness, and exasperation emerged as a group emotion related to policy neglect. In relation to marginalised masculinities, binary positions of emotional/rational and female/male become muddied, which impacted subject positioning in the emotion stories.

The term "disability culture" has been used to describe common (although still highly varied) bonds of experience and resilience shared between persons with disabilities. Sandra Brown (2002) wrote that disability culture entails a set of beliefs, expressions, and artefacts "created by disabled people ourselves to describe our own life experiences" (p. 49). Disabled culture is one of the multiple cultures that a person would belong to. Reeve (2002) argued that the social model of disability has not sufficiently articulated how oppression and social barriers operate at the emotional level through structuring the emotion stories of persons with disability (for example, being hurt by others' reactions and frustration with lack of access). Disability activists and artists have critiqued the kinds of representations made in the media of the emotion stories that disabled people take part in, which are often falsely portrayed as narrow and limited, rather than full and rich (Johnston, 2018).

Participating in Horizons of Emotion Co-Storying Within Musicking Practices

As Small (1998) wrote, musicking involves participating in a musical performance in any capacity (listening, rehearsing, practising, composing, dancing, and so on). We've discussed how musicking is always a situated activity (Stige, 2012), entailing dialogical relationships of mutuality within which people create, shape, and explore identities (Westvall, 2021). As we reflect on a client's emotion story horizons, we ask how musicking plays a role in their emotion stories and how emotion stories play a role in their musicking. We know clients may enter music therapy already engaged in emotion story horizons within musicking practices, as shown in literature focused on children (Niland, 2017; Soley, 2019), teenagers (Laiho, 2009; McFerran et al., 2016), adults, and the elderly (Creech et al., 2013; Hays & Minichiello, 2005).

In Ansdell's (2016) book *How music helps*, he explained how he asked a range of participants (colleagues, friends, and clients) in his context in the United Kingdom about their relationships with music. One participant, David, described his relationship with music as central and fundamental. Susanna explained that it was a personal and "very deep relationship"; Adam described music as a friend and companion. Music therapists may encounter clients whose relationship with music has been troubled or damaged somehow, resulting in ambivalence, sadness, loss, resistance, competitiveness, and unhappy memories. Both health and illness can be intertwined within our relationships with music.

Participating in Horizons of Emotion Co-Storying Within Health and Illness Practices

The interpretive paradigms within which we make sense of health and illness shape emotion stories (Blaxter, 2010; Nwoye, 2015). For example, Arteaga Pérez (2020) looked at the emotional texture of colorectal cancer treatments in the United Kingdom and how the emotion story of "not getting upset in front of each other" functioned to preserve relationships between patients, their relatives, and health professionals. In an Australian study on the diagnosis of childhood deafness, Harris et al. (2021) found that professionals viewed emotions as impairing parents, but parents considered emotions as motivational tools. Navne et al. (2018) observed how emotion stories of avoidance, distancing, and detachment emerged for parents in a neonatal intensive care unit in relation to uncertainty regarding whether their infant would survive. In a study on emotion and diabetes in Kenya, Mendenhall et al. (2020) found that distress stories were associated not only with the physical impact of diabetes but with the social and emotional complexities of diabetes and how these related to social factors (such as financial concerns, lack of neighbourhood safety, past or ongoing interpersonal abuse, childcare and care for the elderly).

Participating in Horizons of Emotion Co-Storying Within Religious and Spiritual Practices

Emotions are often central to religious and spiritual experiences. Religious institutions, beliefs, and practices can actively shape how participants co-story emotions (Potvin & Argue, 2014; Vishkin, 2021). Religion (which can also be viewed as a cultural system) offers spiritual narrative identities that value particular emotions and ways of regulating emotions (Schnitker et al., 2017). Beliefs about the controllability of emotions also vary within different religions (for example, emotions can be viewed as controllable through prescribing what individuals should feel and what they should not feel) (Cohen & Rozin, 2001). Sometimes this can take place quite explicitly as specific emotions and expressions of emotions are valued (such as joy), and others are discouraged (such as hate) (Vishkin, 2021).

In their article named after a Christian song titled, *You shall go forth with joy*, Carlson et al. (2021) described how people who identify as being more religious report having greater quality of life and coping better with adversity, but they argued that this may not just relate to what individuals feel, but also to how they interpret events, and what emotions they aspire to experience as members of their faith, which they found to be at least partly the case. In research by Vishkin et al. (2020), people across several religions who reported being more religious desired specific emotions that strengthened their foundational religious beliefs (for example, gratitude and awe) and fewer emotions that centred the self (for example, pride).

Emotion regulation involves meaning-making strategies such as cognitive reappraisal, a nuanced process that can leverage a broad range of meaning frameworks, including religious ones, for example, reinterpreting a tough experience as having deeper meaning. A person may feel as if they are exercising emotion regulation as an "inner process" (although this is still a form of co-action as it's an inextricably social act), but emotion regulation can also be overtly extrinsic. In collectivist cultures, religious and spiritual traditions may promote regulation through sharing emotion stories and communal rituals that shape group emotions (Vishkin, 2021). Due to the central role musicking plays in many religions (Demmrich, 2020), a client's emotion story horizon may involve an intricate intertwining of religious and musical threads.

The Music Therapist's Story Horizons

As therapists, we enter therapeutic relationships with clients already as participants in multiple emotion stories within our own lives, from interpersonal ones playing out in our close relationships to sociocultural stories, in which we may find ourselves complicit in confluences that affirm specific stories. I bring my set of concepts into my therapeutic presence, acts and interpretations, acquired through my family, culture, education, and professional discourse. My task is to expand my repertoire of concepts so that I can practice in more flexible ways, be critically aware of my limitations, and engage with humble situated awareness of how my story horizons are influencing the therapeutic scenario as I co-story with clients. In other words, I need to engage in relational empathy.

Micro-Stories

Emotion stories emerge in moment-to-moment micro-interactions in music therapy sessions. Let's examine a vignette to explore what a process of situated awareness of emotion co-storying at a micro-level could entail. Pavlicevic (1999) began her book *Music therapy: Intimate notes* with a description of a client walking into a room full of musical instruments and being greeted by a music therapist, who invited them to choose an instrument. The client picked up two mallets, approached a glockenspiel, and gave it a tap. Finding this quite enjoyable and intriguing, they tapped a bit more, trying to create a tune, although it didn't seem to work. Noticing a drum, the client walked over to explore the sounds it could make. They then became aware that the music therapist was playing the piano, accompanying them. The client's drumming grew louder and more robust, and before they knew it, they were feeling drawn along inside an expansive sound that seemed to be all over the room. Hearing the fullness and strength of the piano sound, the client hit the drum harder and harder, feeling excited: this is fast, loud, and fun! Suddenly, the client had a feeling of wanting to burst (into laughing? maybe into crying?). They quivered, feeling as if the sound was inside them and around them. It seemed as if there was colour in the room, perhaps dark maroon, and they felt this colour as if it was moving down their back. Their body felt warm around their stomach, and the colour slowed down, spreading and getting broader. They felt a bit heavier and slower after all the speeding, and they wanted to cry quietly. Then there was silence in the room. The client wondered what had happened and who this person at the piano was, who seemed rather good at playing it. After offering this description, Pavlicevic wrote that this client might be wondering what music therapy is all about after this first session, while "knowing there is a feeling about it, a colour, a heat, some 'thing' that seems to take over and make you feel feelings that were not there before" (p. 10).

We explored earlier that situated awareness of emotions as co-storying could be articulated verbally with emotion concept labels. However, this is not the only way to gain and explain our awareness. In this vignette, we see a different kind of situated awareness of an unfolding emotion story, as a feeling, as colours, as heat, and as the expansiveness of sound. During the middle of the vignette, the client became immersed in the story itself. In the end, though, they become aware again of their co-storyteller at the piano. There is a glimmer of relational empathy: "a story unfolded between us, and it had a life of its own."

Let's look at another example and use the questions listed in the section above to see how a process of relational empathy could unfold. I draw this vignette from a music therapy process I facilitated with teenagers. There was a moment in one group where a member, I've used the pseudonym Ammaarah to protect her identity, shared with the group how heartbroken she felt because her relationship with her boyfriend had recently ended. I suggested we could sing a vocal improvisation as a group to surround Ammaarah with care and comfort musically. As we vocalised smooth, flowing sounds, another member, whom I've called Melissa, tried to join in but then stopped singing and turned away.

Turning back again, she smiled and started singing once more, this time in a high-pitched voice, "We love you." Her facial expression communicated some embarrassment. I suggested that we didn't need to find any words if that felt uncomfortable; we could just hum, perhaps. Melissa started tapping softly on her djembe, but just as her rhythm aligned with the others' singing, she pulled back again and stopped playing. My reflection at the time was that Melissa seemed to struggle to step into the painful emotion story (perhaps because this story interacted with others that were part of her emotion story horizon). I attempted gently to invite her to try, by saying, "Stay with it, Melissa. It's okay; let's keep singing." Melissa began to hum. Another member, Natalie, responded with some tears in her eyes, saying, "I hear it, I hear it" (dos Santos, 2019).

We can already see glimpses of the importance of first exploring the horizon of emotion stories that each of the participants in this group was part of. With our current focus on micro-story, though, let's see how relational empathy may have been playing out in this moment.

What are the emotion stories about here? Ammaarah used the concept "heartbroken," and, in my reflection on the emotion story created in the improvisation, I referred to it as a "painful" story. Heartbreak over the ending of a romantic relationship was performed here as co-action. Ammaarah verbally and nonverbally "did" her heartbreak in a way that was recognisable to the other group members and me as the music therapist. I attempted to respond by confirming the validity of her authentic emotion performance. The quality of the vocal improvisation ("smooth" and "flowing") was another form of co-action as we did the emotion together, explicitly enacting it as a story between us, instead of constructing it as a sensation within Ammaarah that she would need to cope with alone. I suggested an expression of the story we could co-create musically between us, simultaneously as an act of affirming Ammaarah's heartbreak and introducing some calm, security, and comfort. As Melissa leaned in and out of the heartbreak story, another emotion story (that I interpreted using the concept of embarrassment) emerged and intertwined. Melissa was selfing as one positioned as uncomfortable in the heartbreak story.

What's the scenario? There were multiple emotion stories at play here (that were each co-storied and impacting one another): a heartbreak story, a comfort story and an embarrassed story (at least). We see how the participants co-regulated at times; their emotion stories bumped up against each other; yet, at other times (primarily through the music), some stories acted as a current pulling others along. This is a relatively flexible scenario as it contained a few different emotion stories pulling and pushing against each other while still being held together in the music.

What's the setting of these emotion stories? The emotion stories within this music therapy process were broadly situated within an under-resourced school in an area where aggression between school learners is commonplace (Louw, 2013). This is a confluence where aggression had frequently come to make sense. The participants in this micro-story included all the group members and the music therapist. It was also about Ammaarah's ex-boyfriend. It took place at a stage in the therapeutic process where the group members were starting to trust one another and were willing to participate in selfing in more vulnerable ways. The music therapy confluence had become one where sharing about heartbreak made sense, and supporting one another within such emotion stories became meaningful.

What are the events in these emotion stories? The description of this moment in a session included several events. For example, Ammaarah shared she felt heartbroken; I suggested that we improvise vocally to care for her; the group members created soft, smooth, and flowing vocalisations; Melissa's shifted between joining and retreating, and so on. We could tease these events in the scenario out and organise them into the separate stories (heartbreak; calm; embarrassment) by examining the words, quality of utterances, musical expressions, body language, invitations and responses, flow and resistance, synchrony and dyssynchrony, the links between them, and the patterns they create as processes of participatory sense-making regarding navigating heartbreak, supporting, and being supported.

What happened at the end of this micro-story? When Melissa aligned her tentative drumming with the group members' singing and then pulled back again just as she started to feel drawn back into the heartbreak story, I attempted to help her remain within it gently, and she started softly humming. Natalie then commented, "I hear it, I hear it," with some tears in her eyes. The group remained in this gentle vocal flow until the improvisation gradually ended naturally. We then sat in silence for a moment.

How are we telling these emotion stories (how are we representing this scenario)? As a group, we co-created the heartbreak story in the context of the session. Ammaarah did the emotion of heartbreak (through her words and body language), and the group responded by co-constructing the story through recognition and validation. The story was then co-constructed further through musicking, as the other storylines also interweaved. One phenomenon we can explore here is positioning. Let's take the example of Melissa. She alternated between positioning herself inside and outside the story. She may have experienced the story calling her in and pushing her out (perhaps the emotion stories she brought with her from her story horizon were repelling this current micro-story to some degree). The group told this story relatively flexibly. I was working towards sustaining it, even as Melissa's actions held the potential to derail it (or to lead it somewhere else).

Were we reflecting on the emotion stories (and the scenario as a whole) as we co-created them? Ammaarah chose to discuss her breakup experience with the group so that it could become part of the co-action of the therapy process. I invited the group into situated awareness of the emotion story. As music therapists, we can create opportunities that afford relational empathy and so too can our clients.

Story Arcs

Many emotion micro-stories link up to form the story arc(s) of a therapeutic process with a client, group, or community. We can "zoom in" and "zoom out." As we plot a therapy process, we are knitting together the significant emotion micro-stories that have emerged along the way.

References

Ansdell, G. (2016). *How music helps in music therapy and everyday life.* Routledge.

Arteaga Pérez, I. (2020). Emotion work during colorectal cancer treatments. *Medical Anthropology*, 1–13.

Barrett, L. F. (2017). *How emotions are made: The secret life of the brain.* Houghton Mifflin Harcourt.

Berscheid, E., & Ammazzalorso, H. (2001). Emotional experience in close relationships. In G. Fletcher & M. Clark (Eds.), *Blackwell handbook of social psychology: Interpersonal processes*, (pp. 308–330). Blackwell.

Blaxter, M. (2010). *Health.* Polity Press.

Boiger, M., & Mesquita, B. (2012). The construction of emotion in interactions, relationships, and cultures. *Emotion Review, 4*(3), 221–229.

Brown, B. (2021). *Atlas of the heart*. Random House.

Brown, J. M. (2002). Towards a culturally centered music therapy practice. *Voices: A world forum for music therapy, 2*(1).

Carlson, S. J., Levine, L. J., Lench, H. C., Flynn, E., Carpenter, Z. K., Perez, K. A., & Bench, S. W. (2021). You shall go forth with joy: Religion and aspirational judgments about emotion. *Psychology of Religion and Spirituality, Advance Online Publication.* https://doi.org/10.1037/rel0000327

Cohen, A. B., & Rozin, P. (2001). Religion and the morality of mentality. *Journal of Personality and Social Psychology, 81*(4), 697–710.

Creech, A., Hallam, S., Varvarigou, M., McQueen, H., & Gaunt, H. (2013). Active music making: A route to enhanced subjective well-being among older people. *Perspectives in Public Health, 133*(1), 36–43.

Demmrich, S. (2020). Music as a trigger of religious experience: What role does culture play? *Psychology of Music, 48*(1), 35–49.

dos Santos, A. (2019). Group music therapy with adolescents referred for aggression. In K. McFerran, P. Derrington, & S. Saarikallio (Eds.), *Handbook of music, adolescents, and wellbeing* (pp. 15–24). Oxford University Press.

Esin, C. (2011). Narrative analysis approaches. In N. Frost (Ed.), *Qualitative research methods in psychology: Combining core approaches* (pp. 92–118). Open University Press.

Gergen, K. (2009). *Relational being: Beyond self and community.* Oxford University Press.

Harris, C., Hemer, S., & Chur-Hansen, A. (2021). Emotion as motivator: Parents, professionals and diagnosing childhood deafness. *Medical Anthropology, 40*(3), 254–266.

Hays, T., & Minichiello, V. (2005). The meaning of music in the lives of older people: A qualitative study. *Psychology of Music, 33*(4), 437–451.

Johnston, K. (2018). Great reckonings in more accessible rooms: The provocative reimaginings of disability theatre. In B. Hadley & D. McDonald (Eds.), *The Routledge handbook of disability arts, culture, and media* (pp. 19–35). Routledge.

Kappas, A. (2013). Social regulation of emotion: Messy layers. *Frontiers in Psychology, 4*, 51.

Koenig, J. (2021). *The dictionary of obscure sorrows*. Simon and Schuster.

Labov, W. (1972). *Language in the inner city*. Blackwell.

Laiho, A. (2009). The psychological functions of music in adolescence. *Nordic Journal of Music Therapy, 13*(1), 47–63.

Louw, W. (2013). *Community-based educational programmes as support structures for adolescents within the context of HIV and AIDS* (Doctoral dissertation). University of Pretoria.

Lutz, C., & White, G. M. (1986). The anthropology of emotions. *Annual Review of Anthropology, 15*, 405–436.

McFerran, K. S., Garrido, S., & Saarikallio, S. (2016). A critical interpretive synthesis of the literature linking music and adolescent mental health. *Youth & Society, 48*, 521–538.

Mendenhall, E., Musau, A., Bosire, E., Mutiso, V., Ndetei, D., & Rock, M. (2020). What drives distress? Rethinking the roles of emotion and diagnosis among people with diabetes in Nairobi, Kenya. *Anthropology & Medicine, 27*(3), 252–267.

Meth, P. (2009). Marginalised men's emotions: Politics and place. *Geoforum, 40*(5), 853–863.

Navne, L., Svendsen, M., & Gammeltoft, T. (2018). The attachment imperative: Parental experiences of relation-making in a Danish neonatal intensive care unit. *Medical Anthropology Quarterly, 32*(1), 120–137.

Niland, A. (2017). Singing and playing together: A community music group in an early intervention setting. *International Journal of Community Music, 10*(3), 273–288.

Nwoye, A. (2015). African psychology and the Africentric paradigm to clinical diagnosis and treatment. *South African Journal of Psychology, 45*(3), 305–317.

Orange, D. M. (2008). Recognition as: Intersubjective vulnerability in the psychoanalytic dialogue. *International Journal of Psychoanalytic Self Psychology, 3*(2), 178–194.
Pavlicevic, M. (1999). *Music therapy: Intimate notes.* Jessica Kingsley.
Perel, E. (2018). In search of erotic intelligence. In A. R. Ben-Shahar, L. Lipkies, & N. Oster (Eds.), *Speaking of bodies* (pp. 49–55). Routledge.
Potvin, N., & Argue, J. (2014). Theoretical considerations of spirit and spirituality in music therapy. *Music Therapy Perspectives, 32*(2), 118–128.
Reeve, D. (2002). Negotiating psycho-emotional dimensions of disability and their influence on identity constructions. *Disability & Society, 17*(5), 493–508.
Riessman, C. (2008). *Narrative methods for the human sciences.* Sage.
Schnitker, S. A., Houltberg, B., Dyrness, W., & Redmond, N. (2017). The virtue of patience, spirituality, and suffering: Integrating lessons from positive psychology, psychology of religion, and Christian theology. *Psychology of Religion and Spirituality, 9*(3), 264.
Schoebi, D., & Randall, A. K. (2015). Emotional dynamics in intimate relationships. *Emotion Review, 7*(4), 342–348.
Seu, I. B., & Cameron, L. (2013). Empathic mutual positioning in conflict transformation and reconciliation. *Peace and Conflict: Journal of Peace Psychology, 19*(3), 266.
Shaver, P. R., Collins, N., & Clark, C. L. (1996). Attachment styles and internal working models of self and relationship partners. In G. Fletcher & J. Fitness (Eds.), *Knowledge structures in close relationships: A social psychological approach* (pp. 25–62). Lawrence Erlbaum.
Shields, S. A. (2000). Thinking about gender, thinking about theory: Gender and emotional experience. In A. Fischer (Ed.), *Gender and emotion: Social psychological perspectives* (pp. 3–22). Cambridge University Press.
Small, C. (1998). *Musicking: The meanings of performing and listening.* Wesleyan University Press.
Soley, G. (2019). The social meaning of shared musical experiences in infancy and early childhood. In S. Young & B. Ilari (Eds.), *Music in early childhood* (pp. 73–85). Springer.
Stige, B. (2012). Health musicking: A perspective on music and health as action and performance. In R. MacDonald, G. Kreutz, & L. Mitchell (Eds.), *Music, health, and wellbeing* (pp. 183–195). Oxford University Press.
Vishkin, A. (2021). Variation and consistency in the links between religion and emotion regulation. *Current Opinion in Psychology, 40*, 6–9.

Vishkin, A., Schwartz, S., Ben-Nun Bloom, P., Solak, N., & Tamir, M. (2020). Religiosity and desired emotions: Belief maintenance or prosocial facilitation? *Personality and Social Psychology Bulletin, 46*(7), 1090–1106.
Watt Smith, T. (2016). *The book of human emotions.* Little Brown Spark.
Westvall, M. (2021). *Participatory music-making in diasporic contexts.* Danish Musicology Online.

19

Responding by Affirming and Changing Emotion Stories

Situated awareness requires acknowledging that I am a participant in an emotion story. I cannot examine the event objectively from the "outside." *Awareness* calls me to notice that I'm in the story, without getting lost in it. I (and preferably we) can explore it from the inside, ask questions, value what is generative about it, and possibly wonder how it could be different if it has degenerative qualities. We can be a team exploring an anger story together, for example, instead of remaining within separate, isolated emotional worlds while one person tries to understand what is happening inside the "inner world" of the other.

Forms of Relational Flow

Gergen (2009) distinguished between generative and degenerative relational flow. Degenerative relational flow is corrosive and shuts down ongoing co-storying ("How could you possibly feel that way?" "I just don't care how you feel anymore"). Generative relational flow brings greater vitality to relations and opens up enriching new potentials. We could draw on Seu and Cameron's (2013) idea of empathic mutual

A. dos Santos, *Empathy Pathways*,
https://doi.org/10.1007/978-3-031-08556-7_19

positioning within reconciliation here, where we position each other as having benevolent intent, even when in conflict. They identify three functions of empathic mutual positioning. First, participants allow for expanded positions for themselves and others. We are allowed to be complex multi-beings, and our stories and co-storying can change, grow, and take new directions. Second, we embrace, accept, recognise, and respect differences because we acknowledge that we can be conflicted and our needs can be asymmetric as multi-beings. Third, we reflexively allow the scenario to hold multi-layered stories of wounding and woundedness that do not negate each other. Seu and Cameron explained that empathic mutual positioning is formative, not merely facilitative, because identity forms in a dialogical process that is always relational. We take on new identities in new kinds of relationships.

Participating in emotion co-storying that fuels generative relational flow can be reparative for a client. Simon Procter (2011) offered an example of a micro-story in a session with a client called John. As John drummed, Simon felt a limited sense of contact with him. Even though Simon played with him on the piano, continuously seeking to catch his attention and elicit connection, John drummed without any responsiveness to Simon. After about 15 minutes, as Simon shaped his piano playing in the rhythm of a waltz, he suddenly sensed a shift in the quality of John's beating as his drumming started to move along in 3/4 time. John lifted his head, and he made eye contact with Simon, grinning. A sense of fun emerged between them. Simon increased the tempo, and John followed as they moved faster and faster to a bursting ending where both of them laughed.

This wasn't a case of no co-action shifting to co-action. In Chapter 16, we discussed how all action is co-action. Limiting contact with another person is still a form of co-action. As participants in the co-action of the session, John and Simon were initially selfing as closed off and seeking, respectively. Perhaps (and here I'm just speculating) John was initially taking part in an emotion story of glumness or alienation and Simon in an emotion story of focus, liveliness, or even anxiety. Gradually, we see these stories blending and moving into cheerful, energised playfulness. Proctor explained how the musical norm of the waltz offered a framework within which they could interact, perceive their interaction, and evoke memories of stories within their horizons that were reciprocal, trusting, and enjoyable, which then prepared them to risk trusting another person another time.

In relation to her music therapy work with families in a children's psychiatric unit, Jones and Oldfield (1999) presented a description of a therapeutic process with a boy (also) called John, as written from the perspective of John's mother Anna who attended music therapy sessions with him. John was diagnosed with ASD and severe learning difficulties. Initially, John screamed whenever the music therapist offered him anything and refused to sit down. After some time, Anna explained, "John's enthusiasm and pleasure in music-making were so intense that it was impossible not to feel happy myself, especially when he started to share his enjoyment with me" (p. 171). Within a constituent ontology, we might frame this as emotional contagion. Along our current pathway of relational empathy, however, as our focus has shifted to how emotions are always between us as stories that emerge through our relationship, we can see this as a moment of generative co-storying that became available to John and Anna. We see Anna showing relational empathy in the situated awareness she highlighted in her description.

Relationships can be a profound source of psychological pain at the end of life. A dying person may experience anticipatory grief in relation to the loss of their spouse, children, friends, and future meaningful life events with those they are close to. They may worry about who will care for those they are responsible for. They may also be concerned about strained relationships and "unfinished business." Clements-Cortés (2010) conducted music therapy processes in palliative care settings to enhance relationship completion. Human suffering can include threats to one's relationships, one's notion of self, and threats to one's relationship with a source of transcendent meaning. Serious illness can challenge and change all these relationships. Clements-Cortes presented the case of Sharon, a 40-year-old woman dying from breast cancer. Sharon articulated overwhelming anxiety about dying and was grieving the loss of her role as a mother to her children. She felt she couldn't share these concerns with her family and felt isolated. When she'd previously gone into remission, her family attempted to construct everything as having returned to "normal" and performed a story of happiness, never speaking of Sharon's illness or prognosis. Within music therapy, Sharon could engage in selfing as one who was emotionally struggling. Through the supportive and trusting relationship that she developed with the music therapist, she expressed her desire to engage in and discuss her death with her husband and children. She wished to take part in grief co-storying as a generative process with her family. Through music therapy, she gained the courage to initiate this and could bring about the closure that she sought.

We create emotional stories between us, changing or abandoning ones that don't serve us. Gergen (2009) suggested ways of shifting how we are co-storying. In the light of co-action, something becomes real only through collaborative confirmation. Think of a situation where one partner shouts at their spouse, who then yells back, "Why are you so angry?"; and then compare that to a situation where the spouse might reply, "I think we're both very tense at the moment. Otherwise, we wouldn't be treating each other like this." Here the anger was redefined as tension (through drawing on a different emotion concept), and a different relational trajectory could be invited. Some meta-moves could be employed, as the couple steps back to reflect on the conversation itself: "What are we doing? Do we really want to treat each other like this?" or "How important is what we're fighting about, really?" "Is there a way we could look at this problem differently?" The pair could also shift their emotional performance. The second spouse might reply, "I feel so sad that we are hurting each other like this." A deviation from their conventional pattern of exchange could invite new forms of co-storying. Gergen also referred to the "theatrical move": "We're really making a mess of this conversation. How about we start over?".

Music Therapy as Relational Recovery

Relational therapists refuse to see their clients as self-contained individuals who need "fixing" (DeYoung, 2014). The problem doesn't lie "inside" the client. We're exploring together how emotional difficulties are actions in relationships and how co-storying could happen differently. Change happens in complex, non-linear, systemic ways. In DeYoung's explanation of relational therapy, she referred to a hypothetical client called Jane and encouraged the question, "What's wrong for Jane" rather than "What's wrong with Jane." The "wrong" that Jane brings to therapy is not inside her, to be rooted out or cured. What's wrong for Jane "is entangled with everything Jane knows and feels about being in the world—especially about being with others in the world" (p. 18). DeYoung also gave an example of Ben, who thought there was something wrong with him. In response to his description of feeling shame, DeYoung suggested that she could have replied by saying: I don't think the problem you have with feeling stupid is just inside you. I think it comes up when things happen…between you and other people. Things keep happening; they lead to the same old bad feelings; and then you think there's something really wrong with you.

As psychological problems are self-in-relation problems, therapy is a process of self-with-other performative change. A relational therapist would work with clients to explore how their relational world works. Change occurs within the relationships with another person (which may result in new ways of selfing). New life meanings can be actively created that expand the capacity for engagement and connection. DeYoung (2014) described empathy as "a system of mutual cues and responses that regulate each participant's experience of self and the other in the system" (p. 50).

Music therapists frequently work with clients experiencing some form of isolation, and for whom opportunities for co-storying with others in the here-and-now may be few and far between. Individuals may be isolated from their social context (perhaps in care facilities). They may be isolated through conditions that separate them physically and mentally from the social world, through stigma, or because social contact is highly anxiety-provoking. Isolation can come about through social circumstances such as migration, family separation, economic struggles, and family challenges such as childcare (Davidson, 2004).

Pavlicevic (1999) offered a case study of work between Oksana, a music therapist, and Daniel, a 6-year-old with cerebral palsy, epilepsy, learning difficulties, and no speech. In the first three sessions, nothing happened. Daniel was detached; he didn't make any eye contact and didn't appear to respond to any musical sounds. When he offered some short vocalisations in session four, Oksana listened carefully to the quality, pitch, and rhythm. She vocalised with him, adapting, matching, and extending Daniel's sounds. Daniel's vocalisations intensified, growing in range and flexibility, and the two improvised together for ten minutes. Oksana was influenced by Daniel's sounds and how he made them (even when they were tiny and haphazard at first). Daniel was influenced by Oksana as he responded to the extensions that she offered, increasing the diversity of his sounds to retain musical and interpersonal connection. Pavlicevic reflected on how "Daniel and Oksana together enter[ed] into emotional meaning that [was] exclusive to them: to who each of them [was], and to each of them in relation to one another" (p. 21). This may have been an uncommon experience for Daniel.

When I'm in a relational encounter, I can step back to reflect on the emotion co-storying between us as a way of declaring that this relationship matters. A relationship is the core of most therapeutic change, and, therefore, relational empathy can be understood as indispensable. Also, helping clients enhance their relational empathy is a way of affording them an opportunity to become more empowered to construct increasingly generative relationships. Gergen (2009) characterised therapy as a process of restoring viable relations for clients.

We see a focus on relationships in approaches like restorative justice, where the emphasis is placed on repairing harm and resolving conflict. Those who have suffered harm are offered opportunities to have this acknowledged and receive amends. Those who have caused harm are given opportunities to recognise the impact of their actions and to make reparations. Besides restoring the well-being of individuals, restorative justice seeks to restore the well-being of communities and prevent further harm from being done (Liebmann, 2007).

Perhaps you are engaging in relational empathy, but the person you are co-storying with refuses to do the same (which is not the act of a discrete individual who is showing resistance; this is a relational story too). In some ways, even when one person shifts their engagement with the story, it can significantly change the co-storying as a whole.

Relational Responses to Burnout

The emotional world is between us, so we're already always in the emotion story. Relational empathy may offer a way to notice when this is generative, identify when it isn't, and make some choices when we need to change it. In terms of the music therapist's experience, one may wonder whether framing emotions as co-storying means that one's chances of developing burnout are, therefore, high. However, relational reflexivity is a way of not getting lost in the co-storying. Relational empathy may, in fact, serve as protective against getting drawn into burnout.

Crucially, however, we are not necessarily thinking about minimising burnout as an individual task within this pathway. As Reynolds (2011, 2012) argued, resisting burnout is a process that is situated in collective sustainability. The problem of burnout is not inside individual heads or hearts, but in the real world, where there is injustice. Reynolds (2011) explained that the people she worked alongside didn't burn her out (instead, they challenged, inspired, and transformed her). What did harm her was the injustice and indignity suffered by her clients, as well as frustration at her inability to change the unjust structures within which they lived and struggled.

As a healthcare practitioner, no amount of "self-care" will address the poverty, violence, marginalisation, and threats to basic dignity that our clients struggle with. As suffering is rooted in society, our response needs to be collective and relational, embracing solidarity, co-creation, and interconnectedness. Reynolds championed collective ethics and connective practices of resisting oppression. As healthcare workers, community workers, clients, families, communities, we are meant to do this work together. To sustain our practice with aliveness, spirited presence, and connectedness, and not to become burned out, requires us to remain fully and relationally engaged. It often demands creativity to stay connected across difference and conflict and amidst limited resources. We do more than survive our work; we are changed through doing it alongside our clients.

References

Clements-Cortés, A. (2010). The role of music therapy in facilitating relationship completion in end-of-life care. *Canadian Journal of Music Therapy, 16*(1).

Davidson, J. (2004). What can the social psychology of music offer community music therapy? In M. Pavlicevic & G. Ansdell (Eds.), *Community music therapy* (pp. 114–128). Jessica Kingsley.

DeYoung, P. A. (2014). *Relational psychotherapy: A primer.* Routledge.

Gergen, K. (2009). *Relational being: Beyond self and community*. Oxford University Press.

Jones, A., & Oldfield, A. (1999). Sharing sessions with John. In J. Hibben (Ed.), *Inside music therapy: Client experiences* (pp. 168–173). Barcelona Publishers.

Liebmann, M. (2007). *Restorative justice: How it works.* Jessica Kingsley.

Pavlicevic, M. (1999). *Music therapy: Intimate notes.* Jessica Kingsley.

Procter, S. (2011). Reparative musicing: Thinking on the usefulness of social capital theory within music therapy. *Nordic Journal of Music Therapy, 20*(3), 242–262.

Reynolds, V. (2011). Resisting burnout with justice-doing. *International Journal of Narrative Therapy & Community Work, 4*, 27–45.

Reynolds, V. (2012). An ethical stance for justice-doing in community work and therapy. *Journal of Systemic Therapies, 31*(4), 18–33.

Seu, I. B., & Cameron, L. (2013). Empathic mutual positioning in conflict transformation and reconciliation. *Peace and Conflict: Journal of Peace Psychology, 19*(3), 266.

20
Conclusion

Thinking deeply about our work as music therapists is an act of compassion. It's a way of taking our clients seriously and affording their processes due respect. We immerse ourselves in contemplation as a profound act of care. This is why theory matters in music therapy. As bell hooks (1991) wrote, the theory is a location for healing.

How Might We Use These Four Pathways?

As we journeyed along the four empathy pathways in this book, we encountered the main stepping stones summarised in Table 20.1.

Perhaps one pathway felt the most sense-filled to you, the most relatable, or the most helpful as a thinking tool within your practice, community, classroom, or research study. Alternatively, at this point, you may be thinking to yourself that these four pathways of empathy merely offer different terminology for essentially the same process. Maybe we are just inconsequentially playing around with different metaphors?

A. dos Santos, *Empathy Pathways*,
https://doi.org/10.1007/978-3-031-08556-7_20

Table 20.1 Four pathways of empathy

	Insightful empathy	Translational empathy	Empathising assemblages	Relational empathy
STANCE	Receptivity	Honouring opacity	Always already entangled	Recognising that we are selfing through relationships
AWARENESS	Multifaceted perceiving	Translational processes	Mapping relationships	Emotions as co-storying
FEELING	Sharing emotions	Responsive withness	Affective attunement	Layers of co-storying
MEANING	Understanding emotions	Negotiating meanings		
RESPONSE	Responding with action	Accompaniment	Enhancing response-ability	Affirming and changing emotion stories

Metaphors do matter, however. Metaphors operate linguistically and at a conceptual level through a process called metaphoric transfer (Landau et al., 2011).

Participants in a study by Williams and Bargh (2008), for example, explained the personality of someone they encountered as "warmer" after they'd held a warm beverage. People who took part in a study by Meier et al. (2007) were better able to locate God-related terms near the top of a computer screen as they conceptually associated God as being "up" in heaven, and this enabled them to find words quicker when they were higher up on the screen. After watching a series of expanding squares, people who took part in a study by Landau et al. (2011) reported greater feelings of self-actualisation than those who had watched a series of contracting squares. Schlegel et al. (2012) also studied metaphoric transfer in relation to the metaphors people used to understand themselves. Holding a self-discovery metaphor had a more positive influence on people's search for meaning compared to a self-creation metaphor. The metaphors we use alter what we notice, how we behave, and what we create (Hausman, 1991). These four pathways of empathy may elicit imaginative metaphoric journeys, but, crucially, they also invite different action. The four perspectives on empathy presented in this book offer thinking tools to open up multiple ways of encountering the world, engaging in relationships, and understanding ourselves and others. As Reddekop (2014) wrote, when we change our ontology, we change our thinking, and our mode of being and doing in the world.

Perhaps you have travelled carefully down all four pathways and are now feeling a bit confused: each one holds some value, and yet they seem incompatible in their underpinnings, so what does that mean for practice? Can we blend these empathic strategies? Can we walk along different pathways in different contexts?

Perhaps we could take some inspiration from researchers. When faced with the challenge of incompatible ontological foundations, mixed methods researchers respond with pragmatism. They ask, "What will work best here and now to address the question that seems most pressing?" If we took this kind of approach to select which empathy pathway to walk along, it would justify travelling down different pathways in response to particular clients' worldviews, needs, and strengths in specific places and spaces.

When seeking to examine a phenomenon from multiple perspectives (positivist-oriented), qualitative researchers like Lincoln and Guba (1985) developed the approach of triangulation to look upon a singular phenomenon from various angles. This is like the story of many people standing around an elephant with blindfolds on, where one feels the tail and thinks it's a rope, the other feels the side and thinks it's a wall, and so on. Only when they pool their insights, do they realise that a (hopefully calm) elephant stands before them. Triangulation would invite us to see the four empathy pathways as facets of a similar phenomenon that we could explore from different angles (although one might argue that the vastly different ontological foundations of the different types of empathy would make this strategy tricky to justify).

Richardson (2000) argued that postmodernist qualitative researchers don't triangulate; they crystallise. The image of a crystal reminds us that we're not examining a fixed, rigid two-dimensional object; we're exploring something—empathy in this case—that has a potentially endless variety of shapes, dimensions, and angles of approach. Paradoxically, we know that there's always more to know. Ellingson (2011) described crystallisation as combining multiple genres of representation and forms of analysis into a coherent text or set of related texts. For example, one may combine rigorous conventional forms of data analysis and creative (re)presentation within the same study to build a rich, open (yet still partial) account of a phenomenon. The process problematises its own construction, highlights researchers' positionality and vulnerabilities, acknowledges the social and relational construction of meaning, and reveals how knowledge claims are always indeterminate. When we resist dichotomies, we open ourselves up to a nuanced range of possibilities.

In his book *How our lives become stories,* Eakin (2019) argued for a comprehensive model of the self that includes both the physical body and the body of relational encounters. He drew on the world of Neisser (1988), who attempted to distinguish between five types of self-knowledge (which are not usually experienced as distinct, however). Neisser described the ecological self (the person who is here in this place, participating in this activity), the interpersonal self (who is engaged in social interaction with another person), the extended self (of memory and anticipation), the private self (whose conscious experiences are not available to others), and the conceptual self (described through notions of social roles, personality types, theories of body and mind, and perceptions of the self as a category). Eakin reflected on the strengths of this model as avoiding the mind/body split, resisting the concept of a unified self and not privileging any of the five types of self-knowledge. He argued that it's time to discard restrictive notions of the self and the subject, instead opening the way to broader, inclusive, and flexible approaches. If we elected to see the self in such an inclusive way, for example, then the four pathways of empathy explored in this book allow us to move fluidly within such a multifaceted approach.

Hope in Action

Hope, just like empathy, has always been defined in different and sometimes contradictory ways within various contexts (Fagiano, 2019). While this can generate confusion, it also creates opportunities. Multifaceted experiences of empathy offer different practices of hope in action. We are always hoping for a better understanding of one another, better connection, a more response-able world, and healthier relationships. As music therapists, we continue to critically (re)consider the values, beliefs, and practices that sustain our work and our participation in the world at large as we seek to create a more inclusive and kinder one.

This book has brought me hope as I've written it. While writing, I found myself in certain encounters both in music therapy sessions and in life as a whole where insightful empathy—the one I'd been taught was the only empathy pathway—felt challenging to walk (because I struggled to understand the other person, because I felt hurt, because I felt limited in my narrow life experience and understanding, because working within an individualistic perspective seemed insufficient for serving a social justice agenda, or because I had a nagging feeling that musicking was accomplishing so much more). I decided to lean into those questions with honesty and vulnerability instead of trying to push myself into a box of what I *should* be feeling or what I *should* be able to understand. I also decided not to judge myself when empathy felt difficult, but to approach the situation I was in with curiosity.

As humans, we tend to seek empathic connection, even when it's hard or confusing and even when it costs us greatly. As Shaddock (2000) asserted, "the underlying motivation for all people is the desire for a reliable emotional connection" (p. 2). We need each other. Through exploring all four empathy pathways, I discovered that there are always other ways "in" to empathy because the concept is more expansive than we've given it credit for. There is always hope for connection. As music therapists especially, our daily lived experience of connection through musical relationships offers us a deep well of insight to further this conversation.

References

Eakin, P. J. (2019). *How our lives become stories*. Cornell University Press.

Ellingson, L. (2011). *Engaging crystallisation in qualitative research*. Sage.

Fagiano, M. (2019). Relational empathy as an instrument of democratic hope in action. *The Journal of Speculative Philosophy, 33*(2), 200–219.

Hausman, C. R. (1991). Language and metaphysics: The ontology of metaphor. *Philosophy & Rhetoric, 24*(1), 25–42.

hooks, b. (1991). *Theory as liberatory practice*. Yale JL & Feminism, 4, 1.

Landau, M. J., Vess, M., Arndt, J., Rothschild, Z. K., Sullivan, D., & Atchley, R. A. (2011). Embodied metaphor and the "true" self: Priming entity expansion and protection influences intrinsic self-expressions in self-perceptions and interpersonal behavior. *Journal of Experimental Social Psychology, 47*(1), 79–87.

Lincoln, Y. S., & Guba, E. G. (1985). Establishing trustworthiness. *Naturalistic Inquiry, 289*(331), 289–327.

Meier, B. P., Hauser, D. J., Robinson, M. D., Friesen, C. K., & Schjeldahl, K. (2007). What's "up" with God? Vertical space as a representation of the divine. *Journal of Personality and Social Psychology, 93*, 699–710.

Neisser, U. (1988). Five kinds of self-knowledge. *Philosophical Psychology, 1*(1), 35–59.

Reddekop, J. (2014). *Thinking across worlds: Indigenous thought, relational ontology, and the politics of nature; Or, if only Nietzsche could meet a Yachaj* [Doctoral dissertation, University of Western Ontario]. Electronic Thesis and Dissertation Repository. 2082.

Richardson, L. (2000). Writing: A method of inquiry. In N. K. Denzin & Y. S. Lincoln (Eds.), *Handbook of qualitative research* (2nd ed., pp. 923–943). Sage.

Schlegel, R. J., Vess, M., & Arndt, J. (2012). To discover or to create: Metaphors and the true self. *Journal of Personality, 80*(4), 969–993.

Shaddock, D. (2000). *Contexts and connections: An intersubjective systems approach to couples therapy.* Basic Books.

Williams, L., & Bargh, J. (2008). Experiencing physical warmth promotes interpersonal warmth. *Science, 322*(5901), 606–607.

Appendix: Empathy Pathways

Reflection Sheets for Music Therapists

A. dos Santos, *Empathy Pathways*,
https://doi.org/10.1007/978-3-031-08556-7

1. Insightful Empathy

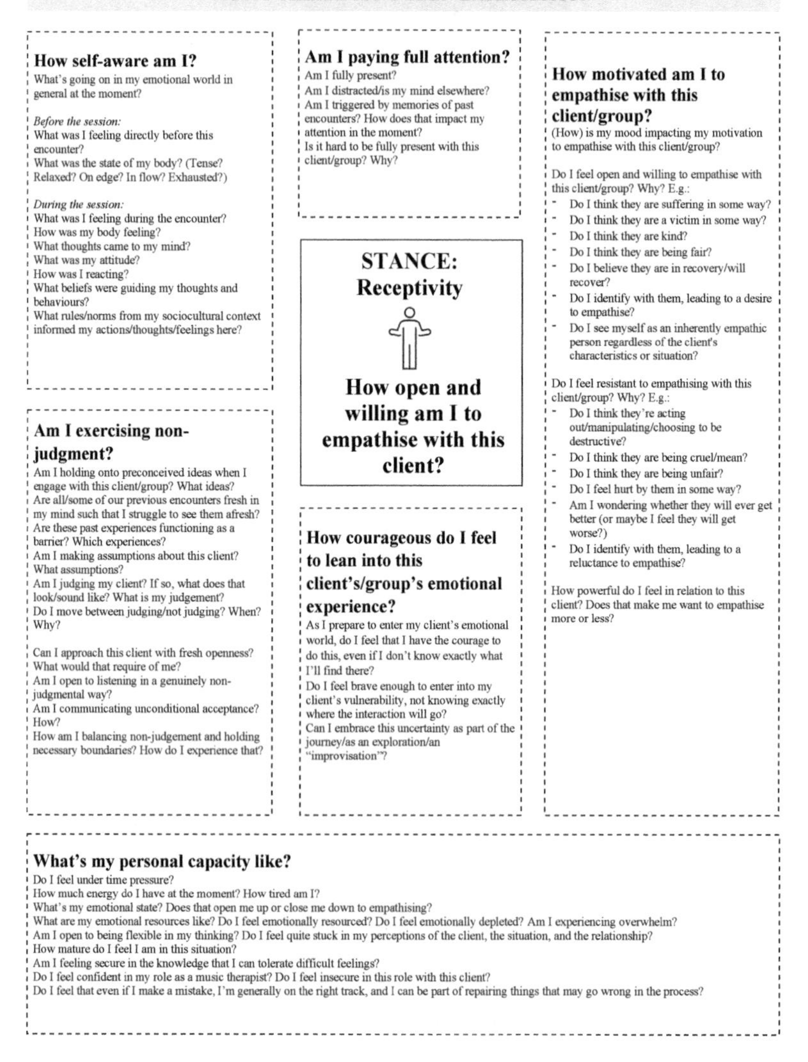

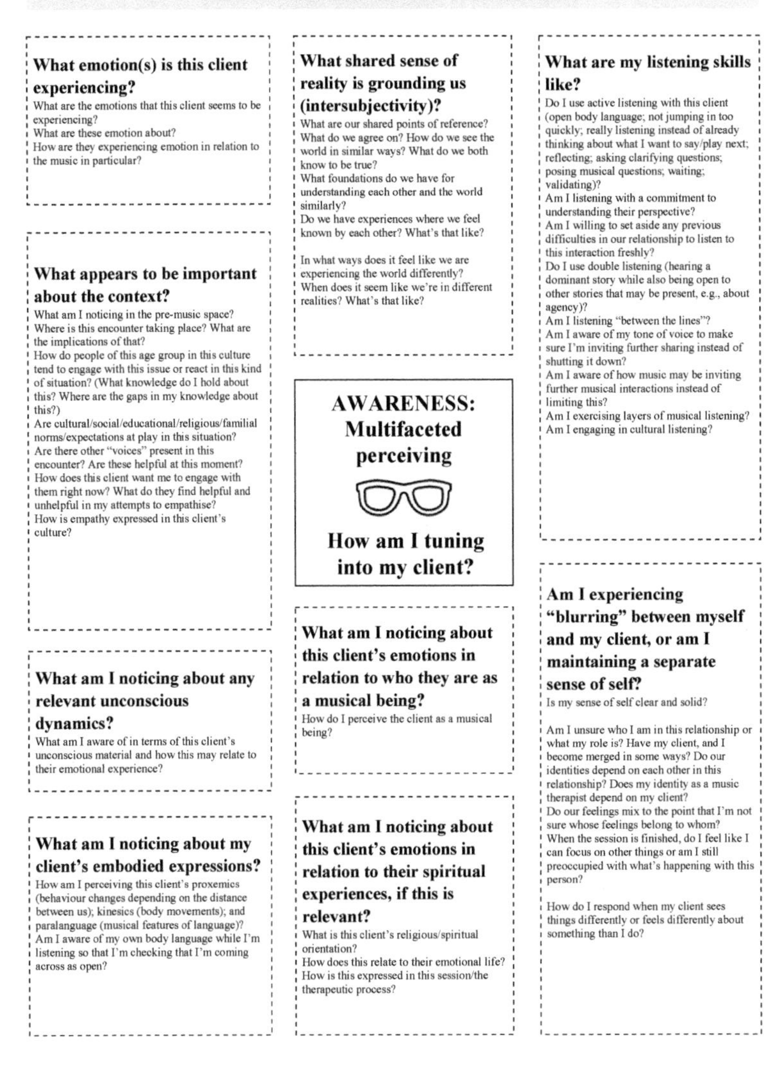
INSIGHTFUL EMPATHY
Purposefully sharing and understanding another's emotions
MUSIC THERAPIST'S SELF-EXPLORATION
What emotion(s) is this client experiencing?
What are the emotions that this client seems to be experiencing? What are these emotion about? How are they experiencing emotion in relation to the music in particular?
What appears to be important about the context?
What am I noticing in the pre-music space? Where is this encounter taking place? What are the implications of that? How do people of this age group in this culture tend to engage with this issue or react in this kind of situation? (What knowledge do I hold about this? Where are the gaps in my knowledge about this?) Are cultural/social/educational/religious/familial norms/expectations at play in this situation? Are there other "voices" present in this encounter? Are these helpful at this moment? How does this client want me to engage with them right now? What do they find helpful and unhelpful in my attempts to empathise? How is empathy expressed in this client's culture?
What am I noticing about any relevant unconscious dynamics?
What am I aware of in terms of this client's unconscious material and how this may relate to their emotional experience?
What am I noticing about my client's embodied expressions?
How am I perceiving this client's proxemics (behaviour changes depending on the distance between us); kinesics (body movements); and paralanguage (musical features of language)? Am I aware of my own body language while I'm listening so that I'm checking that I'm coming across as open?
What shared sense of reality is grounding us (intersubjectivity)?
What are our shared points of reference? What do we agree on? How do we see the world in similar ways? What do we both know to be true? What foundations do we have for understanding each other and the world similarly? Do we have experiences where we feel known by each other? What's that like?
In what ways does it feel like we are experiencing the world differently? When does it seem like we're in different realities? What's that like?
AWARENESS: Multifaceted perceiving
How am I tuning into my client?
What am I noticing about this client's emotions in relation to who they are as a musical being?
How do I perceive the client as a musical being?
What am I noticing about this client's emotions in relation to their spiritual experiences, if this is relevant?
What is this client's religious/spiritual orientation? How does this relate to their emotional life? How is this expressed in this session/the therapeutic process?
What are my listening skills like?
Do I use active listening with this client (open body language; not jumping in too quickly; really listening instead of already thinking about what I want to say/play next; reflecting; asking clarifying questions; posing musical questions; waiting; validating)? Am I listening with a commitment to understanding their perspective? Am I willing to set aside any previous difficulties in our relationship to listen to this interaction freshly? Do I use double listening (hearing a dominant story while also being open to other stories that may be present, e.g., about agency)? Am I listening "between the lines"? Am I aware of my tone of voice to make sure I'm inviting further sharing instead of shutting it down? Am I aware of how music may be inviting further musical interactions instead of limiting this? Am I exercising layers of musical listening? Am I engaging in cultural listening?
Am I experiencing "blurring" between myself and my client, or am I maintaining a separate sense of self?
Is my sense of self clear and solid?
Am I unsure who I am in this relationship or what my role is? Have my client, and I become merged in some ways? Do our identities depend on each other in this relationship? Does my identity as a music therapist depend on my client? Do our feelings mix to the point that I'm not sure whose feelings belong to whom? When the session is finished, do I feel like I can focus on other things or am I still preoccupied with what's happening with this person?
How do I respond when my client sees things differently or feels differently about something than I do?

INSIGHTFUL EMPATHY

Purposefully sharing and understanding another's emotions

MUSIC THERAPIST'S SELF-EXPLORATION

EMOTIONS: Sharing

How am I sharing in this client's emotions?

Do my feelings stem from my own context?

As I reflect on my own emotions:
where do these emotions stem from? (My interpretations of this situation? My life experiences/history/memories/fears? My expectations?)
How are these emotions related to what's happening here with this client?
How may these emotions be more about me than what my client is experiencing?
Where these emotions are not assisting me in my attempts to empathise, how am I managing them?
Once I have noticed them, what do I do?

Am I experiencing emotional contagion?

What feelings have I "caught" from this client?
What emotions am I mirroring?
Is my embodied experience mirroring my client's (e.g., in terms of tension, posture etc.)?
Am I aware of whose feelings belong to whom?
As these feelings arise, am I actively working on regulating them? How?

What phenomenological stage(s) of empathy am I experiencing?

Am I:
(i) Encountering this client's emotions
(ii) Being drawn into their emotions
(iii) Returning to a clear awareness that these are the client's emotions and not my own?

How am I experiencing embodied resonance?

In what ways am I resonating in an embodied way with this client's emotions?
How am I letting go, moving toward or into, resonating, discovering, or discerning, and grasping or taking hol emotions?
How am I sharing emotions through mirroring, synchrony, and entrainment within our embodied interaction?

How accurate is my empathy?

How accurately do I think I'm sharing in their emotion (in terms of what the feeling is, how intensely they are feeling it, and how pleasant/unpleasant the feeling is for them)?
Could I check my accuracy with them?
What may I need to do to "tune in" even more accurately?

How am I experiencing affective empathy?

As I have chosen to step into my client's emotional world, how am I sharing in their feelings with them?
What's my experience of what that actually entails?
What do I feel as I do this?
What am I sensing? What am I noticing?
How am I checking that my feelings remain more appropriate to their situation than mine?

How am I sharing in this client's emotions musically?

Am I sharing emotions through matching, imitating, reflecting, exaggerating, enhancing, synchronising, incorporating, and/or mirroring?
Am I identifying emotions in the musicking and/or experiencing emotions through the musicking?

INSIGHTFUL EMPATHY

Purposefully sharing and understanding another's emotions

MUSIC THERAPIST'S SELF-EXPLORATION

What do I know about this client?

What am I drawing on to understand this client's emotions? For example:
Assessment information?
Their story?
Knowledge from previous sessions?
My experience of them in this moment?
My knowledge of their socio-cultural world?

Which of my own experiences may be helpful to draw on to understand this client's emotions?

Am I drawing on my own experiences to understand what this client is feeling?
Have I been in similar situations to the one they are in now? How did I feel when I was in that situation?
In what ways is it helpful to draw on my own experiences here?

In what ways is this not helpful?
Am I making problematic assumptions as I draw on my own experiences?
Am I over-identifying?

MEANING: Understanding emotions

How am I going about understanding this client's emotions?

What generalised theories am I using?

What theories do I tend to draw on when trying to understand this client's emotions?
Where did I obtain these theories?
Why am I using them here?
Which ones are helpful?
Which ones are unhelpful?
Do I critically reflect on the theories I'm using?

What other information could I draw on to help me understand this client's emotions?

Where are the gaps in my understanding of this client and their experience?
What might help me better understand my client's experience? How can I seek these out?
How can I be proactive in continuously learning more?
Do I want to learn more? Why/why not?

How am I processing complex emotional expressions?

Are my client's emotions particularly complex/confusing/unpredictable?
Am I in a situation where I need advanced knowledge to understand their experiences?
How am I going about attempting to gain that knowledge? How do I feel about this?
Do I need to exercise advanced emotion perspective-taking skills?
How am I doing that?

INSIGHTFUL EMPATHY

Purposefully sharing and understanding another's emotions

MUSIC THERAPIST'S SELF-EXPLORATION

How am I communicating empathy?

How am I communicating empathy to this client in this situation?
Am I thinking about what kinds of communication make sense/feel appropriate to them, or only how I would want someone to communicate empathy with me?
If they communicate not feeling empathised with, what do I do?
If they do feel empathised with, how do I respond? How do I feel?

When I share in this client's emotions, do I sometimes want to withdraw?

When I step into this client's emotional world, do I sometimes struggle with difficult feelings I become aware of (in them and/or in me)?
When this happens, do I want to withdraw? Do I withdraw? How?
How do I feel as I reflect on these questions?

When I share in this client's emotions, do I sometimes respond in ways that could cause some harm?

Is my empathy with this client an act of pathological altruism?
Is showing empathy here predominantly stroking my ego?
Am I fuelling a triangle (taking the victim's side and becoming less empathic towards the perpetrator)?
Is my empathy an act of exploitative advocacy?
Am I predominantly enjoying being the benevolent helper?
Does my empathy promote my client's dependence on me?

Is an expression of empathy sufficient in this moment?

As I communicate empathy to this client, do they seem to find it beneficial (e.g., for connection; dyadic regulation; validation; self-acceptance; belonging; feeling less alone)?
Does empathy alone seem insufficient?

RESPONSE: Action

How am I communicating empathy and responding with action?

Do I experience opposition from this client when I attempt to empathise with them?

As I communicate empathy to this client, do they seem to show resistance, suspicion, or ambivalence?
How do I respond to this?

If more is needed, what compassionate action is required?

Is this empathic presence only the beginning of a necessary/helpful process of action?
If I have realised that communicating empathy is a starting point here and further action is required, how can the information I gained through the process of empathising guide my clinical decision-making going forward?
What compassionate action is necessary?
On what grounds do I think that this is the best response?
How do I feel about proceeding with this action?
How does this client respond as I engage in this action?

Do I show compassion towards myself?

As I have been working through this process of reflecting on my empathy towards this client, have I been showing empathy toward myself?
Am I responding to myself with compassion?
Am I using narrative distancing to take a step back and look at my own process with kindness?
Am I not only giving myself grace, but also kindly engaging in realistic self-critique so that I can grow?
How can I show myself more self-compassion?
What might happen if I did?

How do I step out of the empathic encounter afterwards?

How do I feel after an empathic encounter with this client?
How do I step out of this client's emotions?
How do I intentionally step out of the empathic interaction with this client to nurture my healthy boundaries?
Can I use strategies to regain my footing in my own emotional world as I exit my client's emotional world?
How do I feel about this idea of "exiting" their emotional world after an empathic encounter?
What does that mean for me?
What resources do I need to be able to step away from their emotional experiences well?

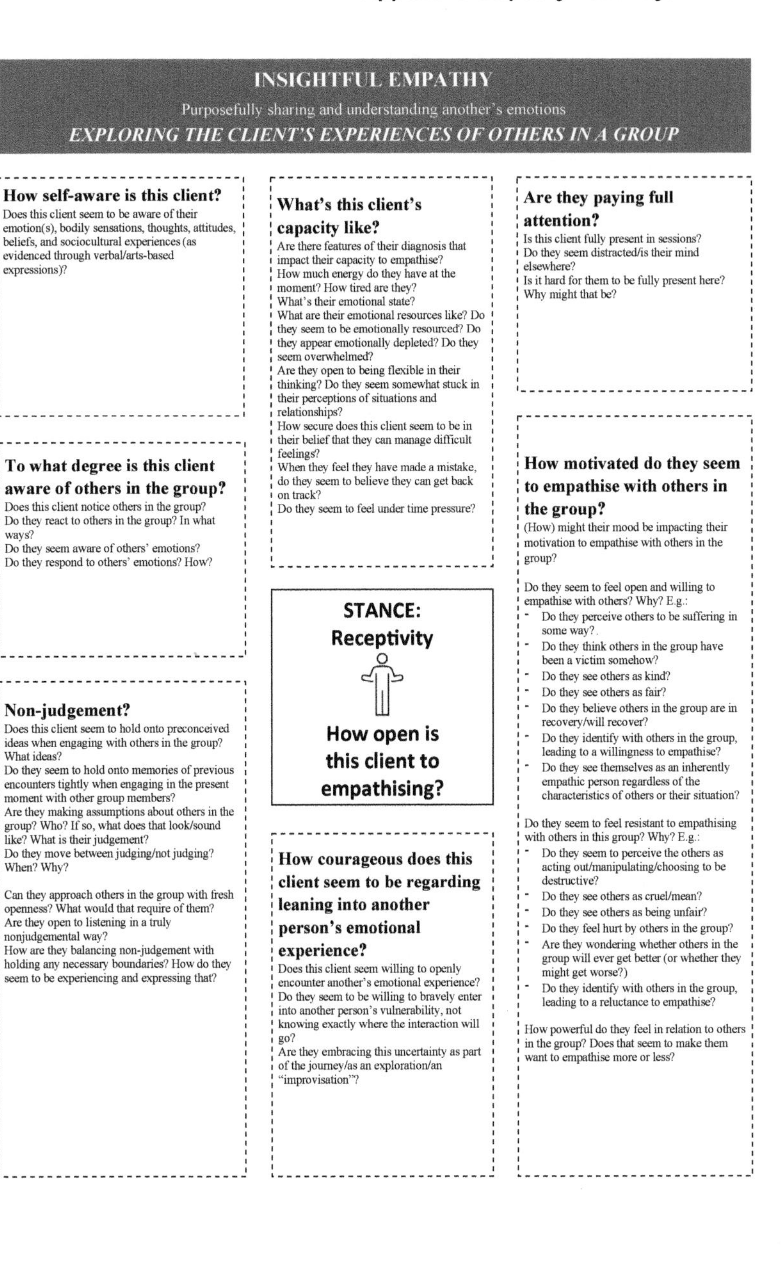

INSIGHTFUL EMPATHY

Purposefully sharing and understanding another's emotions

EXPLORING THE CLIENT'S EXPERIENCES OF OTHERS IN A GROUP

How self-aware is this client?
Does this client seem to be aware of their emotion(s), bodily sensations, thoughts, attitudes, beliefs, and sociocultural experiences (as evidenced through verbal/arts-based expressions)?

To what degree is this client aware of others in the group?
Does this client notice others in the group?
Do they react to others in the group? In what ways?
Do they seem aware of others' emotions?
Do they respond to others' emotions? How?

Non-judgement?
Does this client seem to hold onto preconceived ideas when engaging with others in the group? What ideas?
Do they seem to hold onto memories of previous encounters tightly when engaging in the present moment with other group members?
Are they making assumptions about others in the group? Who? If so, what does that look/sound like? What is their judgement?
Do they move between judging/not judging? When? Why?

Can they approach others in the group with fresh openness? What would that require of them?
Are they open to listening in a truly nonjudgemental way?
How are they balancing non-judgement with holding any necessary boundaries? How do they seem to be experiencing and expressing that?

What's this client's capacity like?
Are there features of their diagnosis that impact their capacity to empathise?
How much energy do they have at the moment? How tired are they?
What's their emotional state?
What are their emotional resources like? Do they seem to be emotionally resourced? Do they appear emotionally depleted? Do they seem overwhelmed?
Are they open to being flexible in their thinking? Do they seem somewhat stuck in their perceptions of situations and relationships?
How secure does this client seem to be in their belief that they can manage difficult feelings?
When they feel they have made a mistake, do they seem to believe they can get back on track?
Do they seem to feel under time pressure?

STANCE: Receptivity

How open is this client to empathising?

How courageous does this client seem to be regarding leaning into another person's emotional experience?
Does this client seem willing to openly encounter another's emotional experience?
Do they seem to be willing to bravely enter into another person's vulnerability, not knowing exactly where the interaction will go?
Are they embracing this uncertainty as part of the journey/as an exploration/an "improvisation"?

Are they paying full attention?
Is this client fully present in sessions?
Do they seem distracted/is their mind elsewhere?
Is it hard for them to be fully present here? Why might that be?

How motivated do they seem to empathise with others in the group?
(How) might their mood be impacting their motivation to empathise with others in the group?

Do they seem to feel open and willing to empathise with others? Why? E.g.:
- Do they perceive others to be suffering in some way?.
- Do they think others in the group have been a victim somehow?
- Do they see others as kind?
- Do they see others as fair?
- Do they believe others in the group are in recovery/will recover?
- Do they identify with others in the group, leading to a willingness to empathise?
- Do they see themselves as an inherently empathic person regardless of the characteristics of others or their situation?

Do they seem to feel resistant to empathising with others in this group? Why? E.g.:
- Do they seem to perceive the others as acting out/manipulating/choosing to be destructive?
- Do they see others as cruel/mean?
- Do they see others as being unfair?
- Do they feel hurt by others in the group?
- Are they wondering whether others in the group will ever get better (or whether they might get worse?)
- Do they identify with others in the group, leading to a reluctance to empathise?

How powerful do they feel in relation to others in the group? Does that seem to make them want to empathise more or less?

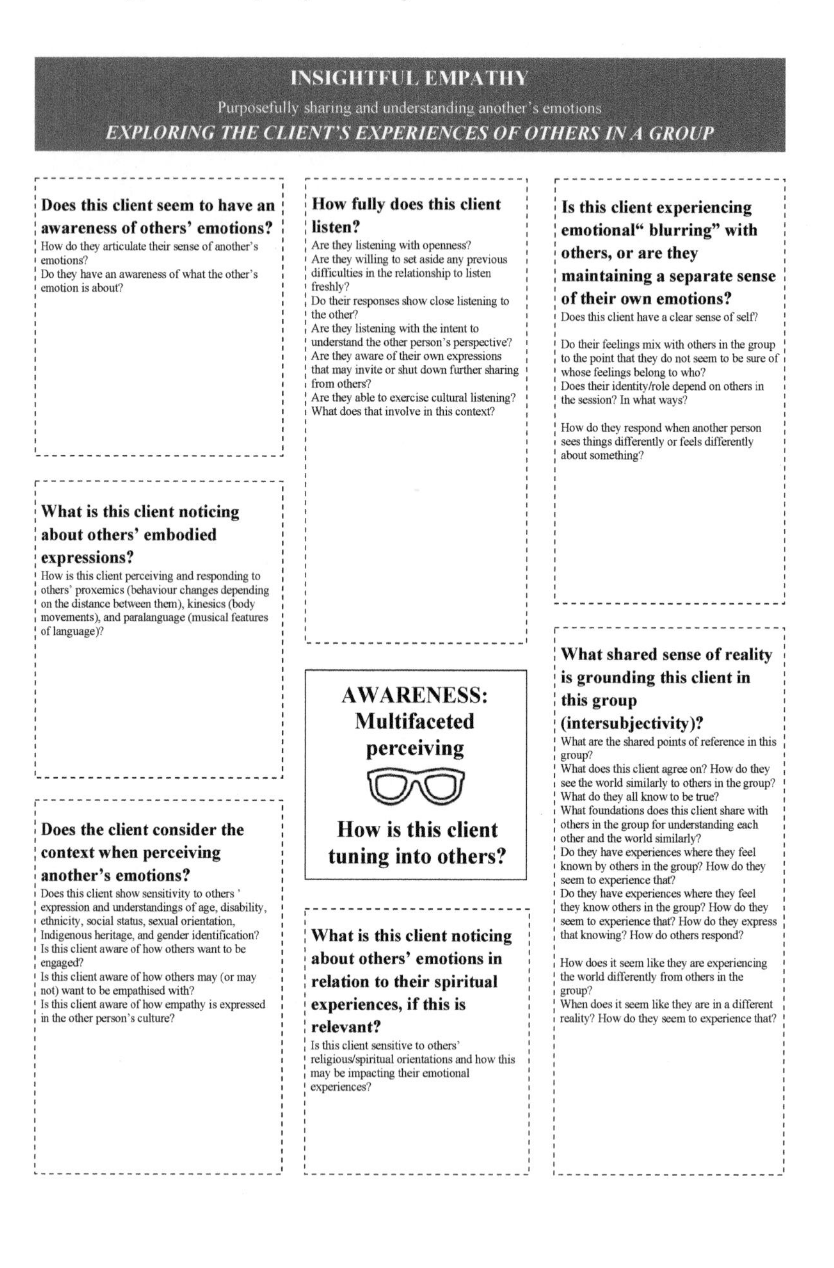

INSIGHTFUL EMPATHY

Purposefully sharing and understanding another's emotions

EXPLORING THE CLIENT'S EXPERIENCES OF OTHERS IN A GROUP

Does this client seem to have an awareness of others' emotions?

How do they articulate their sense of another's emotions?
Do they have an awareness of what the other's emotion is about?

What is this client noticing about others' embodied expressions?

How is this client perceiving and responding to others' proxemics (behaviour changes depending on the distance between them), kinesics (body movements), and paralanguage (musical features of language)?

Does the client consider the context when perceiving another's emotions?

Does this client show sensitivity to others ' expression and understandings of age, disability, ethnicity, social status, sexual orientation, Indigenous heritage, and gender identification?
Is this client aware of how others want to be engaged?
Is this client aware of how others may (or may not) want to be empathised with?
Is this client aware of how empathy is expressed in the other person's culture?

How fully does this client listen?

Are they listening with openness?
Are they willing to set aside any previous difficulties in the relationship to listen freshly?
Do their responses show close listening to the other?
Are they listening with the intent to understand the other person's perspective?
Are they aware of their own expressions that may invite or shut down further sharing from others?
Are they able to exercise cultural listening? What does that involve in this context?

AWARENESS: Multifaceted perceiving

How is this client tuning into others?

What is this client noticing about others' emotions in relation to their spiritual experiences, if this is relevant?

Is this client sensitive to others' religious/spiritual orientations and how this may be impacting their emotional experiences?

Is this client experiencing emotional" blurring" with others, or are they maintaining a separate sense of their own emotions?

Does this client have a clear sense of self?

Do their feelings mix with others in the group to the point that they do not seem to be sure of whose feelings belong to who?
Does their identity/role depend on others in the session? In what ways?

How do they respond when another person sees things differently or feels differently about something?

What shared sense of reality is grounding this client in this group (intersubjectivity)?

What are the shared points of reference in this group?
What does this client agree on? How do they see the world similarly to others in the group?
What do they all know to be true?
What foundations does this client share with others in the group for understanding each other and the world similarly?
Do they have experiences where they feel known by others in the group? How do they seem to experience that?
Do they have experiences where they feel they know others in the group? How do they seem to experience that? How do they express that knowing? How do others respond?

How does it seem like they are experiencing the world differently from others in the group?
When does it seem like they are in a different reality? How do they seem to experience that?

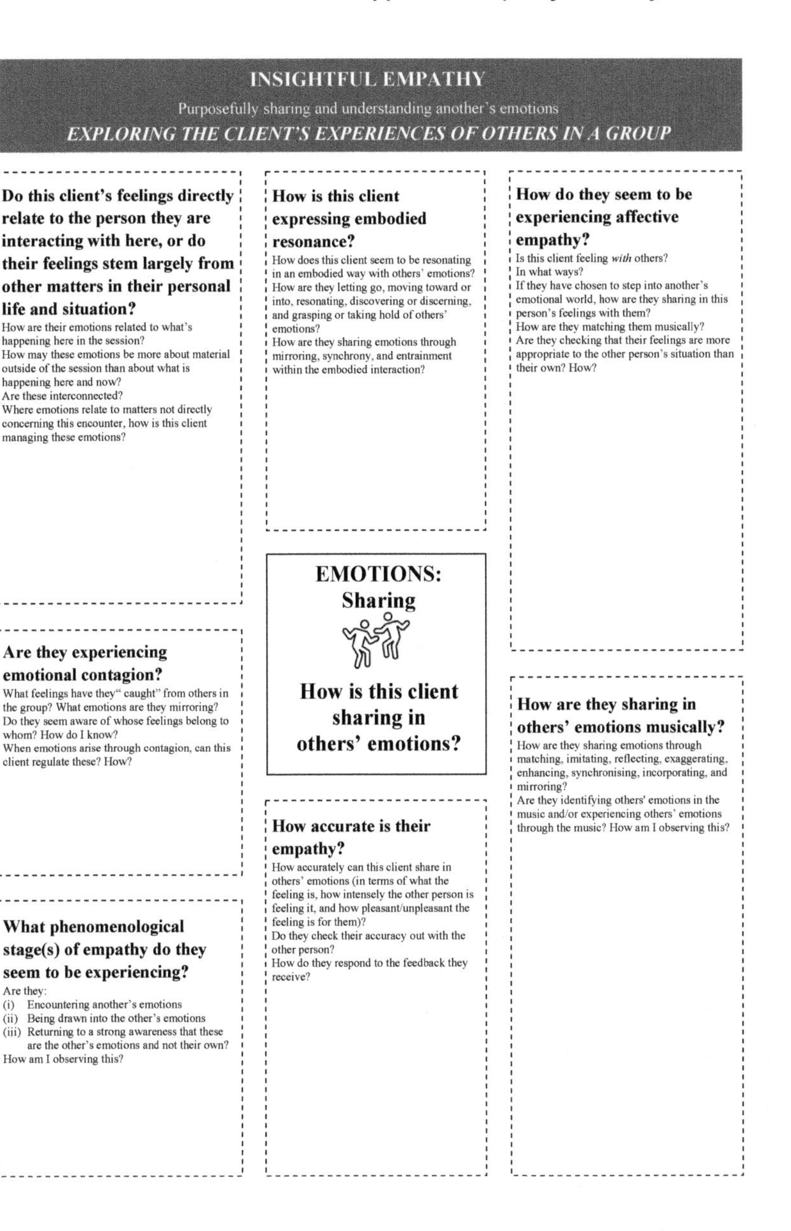

INSIGHTFUL EMPATHY

Purposefully sharing and understanding another's emotions

EXPLORING THE CLIENT'S EXPERIENCES OF OTHERS IN A GROUP

Do this client's feelings directly relate to the person they are interacting with here, or do their feelings stem largely from other matters in their personal life and situation?
How are their emotions related to what's happening here in the session?
How may these emotions be more about material outside of the session than about what is happening here and now?
Are these interconnected?
Where emotions relate to matters not directly concerning this encounter, how is this client managing these emotions?

How is this client expressing embodied resonance?
How does this client seem to be resonating in an embodied way with others' emotions?
How are they letting go, moving toward or into, resonating, discovering or discerning, and grasping or taking hold of others' emotions?
How are they sharing emotions through mirroring, synchrony, and entrainment within the embodied interaction?

How do they seem to be experiencing affective empathy?
Is this client feeling *with* others?
In what ways?
If they have chosen to step into another's emotional world, how are they sharing in this person's feelings with them?
How are they matching them musically?
Are they checking that their feelings are more appropriate to the other person's situation than their own? How?

EMOTIONS: Sharing

How is this client sharing in others' emotions?

Are they experiencing emotional contagion?
What feelings have they" caught" from others in the group? What emotions are they mirroring?
Do they seem aware of whose feelings belong to whom? How do I know?
When emotions arise through contagion, can this client regulate these? How?

How are they sharing in others' emotions musically?
How are they sharing emotions through matching, imitating, reflecting, exaggerating, enhancing, synchronising, incorporating, and mirroring?
Are they identifying others' emotions in the music and/or experiencing others' emotions through the music? How am I observing this?

How accurate is their empathy?
How accurately can this client share in others' emotions (in terms of what the feeling is, how intensely the other person is feeling it, and how pleasant/unpleasant the feeling is for them)?
Do they check their accuracy out with the other person?
How do they respond to the feedback they receive?

What phenomenological stage(s) of empathy do they seem to be experiencing?
Are they:
(i) Encountering another's emotions
(ii) Being drawn into the other's emotions
(iii) Returning to a strong awareness that these are the other's emotions and not their own?
How am I observing this?

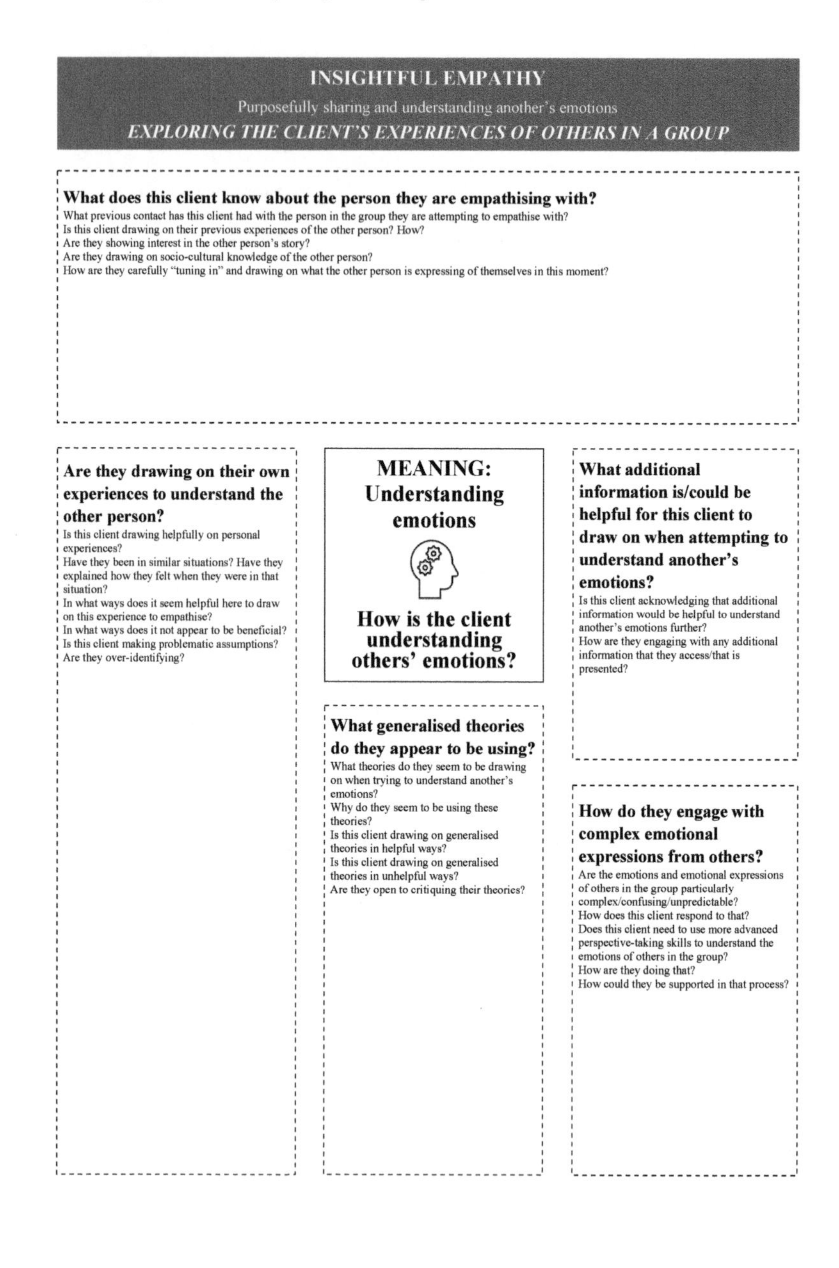
INSIGHTFUL EMPATHY
Purposefully sharing and understanding another's emotions
EXPLORING THE CLIENT'S EXPERIENCES OF OTHERS IN A GROUP
What does this client know about the person they are empathising with?
What previous contact has this client had with the person in the group they are attempting to empathise with?
Is this client drawing on their previous experiences of the other person? How?
Are they showing interest in the other person's story?
Are they drawing on socio-cultural knowledge of the other person?
How are they carefully "tuning in" and drawing on what the other person is expressing of themselves in this moment?
Are they drawing on their own experiences to understand the other person?
Is this client drawing helpfully on personal experiences?
Have they been in similar situations? Have they explained how they felt when they were in that situation?
In what ways does it seem helpful here to draw on this experience to empathise?
In what ways does it not appear to be beneficial?
Is this client making problematic assumptions?
Are they over-identifying?
MEANING: Understanding emotions
How is the client understanding others' emotions?
What generalised theories do they appear to be using?
What theories do they seem to be drawing on when trying to understand another's emotions?
Why do they seem to be using these theories?
Is this client drawing on generalised theories in helpful ways?
Is this client drawing on generalised theories in unhelpful ways?
Are they open to critiquing their theories?
What additional information is/could be helpful for this client to draw on when attempting to understand another's emotions?
Is this client acknowledging that additional information would be helpful to understand another's emotions further?
How are they engaging with any additional information that they access/that is presented?
How do they engage with complex emotional expressions from others?
Are the emotions and emotional expressions of others in the group particularly complex/confusing/unpredictable?
How does this client respond to that?
Does this client need to use more advanced perspective-taking skills to understand the emotions of others in the group?
How are they doing that?
How could they be supported in that process?

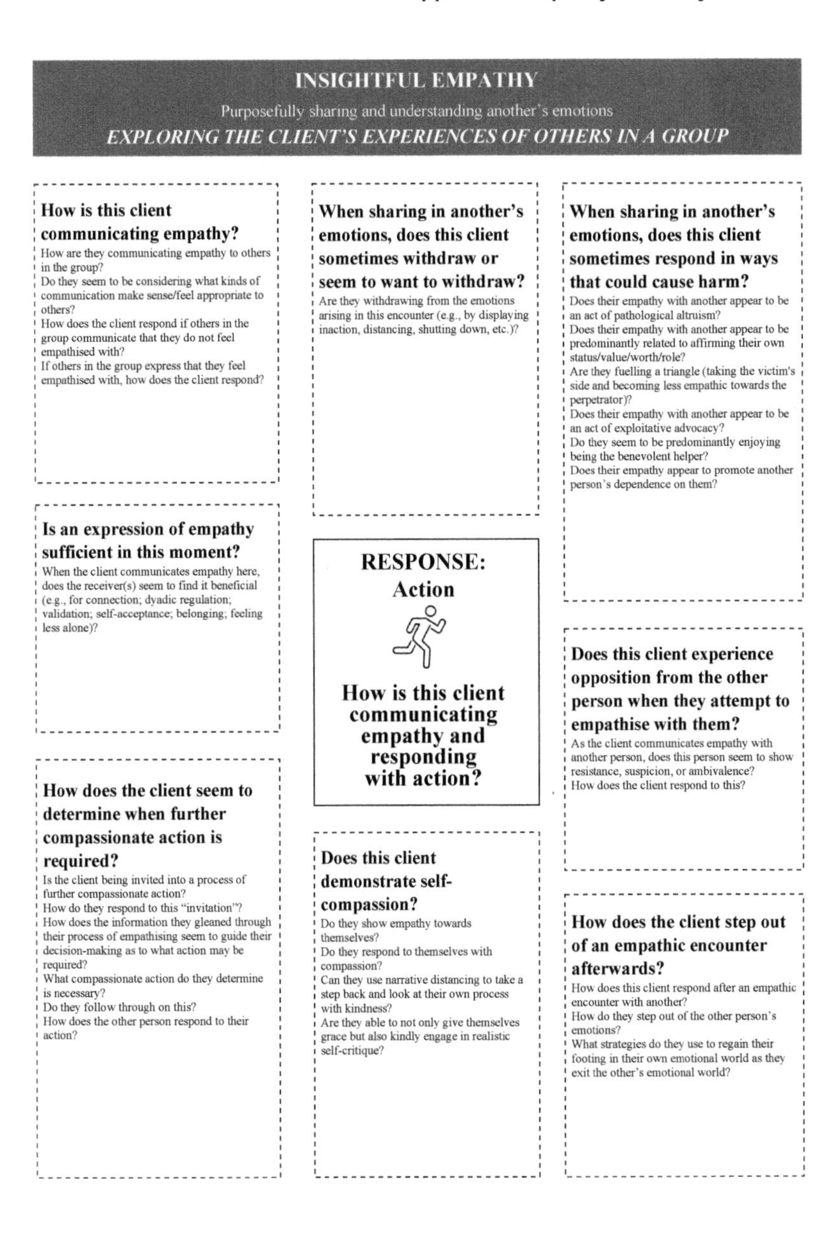

INSIGHTFUL EMPATHY

Purposefully sharing and understanding another's emotions

EXPLORING THE CLIENT'S EXPERIENCES OF OTHERS IN A GROUP

How is this client communicating empathy?

How are they communicating empathy to others in the group?
Do they seem to be considering what kinds of communication make sense/feel appropriate to others?
How does the client respond if others in the group communicate that they do not feel empathised with?
If others in the group express that they feel empathised with, how does the client respond?

Is an expression of empathy sufficient in this moment?

When the client communicates empathy here, does the receiver(s) seem to find it beneficial (e.g., for connection; dyadic regulation; validation; self-acceptance; belonging; feeling less alone)?

How does the client seem to determine when further compassionate action is required?

Is the client being invited into a process of further compassionate action?
How do they respond to this "invitation"?
How does the information they gleaned through their process of empathising seem to guide their decision-making as to what action may be required?
What compassionate action do they determine is necessary?
Do they follow through on this?
How does the other person respond to their action?

When sharing in another's emotions, does this client sometimes withdraw or seem to want to withdraw?

Are they withdrawing from the emotions arising in this encounter (e.g., by displaying inaction, distancing, shutting down, etc.)?

RESPONSE: Action

How is this client communicating empathy and responding with action?

Does this client demonstrate self-compassion?

Do they show empathy towards themselves?
Do they respond to themselves with compassion?
Can they use narrative distancing to take a step back and look at their own process with kindness?
Are they able to not only give themselves grace but also kindly engage in realistic self-critique?

When sharing in another's emotions, does this client sometimes respond in ways that could cause harm?

Does their empathy with another appear to be an act of pathological altruism?
Does their empathy with another appear to be predominantly related to affirming their own status/value/worth/role?
Are they fuelling a triangle (taking the victim's side and becoming less empathic towards the perpetrator)?
Does their empathy with another appear to be an act of exploitative advocacy?
Do they seem to be predominantly enjoying being the benevolent helper?
Does their empathy appear to promote another person's dependence on them?

Does this client experience opposition from the other person when they attempt to empathise with them?

As the client communicates empathy with another person, does this person seem to show resistance, suspicion, or ambivalence?
How does the client respond to this?

How does the client step out of an empathic encounter afterwards?

How does this client respond after an empathic encounter with another?
How do they step out of the other person's emotions?
What strategies do they use to regain their footing in their own emotional world as they exit the other's emotional world?

2. Translational Empathy

TRANSLATIONAL EMPATHY

A quality of presence in which a sense of withness is generated through a situated and productive process of emotion translation moves of expression and response

MUSIC THERAPIST'S SELF-EXPLORATION

How humbly am I approaching this situation and this client?

What do I know about my client's emotional experience?
What areas do I not know about/know less about?
Am I treating my client as an expert on their own experience? In what ways is this helpful? In what ways does this feel difficult?
How open am I to considering my client's perspective even when it differs from mine and even if I don't (fully) understand it?
Do I feel that I need to understand my client's emotional world fully? Why?
Am I open to acknowledging that I may be wrong about what my client is feeling?
In my view, what is at stake if I surrender the desire to gain full insight into my client's emotional world?
What might I gain by letting go of trying to gain full insight into my client's emotional world?
What might they gain if I let go of trying to gain full insight into their emotional world?

Am I displaying continuous curiosity here?

In terms of this client's emotional experience(s), what am I curious about?
Am I intentionally setting aside my preconceived ideas and assumptions as I reflect on what I'm curious about? How?
Am I open to being surprised?
Am I approaching my client playfully and creatively in terms of my thought processes (as opposed to rigidly and through set patterns of exploration)?

STANCE: Honouring opacity

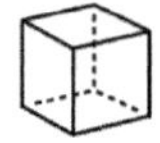

How am I humbly open to embracing what I may not be able to know about the other person?

How willing am I to "sit with" the unknown?

Can I question what I think I know about this client, or do I want to hold onto the certainty of what I think I know?
In what ways am I projecting parts of myself onto this client? As I reflect on that, how do I feel?
How am I seeing this client through the lens of whom I want them to be/think they are, rather than who they actually are? As I reflect on that, how do I feel?
How am I seeing this client according to who they were in the past rather than today? As I reflect on that, how do I feel?
How comfortable or uncomfortable am I with the thought that there are parts of this client that I do not know and perhaps cannot know?
Do I view the idea that this client has parts that I do not know as an invitation to journey with them and potentially learn more, as a barrier, or as a threat?

Am I staying on my side of the "net"?

What am I observing?
As I interpret those observations, am I staying with my reality instead of making assumptions about their reality?

Am I checking that I'm not engaging in damaging forms of othering?

Am I viewing this client as an "other" in any way that may be harmful?
How is this influencing my behaviour?
How do I feel as I reflect on this?
What do I need to do about this?
What feels difficult about doing that? What may be helpful about doing that?

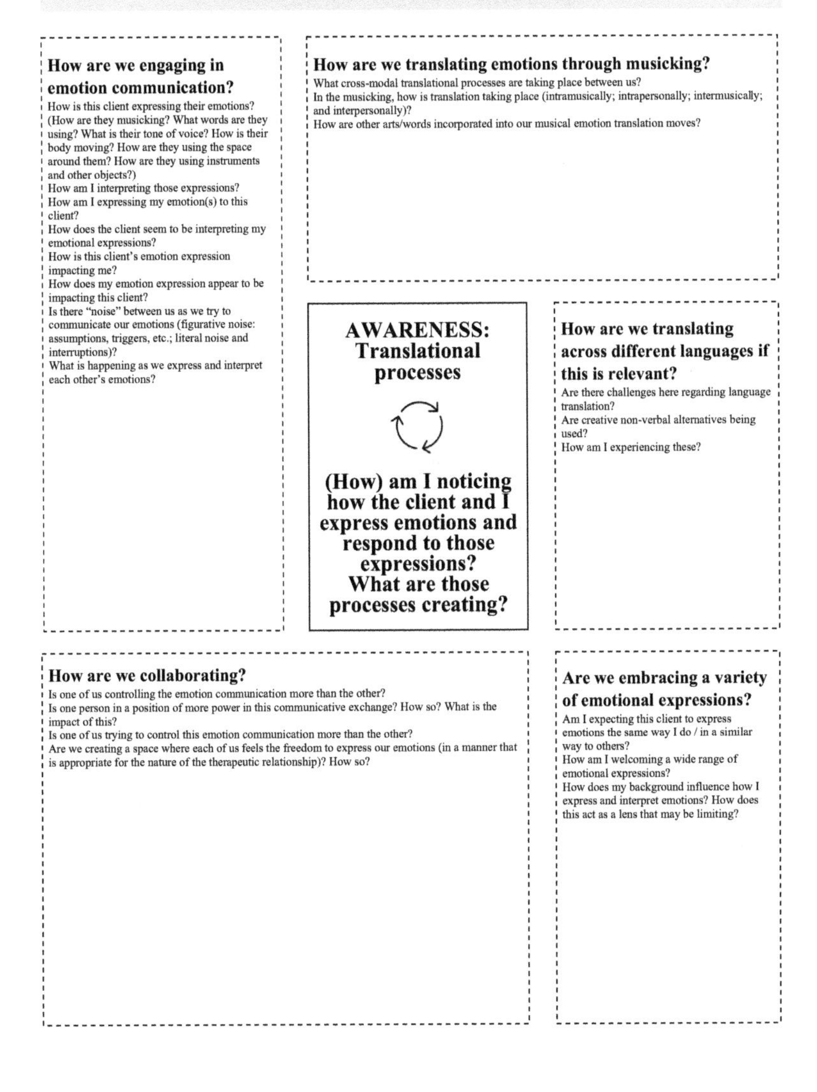
TRANSLATIONAL EMPATHY
A quality of presence in which a sense of withness is generated through a situated and productive process of emotion translation moves of expression and response
MUSIC THERAPIST'S SELF-EXPLORATION
How are we engaging in emotion communication?
How is this client expressing their emotions? (How are they musicking? What words are they using? What is their tone of voice? How is their body moving? How are they using the space around them? How are they using instruments and other objects?)
How am I interpreting those expressions?
How am I expressing my emotion(s) to this client?
How does the client seem to be interpreting my emotional expressions?
How is this client's emotion expression impacting me?
How does my emotion expression appear to be impacting this client?
Is there "noise" between us as we try to communicate our emotions (figurative noise: assumptions, triggers, etc.; literal noise and interruptions)?
What is happening as we express and interpret each other's emotions?
How are we translating emotions through musicking?
What cross-modal translational processes are taking place between us?
In the musicking, how is translation taking place (intramusically; intrapersonally; intermusically; and interpersonally)?
How are other arts/words incorporated into our musical emotion translation moves?
AWARENESS: Translational processes
(How) am I noticing how the client and I express emotions and respond to those expressions? What are those processes creating?
How are we translating across different languages if this is relevant?
Are there challenges here regarding language translation?
Are creative non-verbal alternatives being used?
How am I experiencing these?
How are we collaborating?
Is one of us controlling the emotion communication more than the other?
Is one person in a position of more power in this communicative exchange? How so? What is the impact of this?
Is one of us trying to control this emotion communication more than the other?
Are we creating a space where each of us feels the freedom to express our emotions (in a manner that is appropriate for the nature of the therapeutic relationship)? How so?
Are we embracing a variety of emotional expressions?
Am I expecting this client to express emotions the same way I do / in a similar way to others?
How am I welcoming a wide range of emotional expressions?
How does my background influence how I express and interpret emotions? How does this act as a lens that may be limiting?

TRANSLATIONAL EMPATHY

A quality of presence in which a sense of withness is generated through a situated and productive process of emotion translation moves of expression and response

MUSIC THERAPIST'S SELF-EXPLORATION

How am I a witness?

Am I holding space for this client to feel and express their feelings? How am I doing that?
What does that look/sound like here?
How is the client responding to my attempts to hold space for them?
Do I see myself as a witness?
Am I too entangled in this relationship to be able to step into the role of witness?
How could I try to step into that role?
How am I expressing witnessing?
How am I affirming and validating this client's emotion expressions as I witness them?

How am I taking part in creating a sense of withness in musical and metaphoric ambiguity?

What musical moments of knowing am I experiencing with this client?
What musical moments of not-knowing am I experiencing with this client?
How are we with one another through our experience of vitality affects?
How are we with one another through our use and experience of metaphors?

EMOTIONS: Responsive withness

(How) am I responsively being with this client as they express their emotions?

How am I taking part in creating a sense of withness?

Does this client want me to be offering them a sense of withness right now? Do they want me to be physically present with them? In what way? Do they want me to be emotionally present right now? In what way?
How do I know this?
How do I respond when they don't want me to be with them in their emotional experiences?
What are the benefits of respecting their wish for me not to be with them?
Do I need to stay physically present for safety reasons while finding ways to respect their wishes for emotional distance/separateness?
How do I navigate that?
When the client does desire me to be emotionally with them, am I fully present with them?
What parts of myself am I bringing into this interaction with this client? Why? How is this helpful? How may this not be helpful?
What parts of myself am I not bringing into this interaction with the client? Why? How is this helpful? How may this not be helpful?
How am I dwelling with the client in their pain/struggle, even when I don't fully understand it? How am I expressing that kind of presence to them? What does it look/sound/feel like?
How am I dwelling with the client in their strengths/resilience/creativity/playfulness/curiosity/agency, even when I don't fully understand it?
How am I expressing that kind of presence to the client? What does it look/sound/feel like?

How am I being emotionally responsive?

How am I giving welcoming invitations to this client for emotional expression?
When the client expresses their feelings, how do I respond?
How am I reflecting my understanding of their emotions back to them?
How am I sensitive to changes in their expressions?
What invitations for emotional expression am I receiving from the client?
How am I responding to those invitations?

How am I embracing how we can be both together and separate?

In what ways do we feel a sense of togetherness? How does that feeling come about? How could I describe that sense of "togetherness"?
In what ways do we also have opportunities for separateness/individuality/autonomy?
How do those experiences come about?
How would I describe that aspect of the therapeutic relationship?
How am I experiencing my own separateness?
How am I giving the client space for their own separateness?

TRANSLATIONAL EMPATHY

A quality of presence in which a sense of withness is generated through a situated and productive process of emotion translation moves of expression and response

MUSIC THERAPIST'S SELF-EXPLORATION

Am I engaging in reflexivity?

What assumptions did I bring into this encounter?
How was I feeling before this encounter, and in what ways do my mood, feelings, attitudes, thoughts, and values influence how I am interpreting this client's emotion expressions?
How tired am I, and how does this impact how I interpret this client's emotion expressions?
What are my hopes for this client, and how do these impact how I interpret their emotion expressions?
What else is happening in this context at the moment, and how does this impact the way I am interpreting this client's emotion expressions?

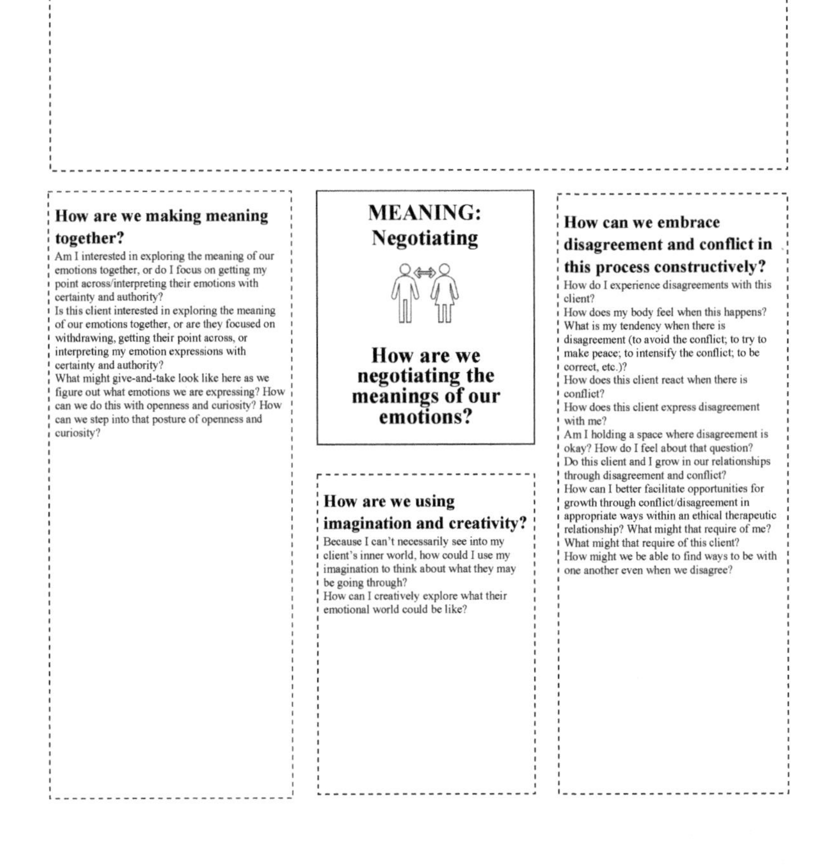

How are we making meaning together?

Am I interested in exploring the meaning of our emotions together, or do I focus on getting my point across/interpreting their emotions with certainty and authority?
Is this client interested in exploring the meaning of our emotions together, or are they focused on withdrawing, getting their point across, or interpreting my emotion expressions with certainty and authority?
What might give-and-take look like here as we figure out what emotions we are expressing? How can we do this with openness and curiosity? How can we step into that posture of openness and curiosity?

MEANING: Negotiating

How are we negotiating the meanings of our emotions?

How are we using imagination and creativity?

Because I can't necessarily see into my client's inner world, how could I use my imagination to think about what they may be going through?
How can I creatively explore what their emotional world could be like?

How can we embrace disagreement and conflict in this process constructively?

How do I experience disagreements with this client?
How does my body feel when this happens?
What is my tendency when there is disagreement (to avoid the conflict; to try to make peace; to intensify the conflict; to be correct, etc.)?
How does this client react when there is conflict?
How does this client express disagreement with me?
Am I holding a space where disagreement is okay? How do I feel about that question?
Do this client and I grow in our relationships through disagreement and conflict?
How can I better facilitate opportunities for growth through conflict/disagreement in appropriate ways within an ethical therapeutic relationship? What might that require of me? What might that require of this client?
How might we be able to find ways to be with one another even when we disagree?

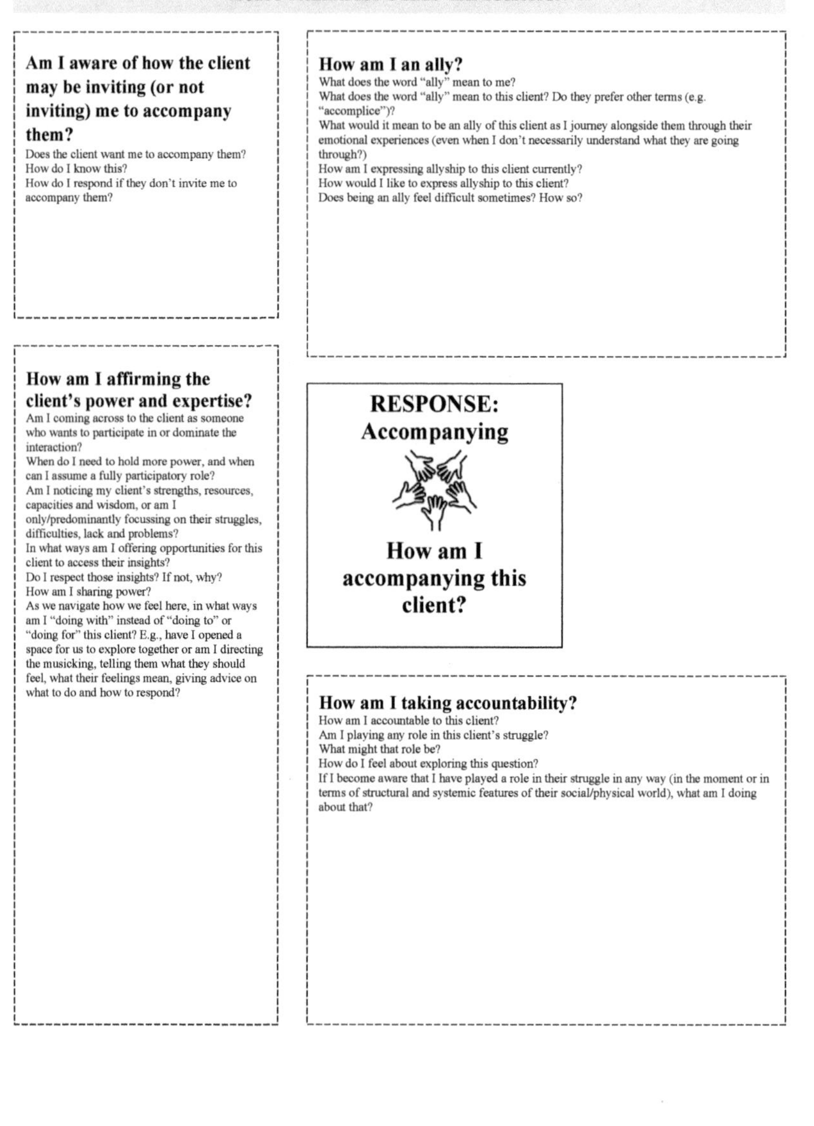
TRANSLATIONAL EMPATHY
A quality of presence in which a sense of withness is generated through a situated and productive process of emotion translation moves of expression and response
MUSIC THERAPIST'S SELF-EXPLORATION
Am I aware of how the client may be inviting (or not inviting) me to accompany them?
Does the client want me to accompany them?
How do I know this?
How do I respond if they don't invite me to accompany them?
How am I an ally?
What does the word "ally" mean to me?
What does the word "ally" mean to this client? Do they prefer other terms (e.g. "accomplice")?
What would it mean to be an ally of this client as I journey alongside them through their emotional experiences (even when I don't necessarily understand what they are going through?)
How am I expressing allyship to this client currently?
How would I like to express allyship to this client?
Does being an ally feel difficult sometimes? How so?
How am I affirming the client's power and expertise?
Am I coming across to the client as someone who wants to participate in or dominate the interaction?
When do I need to hold more power, and when can I assume a fully participatory role?
Am I noticing my client's strengths, resources, capacities and wisdom, or am I only/predominantly focussing on their struggles, difficulties, lack and problems?
In what ways am I offering opportunities for this client to access their insights?
Do I respect those insights? If not, why?
How am I sharing power?
As we navigate how we feel here, in what ways am I "doing with" instead of "doing to" or "doing for" this client? E.g., have I opened a space for us to explore together or am I directing the musicking, telling them what they should feel, what their feelings mean, giving advice on what to do and how to respond?
RESPONSE:
Accompanying
How am I accompanying this client?
How am I taking accountability?
How am I accountable to this client?
Am I playing any role in this client's struggle?
What might that role be?
How do I feel about exploring this question?
If I become aware that I have played a role in their struggle in any way (in the moment or in terms of structural and systemic features of their social/physical world), what am I doing about that?

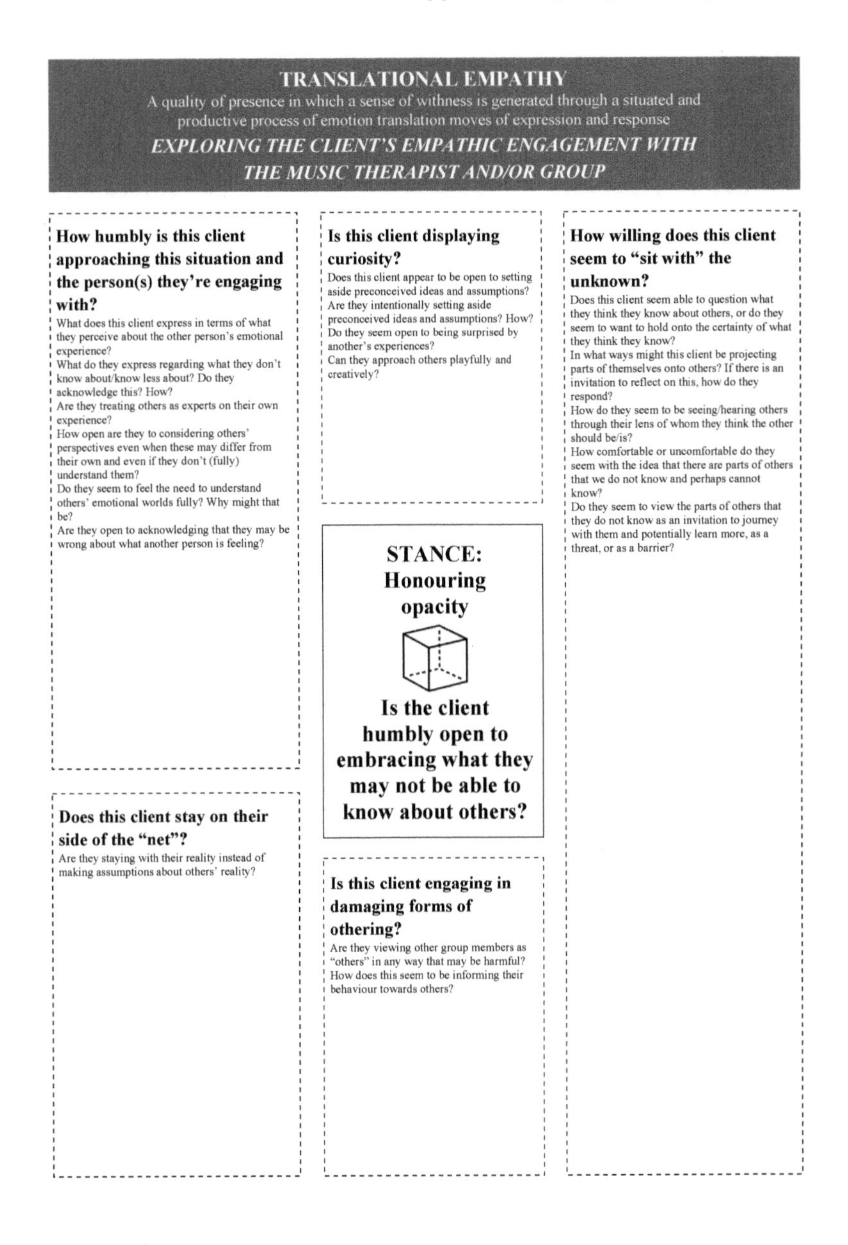

TRANSLATIONAL EMPATHY

A quality of presence in which a sense of withness is generated through a situated and productive process of emotion translation moves of expression and response

EXPLORING THE CLIENT'S EMPATHIC ENGAGEMENT WITH THE MUSIC THERAPIST AND/OR GROUP

How humbly is this client approaching this situation and the person(s) they're engaging with?

What does this client express in terms of what they perceive about the other person's emotional experience?
What do they express regarding what they don't know about/know less about? Do they acknowledge this? How?
Are they treating others as experts on their own experience?
How open are they to considering others' perspectives even when these may differ from their own and even if they don't (fully) understand them?
Do they seem to feel the need to understand others' emotional worlds fully? Why might that be?
Are they open to acknowledging that they may be wrong about what another person is feeling?

Does this client stay on their side of the "net"?

Are they staying with their reality instead of making assumptions about others' reality?

Is this client displaying curiosity?

Does this client appear to be open to setting aside preconceived ideas and assumptions?
Are they intentionally setting aside preconceived ideas and assumptions? How?
Do they seem open to being surprised by another's experiences?
Can they approach others playfully and creatively?

STANCE: Honouring opacity

Is the client humbly open to embracing what they may not be able to know about others?

Is this client engaging in damaging forms of othering?

Are they viewing other group members as "others" in any way that may be harmful?
How does this seem to be informing their behaviour towards others?

How willing does this client seem to "sit with" the unknown?

Does this client seem able to question what they think they know about others, or do they seem to want to hold onto the certainty of what they think they know?
In what ways might this client be projecting parts of themselves onto others? If there is an invitation to reflect on this, how do they respond?
How do they seem to be seeing/hearing others through their lens of whom they think the other should be/is?
How comfortable or uncomfortable do they seem with the idea that there are parts of others that we do not know and perhaps cannot know?
Do they seem to view the parts of others that they do not know as an invitation to journey with them and potentially learn more, as a threat, or as a barrier?

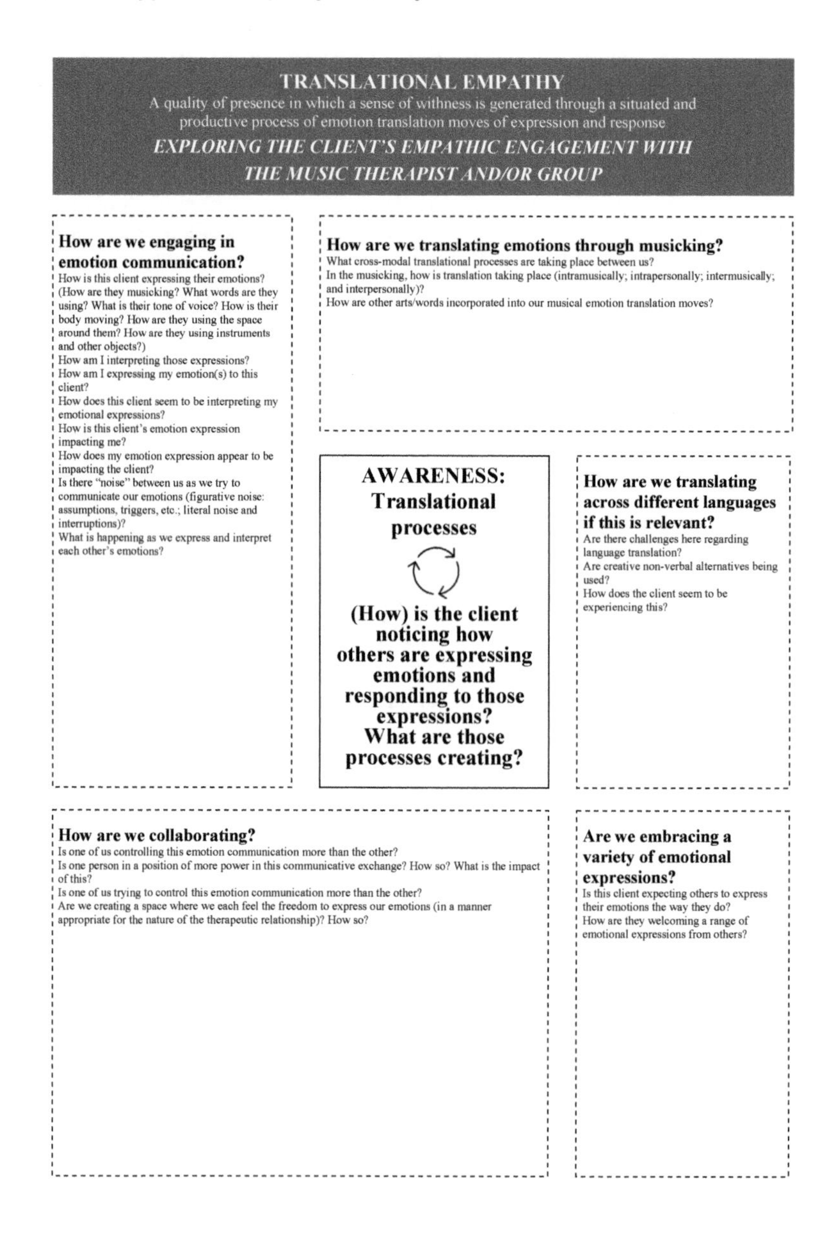
TRANSLATIONAL EMPATHY

A quality of presence in which a sense of withness is generated through a situated and productive process of emotion translation moves of expression and response

EXPLORING THE CLIENT'S EMPATHIC ENGAGEMENT WITH THE MUSIC THERAPIST AND/OR GROUP

How are we engaging in emotion communication?

How is this client expressing their emotions? (How are they musicking? What words are they using? What is their tone of voice? How is their body moving? How are they using the space around them? How are they using instruments and other objects?)
How am I interpreting those expressions?
How am I expressing my emotion(s) to this client?
How does this client seem to be interpreting my emotional expressions?
How is this client's emotion expression impacting me?
How does my emotion expression appear to be impacting the client?
Is there "noise" between us as we try to communicate our emotions (figurative noise: assumptions, triggers, etc.; literal noise and interruptions)?
What is happening as we express and interpret each other's emotions?

How are we translating emotions through musicking?

What cross-modal translational processes are taking place between us?
In the musicking, how is translation taking place (intramusically; intrapersonally; intermusically; and interpersonally)?
How are other arts/words incorporated into our musical emotion translation moves?

AWARENESS: Translational processes

(How) is the client noticing how others are expressing emotions and responding to those expressions? What are those processes creating?

How are we translating across different languages if this is relevant?

Are there challenges here regarding language translation?
Are creative non-verbal alternatives being used?
How does the client seem to be experiencing this?

How are we collaborating?

Is one of us controlling this emotion communication more than the other?
Is one person in a position of more power in this communicative exchange? How so? What is the impact of this?
Is one of us trying to control this emotion communication more than the other?
Are we creating a space where we each feel the freedom to express our emotions (in a manner appropriate for the nature of the therapeutic relationship)? How so?

Are we embracing a variety of emotional expressions?

Is this client expecting others to express their emotions the way they do?
How are they welcoming a range of emotional expressions from others?

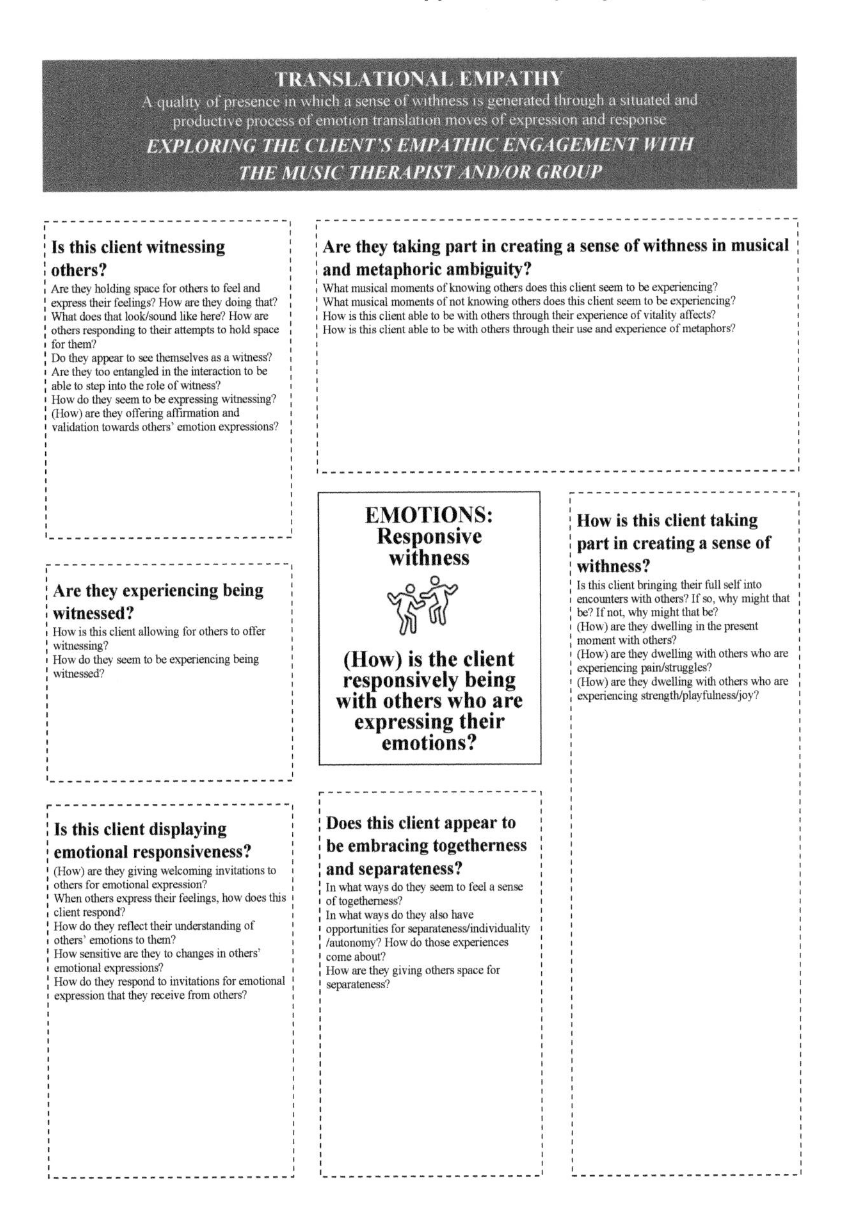
TRANSLATIONAL EMPATHY
A quality of presence in which a sense of withness is generated through a situated and productive process of emotion translation moves of expression and response
EXPLORING THE CLIENT'S EMPATHIC ENGAGEMENT WITH THE MUSIC THERAPIST AND/OR GROUP
Is this client witnessing others?
Are they holding space for others to feel and express their feelings? How are they doing that? What does that look/sound like here? How are others responding to their attempts to hold space for them?
Do they appear to see themselves as a witness? Are they too entangled in the interaction to be able to step into the role of witness?
How do they seem to be expressing witnessing?
(How) are they offering affirmation and validation towards others' emotion expressions?
Are they taking part in creating a sense of withness in musical and metaphoric ambiguity?
What musical moments of knowing others does this client seem to be experiencing?
What musical moments of not knowing others does this client seem to be experiencing?
How is this client able to be with others through their experience of vitality affects?
How is this client able to be with others through their use and experience of metaphors?
EMOTIONS: Responsive withness
(How) is the client responsively being with others who are expressing their emotions?
Are they experiencing being witnessed?
How is this client allowing for others to offer witnessing?
How do they seem to be experiencing being witnessed?
How is this client taking part in creating a sense of withness?
Is this client bringing their full self into encounters with others? If so, why might that be? If not, why might that be?
(How) are they dwelling in the present moment with others?
(How) are they dwelling with others who are experiencing pain/struggles?
(How) are they dwelling with others who are experiencing strength/playfulness/joy?
Is this client displaying emotional responsiveness?
(How) are they giving welcoming invitations to others for emotional expression?
When others express their feelings, how does this client respond?
How do they reflect their understanding of others' emotions to them?
How sensitive are they to changes in others' emotional expressions?
How do they respond to invitations for emotional expression that they receive from others?
Does this client appear to be embracing togetherness and separateness?
In what ways do they seem to feel a sense of togetherness?
In what ways do they also have opportunities for separateness/individuality /autonomy? How do those experiences come about?
How are they giving others space for separateness?

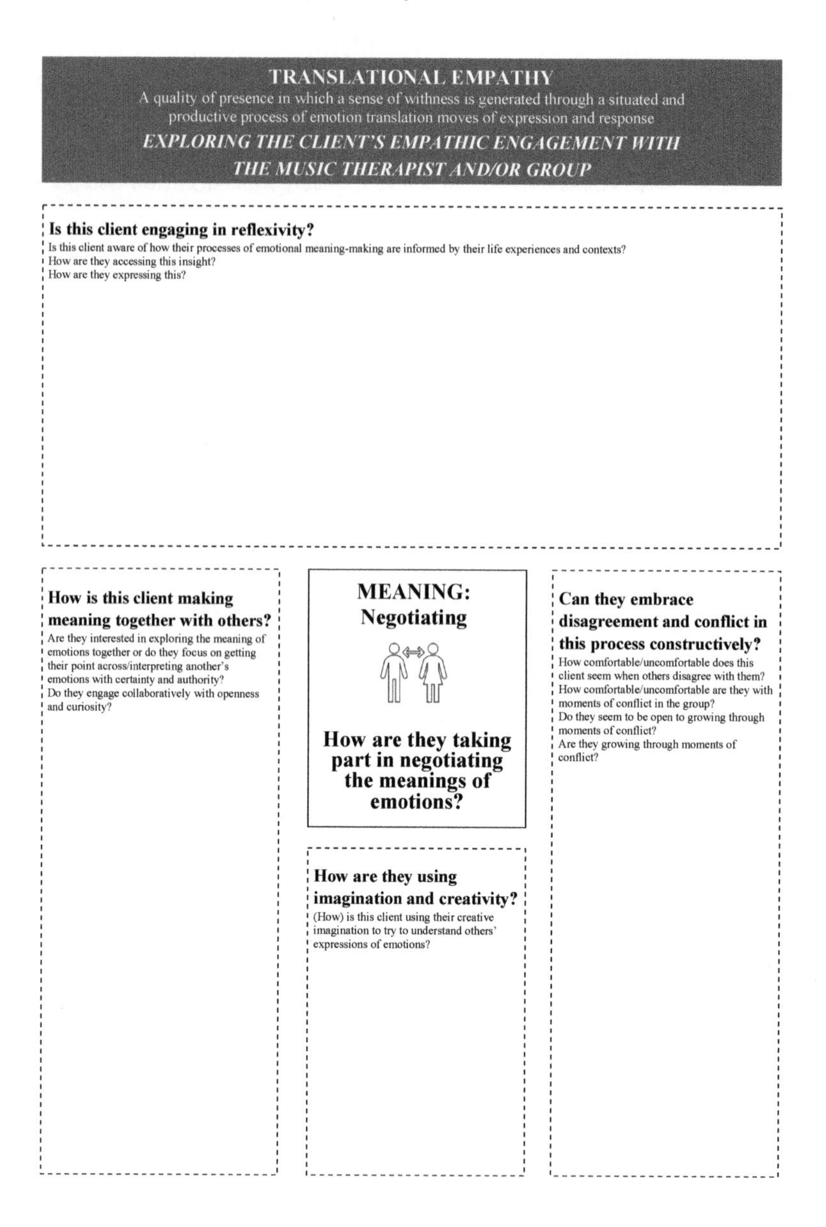
TRANSLATIONAL EMPATHY
A quality of presence in which a sense of withness is generated through a situated and productive process of emotion translation moves of expression and response
EXPLORING THE CLIENT'S EMPATHIC ENGAGEMENT WITH THE MUSIC THERAPIST AND/OR GROUP
Is this client engaging in reflexivity?
Is this client aware of how their processes of emotional meaning-making are informed by their life experiences and contexts?
How are they accessing this insight?
How are they expressing this?
How is this client making meaning together with others?
Are they interested in exploring the meaning of emotions together or do they focus on getting their point across/interpreting another's emotions with certainty and authority?
Do they engage collaboratively with openness and curiosity?
MEANING: Negotiating
How are they taking part in negotiating the meanings of emotions?
Can they embrace disagreement and conflict in this process constructively?
How comfortable/uncomfortable does this client seem when others disagree with them?
How comfortable/uncomfortable are they with moments of conflict in the group?
Do they seem to be open to growing through moments of conflict?
Are they growing through moments of conflict?
How are they using imagination and creativity?
(How) is this client using their creative imagination to try to understand others' expressions of emotions?

TRANSLATIONAL EMPATHY

A quality of presence in which a sense of withness is generated through a situated and productive process of emotion translation moves of expression and response

EXPLORING THE CLIENT'S EMPATHIC ENGAGEMENT WITH THE MUSIC THERAPIST AND/OR GROUP

Is this client aware of how others may be inviting (or not inviting) them to be an accompanier?

How does the client respond to invitations/lack of invitations for accompaniment from others?

In what ways are they an ally?

Where relevant, what role are they playing in expressing allyship?

How are they affirming others' power and expertise?

Do they notice and affirm others' strengths?
Where relevant, how are they actively deconstructing their own power?
How are they sharing power?
In what ways are they "doing with" rather than "doing to" or "doing for"?

RESPONSE: Accompanying

How am I accompanying this client?

In what ways are they taking accountability?

How is this client accountable to others in the group?
How are others accountable to them?

3. An Empathising Assemblage

AN EMPATHISING ASSEMBLAGE
Territorialising an assemblage as mutually affectively response-able
(COLLABORATIVE) EXPLORATION

STANCE:
Always-in-relation

How can we map relationships and enhance mutually affective response-ability here?

RESPONSE:
Enhancing Response-ability

AWARENESS:
Relational Mapping

EMOTION/MEANING:
Affective Attunement

What **things** are part of this situation (as these-becoming-things-always-in-relation)?

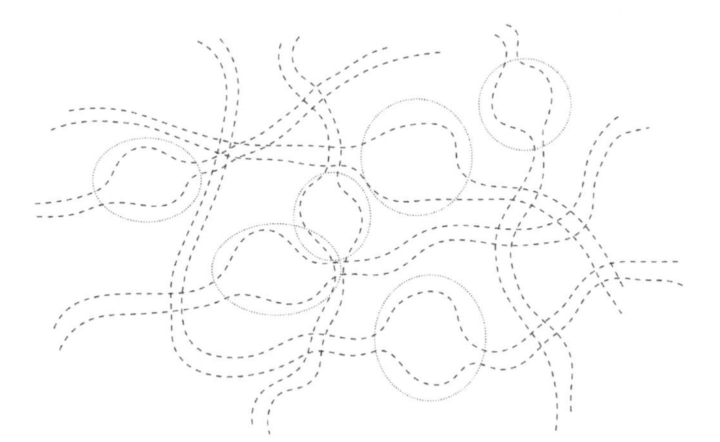

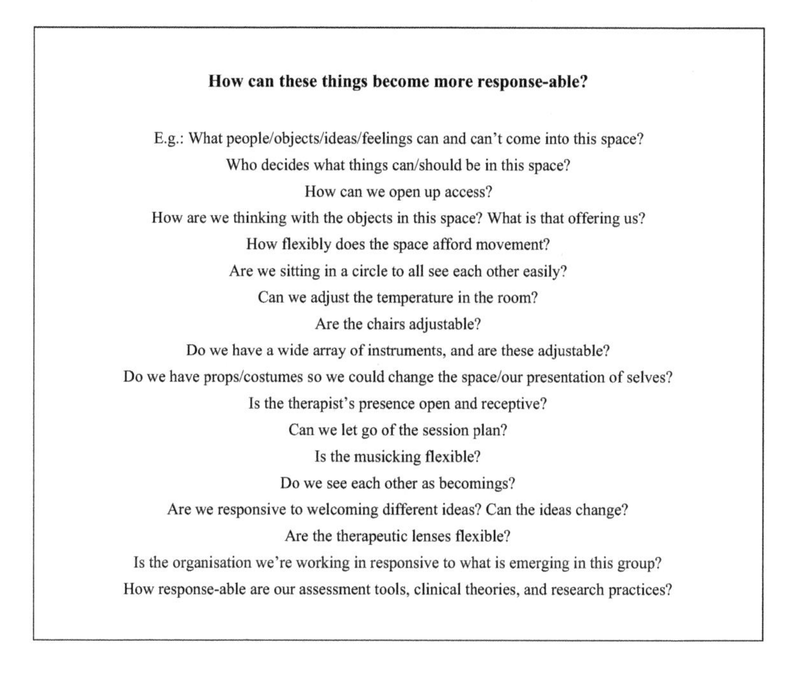

How can these things become more response-able?

E.g.: What people/objects/ideas/feelings can and can't come into this space?
Who decides what things can/should be in this space?
How can we open up access?
How are we thinking with the objects in this space? What is that offering us?
How flexibly does the space afford movement?
Are we sitting in a circle to all see each other easily?
Can we adjust the temperature in the room?
Are the chairs adjustable?
Do we have a wide array of instruments, and are these adjustable?
Do we have props/costumes so we could change the space/our presentation of selves?
Is the therapist's presence open and receptive?
Can we let go of the session plan?
Is the musicking flexible?
Do we see each other as becomings?
Are we responsive to welcoming different ideas? Can the ideas change?
Are the therapeutic lenses flexible?
Is the organisation we're working in responsive to what is emerging in this group?
How response-able are our assessment tools, clinical theories, and research practices?

What **connections** and **pathways** seem important here?

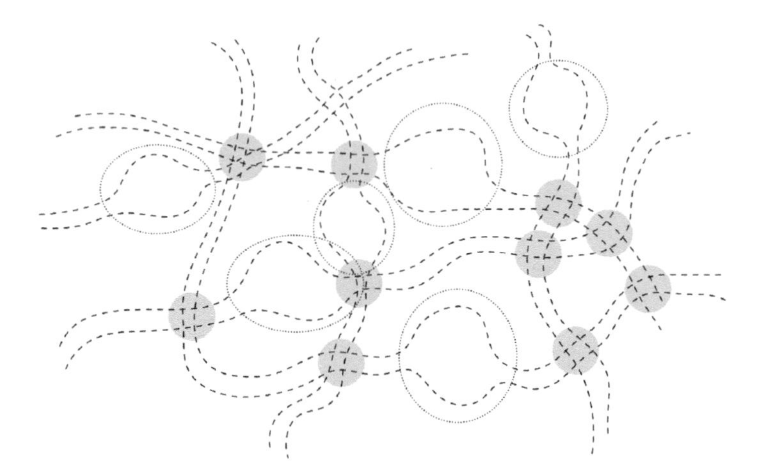

How can the connections/pathways be more response-able?

E.g.: How are people getting to know each other here?
How are they accessing connections with instruments? With genres?
With parts of self? With spaces? With relational roles?
How are the instruments responding to the people?
How is the music therapist changed through interaction with the clients?
Does my music create more or less space for someone else's music?
Do my responses silence others?
Is everyone growing in their capacity to respond here?
Do new ideas emerge between the ideas that are shared here?
Who decides what connections and pathways are possible and what aren't?
What new pathways are opening up here?

What **affects** are flowing here?
How are they flowing?

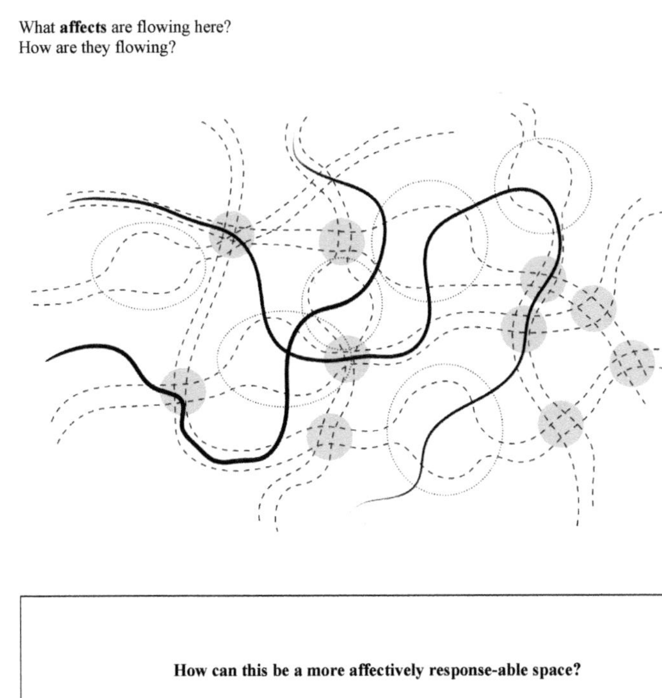

How can this be a more affectively response-able space?

E.g.: Are we open to any vitality affects (in the musicking and other interactions?)
or do we avoid/squash/ignore/push against/fear/punish (some of) them?
Are there predominant vitality affects that mark this space?
Do we all impact the flow of affects here?
How do we respond to the vitality affects that emerge here?
Are some affects getting stuck here (are they getting glued to people or things - the person who is "always…"?)
Who gets to feel and express what/when?
Is it acceptable for certain people to be entangled in particular vitality affects and not others?
Are some affective practices more highly regarded than others (e.g., "shimmering" rather than "sinking")?

How are we shaping vitality affects into emotions (through how we interpret them)?
Who decides? Is this a response-able process?

How are we marking this space (how is it **territorialising**)?

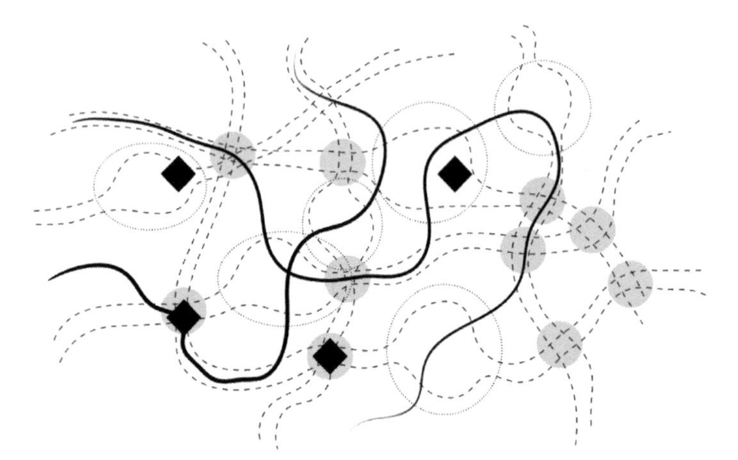

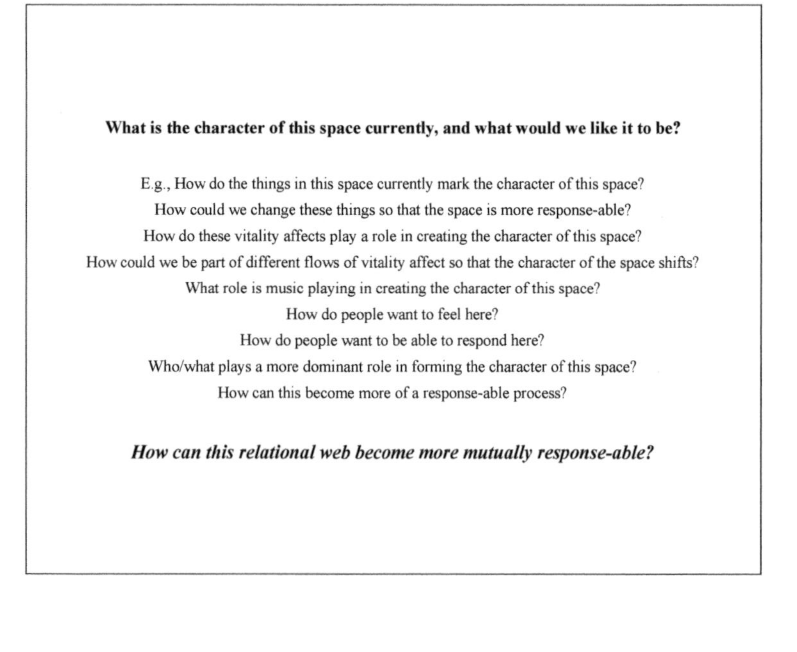

What is the character of this space currently, and what would we like it to be?

E.g., How do the things in this space currently mark the character of this space?
How could we change these things so that the space is more response-able?
How do these vitality affects play a role in creating the character of this space?
How could we be part of different flows of vitality affect so that the character of the space shifts?
What role is music playing in creating the character of this space?
How do people want to feel here?
How do people want to be able to respond here?
Who/what plays a more dominant role in forming the character of this space?
How can this become more of a response-able process?

How can this relational web become more mutually response-able?

4. Relational Empathy

RELATIONAL EMPATHY

Situated awareness of emotion co-storying

(COLLABORATIVE) EXPLORATION

How am I selfing here?

What kind of self do I become when interacting with this client? How am I "selfing" with them? What kind of self does this client become when interacting with me? How are they "selfing" with me?

How do we tend to do emotions here?

How do I tend to "do" emotions with this client? (How do I perform/express my emotion in a recognisable way to them?)
Do they tend to react in a way that affirms my emotion-performance? Or do they respond in a way that invalidates my emotion performance? Can I think of any examples?
How do they "do" their emotion in a recognisable way to me?
Do I tend to react in a way that affirms their emotion-performance? Or do I respond in a way that invalidates their emotion performance? Can I think of any examples?
Why do we "do" our emotions like this? (Are there any norms related to culture, gender, religion, age, etc. that influence us in this manner?)

What tends to unfold emotionally between us in general?

What happens to my feelings when I'm in an encounter with this client? Do they intensify? Do they soften/reduce in intensity? (Do I have some sense of why that may be?)

What seems to happen to the client's feelings when they are in an encounter with me? Do they intensify? Do they soften/reduce in intensity? (Do I have some sense of why that may be?)

How does that story tend to unfold in general?

STANCE: Selfing through relationship

How are we becoming who we are through each other?

AWARENESS of emotions as co-storying

How are emotions unfolding as stories between us?

How can I notice how we are "in" emotions?

When I encounter this client, and I think about what feelings are present, how could I describe what it's like for us to be "in" that feeling? (E.g., rather than "I am frustrated", this would be, "I am in this feeling of frustration" or "I am in this frustration story"; or rather than, "This client seems to be joyful today", this would be, "My client is in a space of joy" or even "We are in a joyful story together".)
What is it like to reframe emotion in this way?

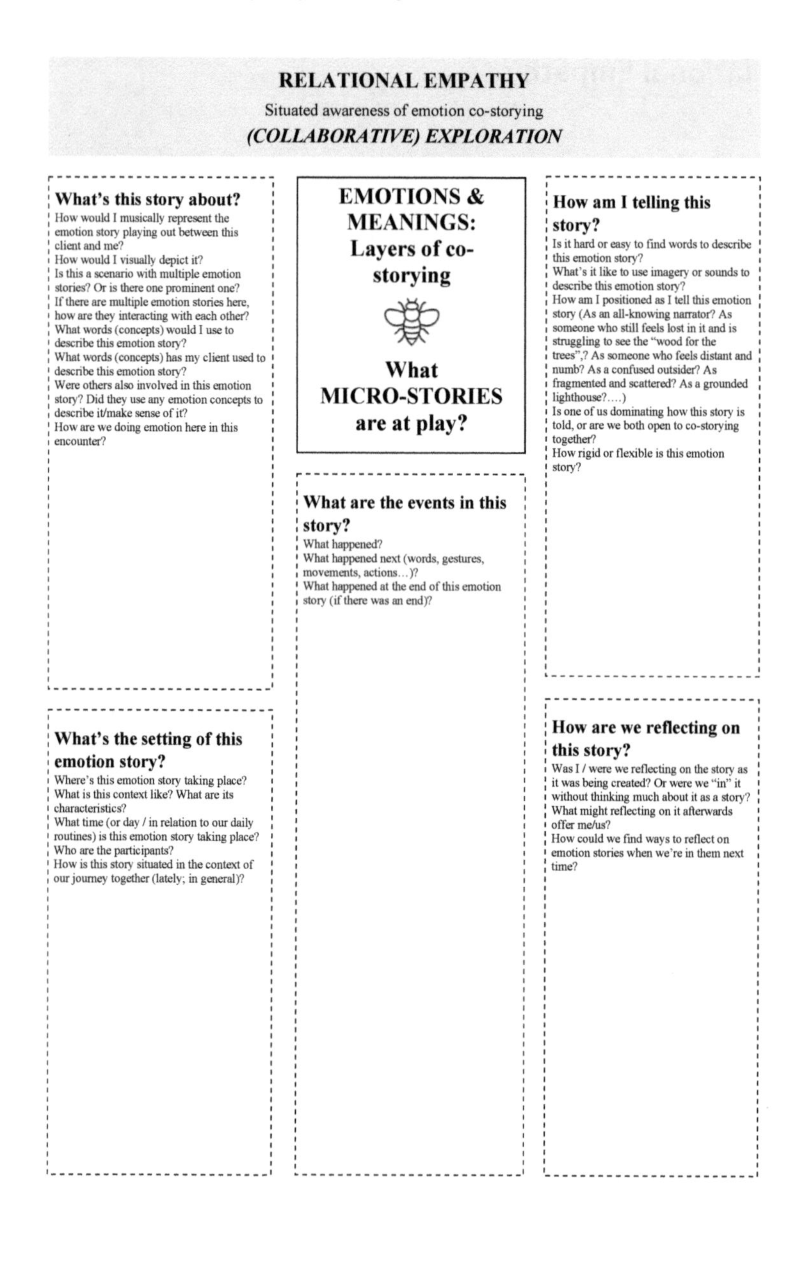
RELATIONAL EMPATHY

Situated awareness of emotion co-storying

(COLLABORATIVE) EXPLORATION

EMOTIONS & MEANINGS: Layers of co-storying

What MICRO-STORIES are at play?

What's this story about?
How would I musically represent the emotion story playing out between this client and me?
How would I visually depict it?
Is this a scenario with multiple emotion stories? Or is there one prominent one?
If there are multiple emotion stories here, how are they interacting with each other?
What words (concepts) would I use to describe this emotion story?
What words (concepts) has my client used to describe this emotion story?
Were others also involved in this emotion story? Did they use any emotion concepts to describe it/make sense of it?
How are we doing emotion here in this encounter?

What's the setting of this emotion story?
Where's this emotion story taking place?
What is this context like? What are its characteristics?
What time (or day / in relation to our daily routines) is this emotion story taking place?
Who are the participants?
How is this story situated in the context of our journey together (lately; in general)?

What are the events in this story?
What happened?
What happened next (words, gestures, movements, actions…)?
What happened at the end of this emotion story (if there was an end)?

How am I telling this story?
Is it hard or easy to find words to describe this emotion story?
What's it like to use imagery or sounds to describe this emotion story?
How am I positioned as I tell this emotion story (As an all-knowing narrator? As someone who still feels lost in it and is struggling to see the "wood for the trees",? As someone who feels distant and numb? As a confused outsider? As fragmented and scattered? As a grounded lighthouse?….)
Is one of us dominating how this story is told, or are we both open to co-storying together?
How rigid or flexible is this emotion story?

How are we reflecting on this story?
Was I / were we reflecting on the story as it was being created? Or were we "in" it without thinking much about it as a story?
What might reflecting on it afterwards offer me/us?
How could we find ways to reflect on emotion stories when we're in them next time?

RELATIONAL EMPATHY

Situated awareness of emotion co-storying

(COLLABORATIVE) EXPLORATION

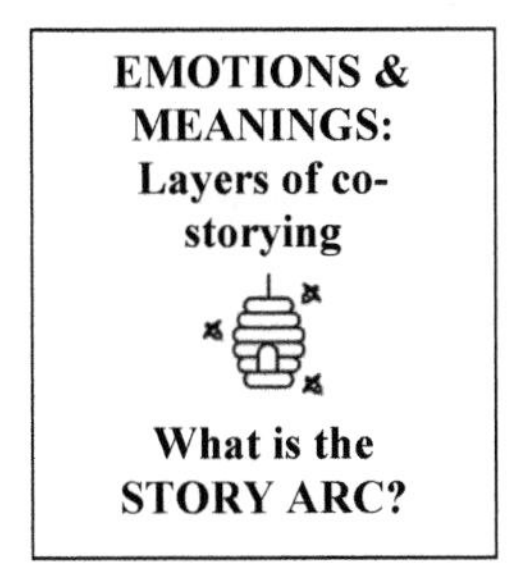

When I look back over the micro-stories of the last while in our therapeutic relationship, what trends/patterns do I see?

RELATIONAL EMPATHY

Situated awareness of emotion co-storying

(COLLABORATIVE) EXPLORATION

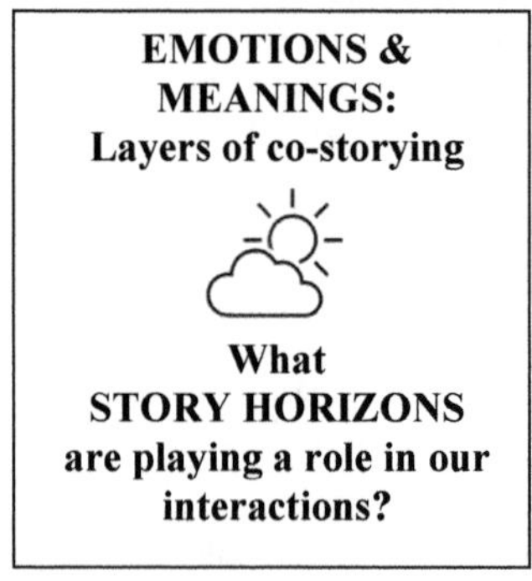

What stories have we brought with us?

What emotion stories do I bring with me into my encounters with this client?
What emotion stories does this client bring with them into their encounters with me?
How do these influence the emotion stories that play out between us in the moment?
What emotion stories are prominent in this context within which we have our sessions?

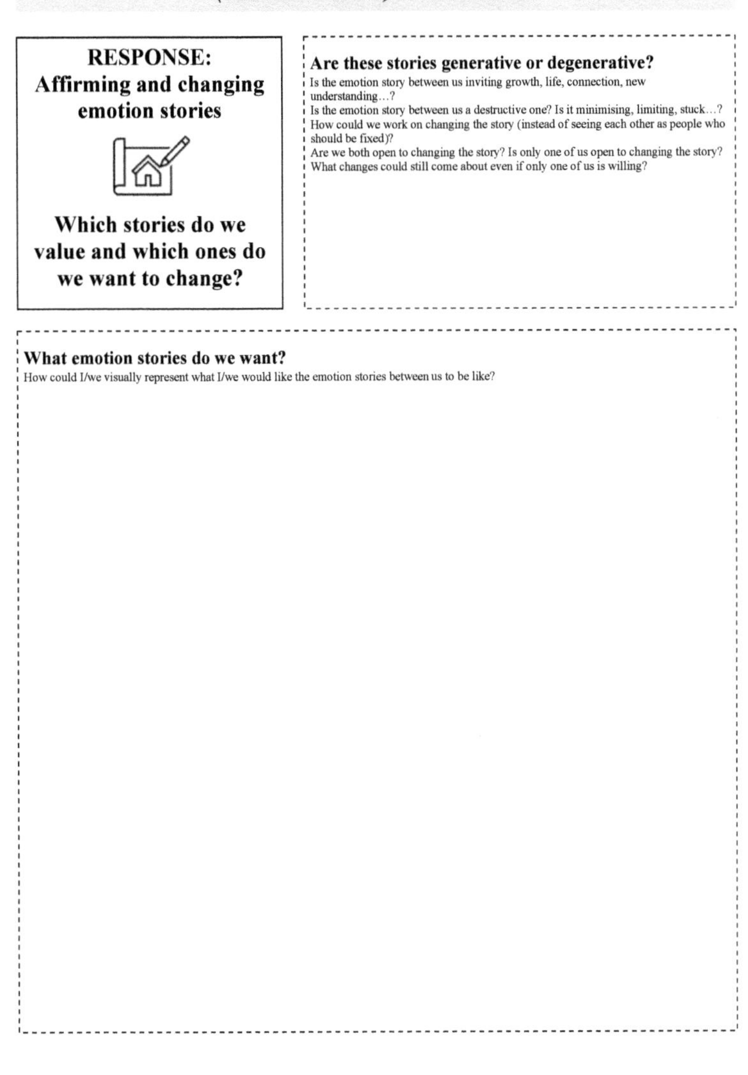

RELATIONAL EMPATHY

Situated awareness of emotion co-storying

(COLLABORATIVE) EXPLORATION

RESPONSE: Affirming and changing emotion stories

Which stories do we value and which ones do we want to change?

Are these stories generative or degenerative?

Is the emotion story between us inviting growth, life, connection, new understanding…?
Is the emotion story between us a destructive one? Is it minimising, limiting, stuck…?
How could we work on changing the story (instead of seeing each other as people who should be fixed)?
Are we both open to changing the story? Is only one of us open to changing the story?
What changes could still come about even if only one of us is willing?

What emotion stories do we want?

How could I/we visually represent what I/we would like the emotion stories between us to be like?

References

Abbott, E. A. (2018). Subjective observation in music therapy: A study of student practicum Logs. *Music Therapy Perspectives, 36*(1), 117–126.

Abrahams, D., & Rohleder, P. (2021). *A clinical guide to psychodynamic psychotherapy*. Routledge.

Abrams, B. (2010). Evidence-based music therapy practice: An integral understanding. *Journal of Music Therapy, 47*(4), 351–379.

Abrams, B. (2018). Understanding humanistic dimensions of music therapy: Editorial introduction. *Music Therapy Perspectives, 36*(2), 139–143.

Adams, D., & Oliver, C. (2011). The expression and assessment of emotions and internal states in individuals with severe or profound intellectual disabilities. *Clinical Psychology Review, 31*(3), 293–306.

Adams, E., Lee, A., Pritchard, C., & White, R. (2010). What stops us from healing the healers: A survey of help-seeking behavior, stigmatisation and depression within the medical profession. *International Journal of Social Psychiatry, 56*(4), 359–370.

Adams, G., Dobles, I., Gómez, L. H., Kurtiş, T., & Molina, L. E. (2015). Decolonizing psychological science: Introduction to the special thematic section. *Journal of Social and Political Psychology, 3*(1), 213–238.

A. dos Santos, *Empathy Pathways*,
https://doi.org/10.1007/978-3-031-08556-7

Adhikari, M. (2005). Contending approaches to coloured identity and the history of the coloured people of South Africa. *History Compass, 3*(1), 1–16.

Adkins, B. (2018). To have done with the transcendental: Deleuze, immanence, Intensity. *The Journal of Speculative Philosophy, 32*(3), 533–543.

Ağören, G. C. (2021). Understanding depressive feelings as situated affections. *Emotion Review*, 17540739211057846.

Agres, K. R., Foubert, K., & Sridhar, S. (2021). Music therapy during COVID-19: Changes to the practice, use of technology, and what to carry forward in the future. *Frontiers in Psychology, 12*, 1317.

Ahmed, S. (2014). *The cultural politics of emotion* (2nd ed.). Edinburgh University Press.

Ahonen, H., & Mongillo Desideri, A. (2014). Heroines' journey-emerging story by refugee women during group analytic music therapy. *Voices: A World Forum for Music Therapy, 14*(1). https://doi.org/10.15845/voices.v14i1.686

Aigen, K. (1999). The true nature of music-centred music therapy theory. *British Journal of Music Therapy, 13*(2), 77–82.

Aigen, K. (2005). *Being in music: Foundations of Nordoff-Robbins music therapy.* Barcelona Publishers.

Aigen, K. (2014). Music-centered dimensions of Nordoff-Robbins music therapy. *Music Therapy Perspectives, 32*(1), 18–29.

Aigen, K. (2015). A critique of evidence-based practice in music therapy. *Music Therapy Perspectives, 33*(1), 12–24.

Aigen, K. (2018). Community music therapy. In G. McPherson & G. Welch (Eds.), *Special needs, community music, and adult learning* (pp. 138–154). Oxford University Press.

Aigen, K. S. (2012). Community music therapy. In G. E. McPherson & G. F. Welch (Eds.), *The Oxford handbook of music education* (Vol. 2, pp. 138–154). Oxford University Press.

Akgün, A. E., Keskin, H., Ayar, H., & Erdoğan, E. (2015). The influence of storytelling approach in travel writings on readers' empathy and travel intentions. *Procedia-Social and Behavioral Sciences, 207*, 577–586.

Al'tman, Y. A., Alyanchikova, Y. O., Guzikov, B. M., & Zakharova, L. E. (2000). Estimation of short musical fragments in normal subjects and patients with chronic depression. *Human Physiology, 26*(5), 553–557.

Aldridge, D. (1989). A phenomenological comparison of the organization of music and the self. *Arts in Psychotherapy, 16*(2), 91–97.

Aldridge, D. (2000). *Spirituality, healing and medicine: Return to the silence.* Jessica Kingsley.

Alifuoco, A. (2017). 'Alive' performance: Toward an immersive activist philosophy. *Performance Philosophy, 3*(1), 126–145.
Alldred, P., & Fox, N. J. (2015). The sexuality-assemblages of young men: A new materialist analysis. *Sexualities, 18*(8), 905–920.
Almanna, A., & Gu, C. (2021). *Translation as a set of frames*. Routledge.
Alvarez, T. T. (2012). Beats, rhymes and life: Rap therapy in an urban setting. In S. Hadley & G. Yancy (Eds.), *Therapeutic uses of rap and hip-hop* (pp. 117–128). Routledge.
American Psychiatric Association. (2000). *Diagnostic and statistical manual of mental disorders* (4th ed., text rev.). American Psychiatric Association.
American Psychiatric Association. (2013). *Diagnostic and statistical manual of mental disorders* (5th ed.). American Psychiatric Association.
Amiel-Houser, T., & Mendelson-Maoz, A. (2015). Against empathy: Levinas and ethical criticism in the 21st century. *Journal of Literary Theory, 8*(1), 199–218.
Amir, D. (1990). A song is born: Discovering meaning in improvised songs through a phenomenological analysis of two music therapy sessions with a traumatic spinal cord injured young adult. *Music Therapy, 9*(1), 62–81.
Anderson, B., Kearnes, M., McFarlane, C., & Swanton, D. (2012). On assemblages and geography. *Dialogues in Human Geography, 2*(2), 171–189.
Ansdell, G. (1995). *Music for life.* Jessica Kingsley.
Ansdell, G. (2003). Community music therapy: Big British balloon or future international trend? In *Community, relationship and spirit: Continuing the dialogue and debate* (pp. 1–13). BSMT Publications.
Ansdell, G. (2014a). *How music helps in music therapy and everyday life.* Ashgate.
Ansdell, A. (2014b). Yes, but, No, but: A contrarian response to Cross. *Psychology of Music, 42*(6) 820–825.
Ansdell, G. (2016). *How music helps in music therapy and everyday life.* Routledge.
Ansdell, G. (2017). Reflection belonging through musicing: Explorations of musical community. In B. Stige, G. Ansdell, C. Elefant, & M. Pavlicevic (Eds.), *Where music helps: Community music therapy in action and reflection* (pp. 41–62). Routledge.
Ansdell, G., & DeNora, T. (2016). *Musical pathways in recovery: Community music therapy and mental wellbeing.* Routledge.
Ansdell, G., & Pavlicevic, M. (2005). Musical companionship, musical community: Music therapy and the process and value of musical communication. In D. Miell, R. MacDonald, & D. Hargreaves (Eds.), *Musical communication* (pp. 193–214). Oxford University Press.

Ansdell, G., & Stige, B. (2015). Community music therapy. In J. Edwards (Ed.), *Oxford handbook of music therapy* (pp. 595–621). Oxford University Press.

Ansdell, G., & Stige, B. (2018). Can music therapy still be humanist? *Music Therapy Perspectives, 36*(2), 175–182.

Apperly, I. (2008). Beyond simulation–theory and theory–theory: Why social cognitive neuroscience should use its own concepts to study "Theory of Mind." *Cognition, 107*(1), 266–283.

Apperly, I. (2011). *Mindreaders: The cognitive basis of "theory of mind."* Psychology Press.

Aragona, M., Kotzalidis, G., & Puzella, A. (2013). The many faces of empathy: Between phenomenology and neuroscience. *Archives of Psychiatry and Psychotherapy, 4*, 5–12.

Archer, M. (2015). The relational subject and the person: Self, agent, and actor. In P. Donati & M. Archer (Eds.), *The relational subject* (pp. 85–122). Cambridge University Press.

Archer, M., & Donati, M. (2015). The plural subject versus the relational subject. In P. Donati & M. Archer (Eds.), *The relational subject* (pp. 33–78). Cambridge University Press.

Arnason, C. (2002). An eclectic approach to the analysis of improvisations in music therapy sessions. *Music Therapy Perspectives, 20*(1), 4–12.

Arnason, C. (2003). Music therapists' listening perspectives in improvisational music therapy. *Nordic Journal of Music Therapy, 12*(2), 124–138.

Arndt, N., & Naudé, L. (2016). Contrast and contradiction: Being a black adolescent in contemporary South Africa. *Journal of Psychology in Africa, 26*(3), 267–275.

Argstatter, H. (2016). Perception of basic emotions in music: Culture-specific or multicultural? *Psychology of Music, 44*(4), 674–690.

Arteaga Pérez, I. (2020). Emotion work during colorectal cancer treatments. *Medical Anthropology*, 1–13.

Atay, A. (2018, March). *A relational ontology for peace* [Keynote lecture]. International Conference on Cultural Studies, Istanbul.

Atwood, G. E. (2012). The abyss of madness and human understanding. *Pragmatic Case Studies in Psychotherapy, 8*(1), 49–59.

Aubé, W., Angulo-Perkins, A., Peretz, I., Concha, L., & Armony, J. L. (2015). Fear across the senses: Brain responses to music, vocalizations and facial expressions. *Social Cognitive and Affective Neuroscience, 10*(3), 399–407.

Aucoin, E., & Kreitzberg, E. (2018). Empathy leads to death: Why empathy is an adversary of capital defendants. *Santa Clara Law Review, 58*, 99–136.

Austin, D. (1996). The role of improvised music in psychodynamic music therapy with adults. *Music Therapy, 14*(1), 29–43.
Austin, D. (2001). In search of the self: The use of vocal holding techniques with adults traumatized as children. *Music Therapy Perspectives, 19*(1), 22–30.
Austin, D. (2010). When the bough breaks: Vocal psychotherapy and traumatized adolescents. In K. Stewaer (Ed.), *Music therapy and trauma: Bridging theory and clinical practice* (pp. 176–187). Satchnote Press.
Aveline, M. (1990). The training and supervision of individual therapists. In W. Dryden (Ed.), *Individual therapy: A handbook* (pp. 434–466). Open University Press.
Avramides, A. (2001). *Other minds.* Routledge.
Ayvazian, A. (2010). Interrupting the cycle: The role of allies as agents of change. In M. Adams, W. J. Blumenfeld, C. R. Castañeda, H. W. Hackman, M. L. Peters, & X. Zúñiga (Eds.), *Readings for diversity and social justice* (2nd ed., pp. 625–628). Routledge.
Bachelor, A. (1988). How clients perceive therapist empathy: A content analysis of "received" empathy. *Psychotherapy, 25*, 227–240.
Baez, S., Manes, F., Huepe, D., Torralva, T., Fiorentino, N., Richter, F., Huepe-Artigas, D., Ferrari, J., Montanes, P., Reyes, P., Matallana, D., Vigliecca, N., Decety, J., & Ibanez, A. (2014). Primary empathy deficits in frontotemporal dementia. *Frontiers in Aging Neuroscience, 6*, article 264.
Baker, F. (2014). *Therapeutic songwriting: Developments in theory, methods, and practice.* Palgrave Macmillan.
Baker, T. A., Clay, O. J., Johnson-Lawrence, V., Minahan, J. A., Mingo, C. A., Thorpe, R. J., Ovalle, F., & Crowe, M. (2017). Association of multiple chronic conditions and pain among older black and white adults with diabetes mellitus. *BMC Geriatrics, 17*(1), 255–259.
Bakhtin, M. (1984). *Problems of Dostoevsky's poetics* (C. Emerson, Trans.). University of Minnesota Press.
Bakker, A. B., Le Blanc, P. M., & Schaufeli, W. B. (2005). Burnout contagion among intensive care nurses. *Journal of Advanced Nursing, 51*(3), 276–287.
Bakker, A. B., Schaufeli, W. B., Sixma, H. J., & Bosveld, W. (2001). Burnout contagion among general practitioners. *Journal of Social and Clinical Psychology, 20*(1), 82–98.
Bal, P. M., & Veltkamp, M. (2013). How does fiction reading influence empathy? An experimental investigation on the role of emotional transportation. *PLoS ONE, 8*(1), e55341.

Bamford, J. M. S., & Davidson, J. W. (2017). Trait empathy associated with agreeableness and rhythmic entrainment in a spontaneous movement to music task: Preliminary exploratory investigations. *Musicae Scientiae, 23*(1), 5–24.

Barad, K. (2003). Posthumanist performativity: Toward an understanding of how matter comes to matter. *Signs: Journal of Women in Culture, 28*(3), 801–831.

Barad, K. (2007). *Meeting the universe halfway: Quantum physics and the entanglement of matter and meaning.* Duke University Press.

Barad, K. (2010). Quantum entanglements and hauntological relations of inheritance: Dis/continuities, spacetime enfoldings, and justice-to-come. *Derrida Today, 3*(2), 240–268.

Barad, K. (2014). Diffracting diffraction: Cutting together-apart. *Parallax, 20*, 168–187.

Barad, K. (2015). Transmaterialities: Trans*/matter/realities and queer political imaginings. *GLQ: A Journal of Lesbian and Gay Studies, 21*(2–3), 387–422.

Barrett, L. F. (2017). *How emotions are made: The secret life of the brain.* Houghton Mifflin Harcourt.

Barrett-Lennard, G. T. (1981). The empathy cycle: Refinement of a nuclear concept. *Journal of Counseling Psychology, 28*, 91–100.

Barsness, R. E. (2017). *Core competencies of relational psychoanalysis: A guide to practice, study and research.* Routledge.

Basch, M. F. (1988). *Understanding psychotherapy: The science behind the art.* Basic Books.

Bateson G. (1991). *A sacred unity: Further steps to an ecology of mind.* Harper Collins.

Batson, C. (2009). These things called empathy: Eight related but distinct phenomena. In J. Decety & W. Ickes (Eds.), *The social neuroscience of empathy* (pp. 3–16). The MIT Press.

Batson, C. (2011). *Altruism in humans.* Oxford University Press.

Batson, C. (2017). The empathy-altruism hypothesis: What and so what? In E. Seppala, E. Simon-Thomas, S. Brown, M. Worline, C. Cameron, & J. Doty (Eds.), *The Oxford handbook of compassion science* (pp. 56–73). Oxford University Press.

Batson, C., Early, S., & Salvarini, G. (1997). Perspective taking: Imagining how another feels versus imagining how you would feel. *Personality and Social Psychology Bulletin, 23*, 751–758.

Batson, C., Klein, T. R., Highberger, L., & Shaw, L. L. (1995). Immorality from empathy-induced altruism: When compassion and justice conflict. *Journal of Personality and Social Psychology, 68*(6), 1042.
Baumeister, J. C., Papa, G., & Foroni, F. (2016). Deeper than skin deep: The effect of botulinum toxin-A on emotion processing. *Toxicon, 118*, 86–90.
Bayram, A. B., & Holmes, M. (2020). Feeling their pain: Affective empathy and public preferences for foreign development aid. *European Journal of International Relations, 26*(3), 820–850.
Beagan, B. L. (2018). A critique of cultural competence: Assumptions, limitations, and alternatives. In C. Frisby & W. O'Donohue (Eds.), *Cultural competence in applied psychology* (pp. 123–138). Springer.
Behrens, G. A. (2012). Use of traditional and nontraditional instruments with traumatized children in Bethlehem, West Bank. *Music Therapy Perspectives, 30*(2), 196–202.
bell, h. (2018). Creative interventions for teaching empathy in the counseling classroom. *Journal of Creativity in Mental Health, 13*(1), 106–120.
Benedetti, F., Bernasconi, A., Bosia, M., Cavallaro, R., Dallaspezia, S., Falini, A., Poletti, S., Radaelli, D., Riccaboni, R., Scotti, G., & Smeralido, E. (2009). Functional and structural brain correlates of theory of mind and empathy deficits in schizophrenia. *Schizophrenia Research, 114*, 154–160.
Benjamin, J. (1988). *The bonds of love*. Pantheon Books.
Benjamin, J. (1988). *The bonds of love: Psychoanalysis, feminism, and the problem of domination.* Pantheon.
Benjamin, J. (1995). *Like subjects, love objects.* Yale University Press.
Benjamin, J. (2004). Beyond doer and done to: An intersubjective view of thirdness. *The Psychoanalytic Quarterly, 73*(1), 5–46.
Bennett, J. (2010). *Vibrant matter: A political ecology of things.* Duke University Press.
Bennett, J. (2015). Creativities in popular songwriting curricula: Teaching or learning? In P. Burnard & L. Haddon (Eds.), *Activating diverse musical creativities: Teaching and learning in higher music education* (pp. 186–199). Bloomington.
Bennett, M. (2001). *The empathic healer: An endangered species.* Academic Press.
Bensimon, M. (2020). Relational needs in music therapy with trauma victims: The perspective of music therapists. *Nordic Journal of Music Therapy, 29*(3), 240–254.
Berger, C., Batanova, M., & Cance, J. D. (2015). Aggressive and prosocial? Examining latent profiles of behavior, social status, machiavellianism, and empathy. *Journal of Youth and Adolescence, 44*, 2230–2244.

Bericat, E. (2016). The sociology of emotions: Four decades of progress. *Current Sociology, 64*(3), 491–513.
Berlant, L. (2004). Introduction: Compassion (and withholding). In L. Berlant (Ed.), *Compassion: The culture and politics of an emotion* (pp. 1–14). Routledge.
Berman, A. (2014). Post-traumatic victimhood and group analytic therapy: Intersubjectivity, empathic witnessing and otherness. *Group Analysis, 47*(3), 242–256.
Bersani, G., Bersani, F. S., Valeriani, G., Robiony, M., Anastasia, A., Colletti, C., Liberati, L., Capra, E., Quartini, A., & Polli, E. (2012). Comparison of facial expression in patients with obsessive-compulsive disorder and schizophrenia using the Facial Action Coding System: A preliminary study. *Neuropsychiatric Disease and Treatment, 8*, 537.
Berscheid, E., & Ammazzalorso, H. (2001). Emotional experience in close relationships. In G. Fletcher & M. Clark (Eds.), *Blackwell handbook of social psychology: Interpersonal processes*, (pp. 308–330). Blackwell.
Biehl, J., & Locke, P. (2010). Deleuze and the anthropology of becoming. *Current Anthropology, 51*(3), 317–351.
Bird-David, N. (1999). "Animism" revisited: Personhood, environment, and relational epistemology. *Current Anthropology, 40*(S1), S67–S91.
Birnbaum, J. (2014). Intersubjectivity and Nordoff-Robbins music therapy. *Music Therapy Perspectives, 32*(1), 30–37.
Bishop, A. (2002). *Becoming an ally: Breaking the cycle of oppression in people* (2nd ed.). Fernwood.
Black, S., Bartel, L., & Rodin, G. (2020). Exit music: The experience of music therapy within medical assistance in dying. *Healthcare, 8*(3), 331–342.
Black, S., & Zimmermann, C., & Rodin, G. (2017). Comfort, connection and music: Experiences of inter-active listening on a palliative care unit. *Music & Medicine, 9*(4), 227–233.
Blacking, J. (1973). *How musical is man?* University of Washington Press.
Blair, R. (2007). Empathic dysfunction in psychopathic individuals. In T. Farrow & P. Woodruff (Eds.), *Empathy in mental illness* (pp. 3–16). Cambridge University Press.
Blandon, A. Y., Calkins, S. D., Keane, S. P., & O'Brien, M. (2008). Individual differences in trajectories of emotion regulation processes: The effects of maternal depressive symptomatology and children's physiological regulation. *Developmental Psychology, 44*, 1110–1123.

Blankenburg, W. (2001). First steps toward a psychopathology of 'common sense' (A. L. Mishara, Trans.). *Philosophy, Psychiatry & Psychology, 8*, 303–315.

Blaxter, M. (2010). *Health*. Polity Press.

Blood, A. J., & Zatorre, R. J. (2001). Intensely pleasurable responses to music correlate with activity in brain regions implicated in reward and emotion. *Proceedings of the National Academy of Sciences, 98*(20), 11818–11823.

Bloom, P. (2017). *Against empathy: The case for rational compassion*. Random House.

Bodner, E., Iancu, I., Gilboa, A., Sarel, A., Mazor, A., & Amir, D. (2007). Finding words for emotions: The reactions of patients with major depressive disorder towards various musical excerpts. *The Arts in Psychotherapy, 34*(2), 142–150.

Bogost, I. (2012). *Alien phenomenology, or, what it's like to be a thing*. University of Minnesota Press.

Boiger, M., & Mesquita, B. (2012). The construction of emotion in interactions, relationships, and cultures. *Emotion Review, 4*(3), 221–229.

Boler, M. (1997). The risks of empathy: Interrogating multiculturalism's gaze. *Cultural Studies, 11*(2), 253–273.

Bolger, L. (2015). Being a player: Understanding collaboration in participatory music projects with communities supporting marginalised young people. *Qualitative Inquiries in Music Therapy, 10*(3), 77–126.

Bolger, L., McFerran, K. S., & Stige, B. (2018). Hanging out and buying in: Rethinking relationship building to avoid tokenism when striving for collaboration in music therapy. *Music Therapy Perspectives, 36*(2), 257–266.

Bolton, G., & Ockenfels, A. (2000). ERC: A theory of equity, reciprocity, and competition. *American Economic Review, 90*, 166–193.

Bonde, L. O. (1997). Music analysis and image potentials in classical music. *Nordic Journal of Music Therapy, 7*(2), 121–128.

Bondi, L., Davidson, J., & Smith, M. (2007). Introduction: Geography's "emotional turn." In L. Bondi, J. Davidson, & M. Smith (Eds.), *Emotional geographies* (pp. 1–18). Ashgate.

Bopp, J., Bopp, M., Brown, L., Lane, P., & Morris, P. (1989). *Sacred tree*. Four Worlds International Institute.

Botvinick, M., Jha, A. P., Bylsma, L. M., Fabian, S. A., Solomon, P. E., & Prkachin, K. M. (2005). Viewing facial expressions of pain engages cortical areas involved in the direct experience of pain. *NeuroImage, 25*, 315–319.

Boyd, C., Hayes, L., Wilson, R., & Bearsley-Smith, C. (2008). Harnessing the social capital of rural communities for youth mental health: An asset-based community development framework. *Australian Journal of Rural Health, 16*(4), 189–193.

Boyd, M. P. (2016). Calling for response-ability in our classrooms. *Language Arts, 93*(3), 226.

Bozalek, V. (2020). Rendering each other capable: Doing response-able research responsibly. In K. Murris (Ed.), *Navigating the postqualitative, new materialist and critical posthumanist terrain across disciplines* (pp. 135–149). Routledge.

Bozalek, V., Bayat, A., Gachago, D., Motala, S., & Mitchell, V. (2018). A pedagogy of response-ability. In V. Bozalek, R. Braidotti, M. Zembylas, & Shefer, T. (Eds.), *Socially just pedagogies in higher education: Critical posthumanist and new feminist materialist perspectives* (pp. 97–112). Bloomsbury Academic.

Brabec de Mori, B., & Seeger, A. (2013). Introduction: Considering music, humans, and non-humans. *Ethnomusicology Forum, 22*(3), 269–286.

Brach. T. (2003). *Radical acceptance: Embracing your life with the heart of a Buddha.* Bantam Books.

Bradford, D., & Robin, C. (2021). *Connect*. Penguin Random House.

Braidotti, R. (2011). *Nomadic theory: The portable Rosi Braidotti.* Columbia University Press.

Braidotti, R. (2013). *The Posthuman.* Polity Press.

Braun, C. (1976). Teacher expectation: Sociopsychological dynamics. *Review of Educational Research, 46*(2), 185–213.

Breithaupt, F. (2019). *The dark sides of empathy*. Cornell University Press.

Brennan, T. (2004). *The transmission of affect.* Cornell University Press.

Bresler, L. (2008). The music lesson. In J. G. Knowles & A. L. Cole (Eds.), *Handbook of the arts in qualitative research* (pp. 225–237). Sage.

Bresler, L. (2013). Cultivating empathic understanding in research and teaching. In B. White & T. Costantino (Eds.), *Aesthetics, empathy and education* (pp. 9–28). Peter Lang.

Brin, D. (2012). Self-addiction and self-righteousness. In B. Oakley, A. Knafo, G. Madhavan, & D. Sloan Wilson (Eds.), *Pathological altruism* (pp. 77–84). Oxford University Press.

British Psychological Society (BPS). (2014). *Understanding psychosis and schizophrenia.* British Psychological Society. https://www1.bps.org.uk/system/files/Public%20files/rep03_understanding_psychosis.pdf

Brons, L. L. (2015). Othering, an analysis. *Transcience, A Journal of Global Studies, 6*(1), 69–90.

Brown, B. (2008). *I thought it was just me (but it isn't): Making the journey from "what will people think?" to" I am enough"*. Avery.
Brown, B. (2021). *Atlas of the heart*. Random House.
Brown, J. M. (2002, March). Towards a culturally centered music therapy practice. *Voices: A world forum for music therapy, 2*(1). https://voices.no/index.php/voices/article/view/1601
Brown, S. E. (2002b). What is disability culture? *Disability Studies Quarterly, 22*(2), 34–50.
Brown, S. (2013). Hullo object! I destroyed you! In L. Bunt, S. Hoskyns, & S. Swami (Eds.), *The handbook of music therapy* (pp. 104–116). Routledge.
Bruneau, E. G., Cikara, M., & Saxe, R. (2017). Parochial empathy predicts reduced altruism and the endorsement of passive harm. *Social Psychological and Personality Science, 8*, 934–942.
Bruscia, K. (1987). *Improvisational models of music therapy*. Thomas.
Bruscia, K. (1998). *The dynamics of music psychotherapy*.
Bruscia, K. (2014a). *Defining music therapy* (3rd ed.). Barcelona.
Bruscia, K. (2014b). *Self-experiences in music therapy education training, and supervision*.
Bruscia, K. E. (2001). A qualitative approach to analyzing client improvisations. *Music Therapy Perspectives, 19*(1), 7–21.
Bubandt, N., & Willerslev, R. (2015). The dark side of empathy: Mimesis, deception, and the magic of alterity. *Comparative Studies in Society and History, 57*(1), 5–34.
Buber, M. (1923/1970). *I and Thou* (W. Kaufmann, Trans.). Touchstone.
Buber, M. (1947/2002). *Between man and man* (R. Smith, Trans.). Routledge.
Buber, M. (1958/2000). *I and thou* (R. Smith, Trans.). Scribner.
Buber, M. (1965). *The knowledge of man*. Harper and Row.
Buber, M. (1969). *A believing humanism: Gleanings*. Simon and Schuster.
Buber, M. (1990). *A believing humanism: My testament* (M. S. Friedman, Trans.). Humanities Press.
Buffone, A. E., & Poulin, M. J. (2014). Empathy, target distress, and neurohormone genes interact to predict aggression for others–even without provocation. *Personality and Social Psychology Bulletin, 40*(11), 1406–1422.
Bullard, E. (2011). Music therapy as an intervention for inpatient treatment of suicidal ideation. *Qualitative Inquiries in Music Therapy, 6*, 75–121.
Bunt, L., & Hoskyns, S. (2002). Practicalities principles of and music basic therapy. In L. Bunt, S. Hoskyns, & S. Swami (Eds.), *The handbook of music therapy* (pp. 19–45). Taylor & Francis.

Burland, K., & Magee, W. (2014). Developing identities using music technology in therapeutic settings. *Psychology of Music, 42*(2), 177–189.

Burnard, P., Hasslen, L., Jong, O., & Murphy, L. (2015). The imperative of diverse and distinctive musical creativities as practices of social justice. In C. Benedict, P. Schmidt, G. Spruce & P. Woodford (Eds.), *The Oxford handbook of social justice in music education* (pp. 357–371). Oxford University Press.

Butler, E. A. (2017). Emotions are temporal interpersonal systems. *Current Opinion in Psychology, 17*, 129–134.

Butler, J. (1993). *Bodies that matter*. Routledge.

Caldwell, C., & Atwijuka, S. (2018). "I see you!"—The Zulu insight to caring leadership. *The Journal of Values-Based Leadership, 11*(1), 13.

Calhoun, L., & Tedeschi, R. (2006). Expert companions: Posttraumatic growth in clinical practice. In L. Calhoun & R. Tedeschi (Eds.), *Handbook of posttraumatic growth* (pp. 291–310). Psychology Press.

Campbell-Sills, L., Barlow, D., Brown, T., & Hofmann, S. (2006). Acceptability and suppression of negative emotion in anxiety and mood disorders. *Emotion, 6*, 587–595.

Camras, L. A., & Witherington, D. C. (2005). Dynamical systems approaches to emotional development. *Developmental Review, 25*, 328–350.

Capan, Z. G., dos Reis, F., & Grasten, M. (2021). The politics of translation in international relations. In Z. G. Capan, F. dos Reis, & M. Grasten (Eds.), *The politics of translation in international relations* (pp. 1–19). Palgrave Macmillan.

Carlson, L. (2016). Encounters with musical others. In L. Carlson & P. Costello (Eds.), *Phenomenology and the arts* (pp. 235–252). Lexington Books.

Carlson, S. J., Levine, L. J., Lench, H. C., Flynn, E., Carpenter, Z. K., Perez, K. A., & Bench, S. W. (2021). You shall go forth with joy: Religion and aspirational judgments about emotion. *Psychology of Religion and Spirituality*, Advance online publication. https://doi.org/10.1037/rel000032

Carpenter, J. M., Green, M. C., & Vacharkulksemsuk, T. (2016). Beyond perspective-taking: Mind-reading motivation. *Motivation and Emotion, 40*, 358–374.

Carr, D. (2000). Emotional intelligence, PSE and self-esteem: A cautionary note. *Pastoral Care in Education, 18*(3), 27–33.

Carrera, P., Oceja, L., Caballero, A., Muñoz, D., López-Pérez, B., & Ambrona, T. (2013). I feel so sorry! Tapping the joint influence of empathy and personal distress on helping behavior. *Motivation and Emotion, 37*(2), 335–345.

Carruthers, P. (2009). How we know our own minds: The relationship between mindreading and metacognition. *Behavioral and Brain Sciences, 32*(2), 121–182.

Cassidy, J., Stern, J. A., Mikulincer, M., Martin, D. R., & Shaver, P. R. (2018). Influences on care for others: Attachment security, personal suffering, and similarity between helper and care recipient. *Personality and Social Psychology Bulletin, 44*(4), 574–588.

Cattell, V., Dines, N., Gesler, W., & Curtis, S. (2008). Mingling, observing, and lingering: Everyday public spaces and their implications for wellbeing and social relations. *Health and Place, 14*(3), 544–561.

Chaitin, J., & Steinberg, S. (2008). "You should know better": Expressions of empathy and disregard among victims of massive social trauma. *Journal of Aggression, Maltreatment & Trauma, 17*(2), 197–226.

Chan, G. (2014). Cross-cultural music therapy in community aged-care: A case vignette of a CALD elderly woman. *Australian Journal of Music Therapy, 25*, 92–102.

Chilisa, B., Major, T. E., & Khudu-Petersen, K. (2017). Community engagement with a postcolonial, African-based relational paradigm. *Qualitative Research, 17*(3), 326–339.

Chinn, D. (2007, October). Reflection and reflexivity. *Clinical Psychology Forum, 178*.

Christensen, J. (2018). *Sounds and the aesthetics of play: A musical ontology of constructed emotions.* Palgrave Macmillan.

Church, J. (2012). Sustainable development and the culture of uBuntu. *De Jure Law Journal, 45*(3), 511–531.

Cialdini, R. B., Brown, S. L., Lewis, B. P., Luce, C., & Neuberg, S. L. (1997a). Reinterpreting the empathy–altruism relationship: When one into one equals oneness. *Journal of Personality and Social Psychology, 73*(3), 481.

Cialdini, R. B., Brown, S. L., Luce, C., Sagarin, B. J., & Lewis, B. P. (1997b). Does empathy lead to anything more than superficial helping? Comment on Batson et al. (1997b). *Journal of Personality and Social Psychology, 73*(3), 510–516.

Ciaunica, A. (2019). The 'meeting of bodies': Empathy and basic forms of shared experience. *Topoi, 38*, 185–195.

Cikara, M., Bruneau, E. G., & Saxe, R. R. (2011). Us and them: Intergroup failures of empathy. *Current Directions in Psychological Science, 20*(3), 149–153.

Clark, A., & Chalmers, D. (1998). The extended mind. *Analysis, 58*, 7–19.
Clark, J. (2000). *Beyond empathy: An ethnographic approach to cross-cultural social work practice.* Unpublished manuscript, Faculty of Social Work, University of Toronto.
Clarke, A. (2005). *Situational analysis.* Sage.
Clarke, A. E. (2016a). Introducing situational analysis. In A. E. Clarke, C. Friese, & R. Washburn (Eds.), *Situational analysis in practice* (pp. 11–76). Routledge.
Clarke, A. E. (2016b). From grounded theory to situational analysis: What's new? Why? How? In A. E. Clarke, C. Friese, & R. Washburn (Eds.), *Situational analysis in practice* (pp. 84–118). Routledge.
Clements-Cortés, A. (2010). The role of music therapy in facilitating relationship completion in end-of-life care. *Canadian Journal of Music Therapy, 16*(1).
Clements-Cortés, A. (2013). Burnout in music therapists: Work, individual, and social factors. *Music Therapy Perspectives, 31*(2), 166–174.
Clohesy, A. (2013). *Politics of empathy: Ethics, solidarity, recognition.* Routledge.
Clough, P. T. (2004). Future matters: Technoscience, global politics, and cultural criticism. *Social Text, 22*(3), 1–23.
Cochran, J. L., & Cochran, N. H. (2015). *The heart of counseling* (2nd ed.). Routledge.
Cohen, A. B., & Rozin, P. (2001). Religion and the morality of mentality. *Journal of Personality and Social Psychology, 81*(4), 697–710.
Cole, J. (2009). Impaired embodiment and intersubjectivity. *Phenomenology and the Cognitive Sciences, 8*(3), 343–360.
Cole, J., & Spalding, H. (2008). *The invisible smile.* Oxford University Press.
Coleman, R., & Ringrose, J. (2013). Introduction: Deleuze and research methodologies. In R. Coleman & J. Ringrose (Eds.), *Deleuze and research methodologies* (pp. 1–22). Edinburgh University Press.
Colle, L., Becchio, C., & Bara, B. G. (2008). The non-problem of the other minds: A neurodevelopmental perspective on shared intentionality. *Human Development, 51*(5–6), 336–348.
Collins, R. (2004). *Interaction ritual chains.* Princeton University Press.
Colombetti, G., & Roberts, T. (2015). Extending the extended mind: The case for extended affectivity. *Philosophical Studies, 172*(5), 1243–1263.
Comings, D. E. (1990). *Tourette syndrome and human behaviour.* Hope Press.
Comte, R. (2016). Neo-colonialism in music therapy: A critical interpretive synthesis of the literature concerning music therapy practice with refugees. *Voices: A World Forum for Music Therapy, 16*(3).

Cooke, C., & Colucci-Gray, L. (2019). Complex knowing: Promoting response-ability within music and science teacher education. In C. Taylor & A. Bayley (Eds.), *Posthumanism and higher education* (pp. 165–185). Palgrave Macmillan.

Cooper, M. (2010). Clinical-musical responses of Nordoff-Robbins music therapists: The process of clinical improvisation. *Qualitative Inquiries in Music Therapy, 5*, 86–115.

Coplan, A. (2011). Will the real empathy please stand up? A case for a narrow conceptualization. *The Southern Journal of Philosophy, 49*, 40–65.

Coplan, A., & Goldie, P. (2011). Introduction. In A. Coplan & P. Goldie (Eds.), *Empathy: Philosophical and psychological perspectives* (pp. ix–xlvii). Oxford University Press.

Corey, G. (2005). *Theory and practice of counselling and psychotherapy* (7th ed.). Brooks/Cole—Thomson Learning.

Corrigall, K., & Schellenberg, E. (2013). Music: The language of emotion. In C. Mohiyeddini, E. Eysenck, & S. Bauer (Eds.), *Handbook of psychology of emotions* (pp. 299–325). Nova Science Publishers.

Couette, M., Mouchabac, S., Bourla, A., Nuss, P., & Ferreri, F. (2020). Social cognition in post-traumatic stress disorder: A systematic review. *British Journal of Clinical Psychology, 59*(2), 117–138.

Country, B., Wright, S., Suchet-Pearson, S., Lloyd, K., Burarrwanga, L., Ganambarr, R., Ganambarr-Stubbs, M., Ganambarr, B., Maymuru, D., & Sweeney, J. (2016). Co-becoming Bawaka: Towards a relational understanding of place/space. *Progress in Human Geography, 40*(4), 455–475.

Creech, A., Hallam, S., Varvarigou, M., McQueen, H., & Gaunt, H. (2013). Active music making: A route to enhanced subjective well-being among older people. *Perspectives in Public Health, 133*(1), 36–43.

Crenshaw, A. O., Leo, K., & Baucom, B. R. (2019). The effect of stress on empathic accuracy in romantic couples. *Journal of Family Psychology, 33*(3), 327.

Cross, I. (2014). Music and communication in music psychology. *Psychology of Music, 42*(6), 809–819.

Cross, I., Laurence, F., & Rabinowitch, T. C. (2012). Empathy and creativity in group musical practices: Towards a concept of empathic creativity. In G. McPherson & G. Welch (Eds.), *The Oxford handbook of music education* (Vol. 2, pp. 337–353). Oxford University Press.

Cross, R. (2010). Recent work on the philosophy of Duns Scotus. *Philosophy Compass, 5*(8), 667–675.

Crossley, J. (2017). *The cyber-guitar system: A study in technologically enabled performance practice* [Doctoral dissertation, University of the Witwatersrand].

Crossley, N. (2010). *Towards relational sociology*. Routledge.

Crossley, N. (2018). Networks, interactions and relations. In F. Depelteau (Ed.), *The Palgrave handbook of relational sociology* (pp. 481–498). Palgrave Macmillan.

Crowe, B. J., & Ratner, E. (2012). The sound design project: An interdisciplinary collaboration of music therapy and industrial design. *Music Therapy Perspectives, 30*(2), 101–108.

Cuff, B., Brown, S., Taylor, L., & Howat, D. (2016). Empathy: A review of the concept. *Emotion Review, 8*(2), 144–153.

Cummings, L. (2016). *Empathy as dialogue in theatre and performance*. Palgrave Macmillan.

Cummins, F. (2009). Rhythm as an affordance for the entrainment of movement. *Phonetica, 66*(1–2), 15–28.

Da Silva, L., Sanson, A., Smart, D., & Toumbourou, J. (2004). Civic responsibility among Australian adolescents: Testing two competing models. *Journal of Community Psychology, 32*(3), 229–255.

Dalla Bella, S., Peretz, I., Rousseau, L., & Gosselin, N. (2001). A developmental study of the affective value of tempo and mode in music. *Cognition, 80*(3), B1–10.

Dallos, R., & Stedmon, J. (2009). Flying over the swampy lowlands: Reflective and reflexive practice. In J. Stedmon & R. Dallos (Eds.), *Reflective practice in psychotherapy and counselling* (pp. 1–22). Open University Press.

Daly, A. (2016). *Merleau-Ponty and the ethics of intersubjectivity*. Palgrave Macmillan.

Datta, R. (2018). Traditional storytelling: An effective Indigenous research methodology and its implications for environmental research. *AlterNative: An International Journal of Indigenous Peoples, 14*(1), 35–44.

Daveson, B., O'Callaghan, C., & Grocke, D. (2008). Indigenous music therapy theory building through grounded theory research: The developing indigenous theory framework. *The Arts in Psychotherapy, 35*(4), 280–286.

Davidson, J. (2004). What can the social psychology of music offer community music therapy? In M. Pavlicevic & G. Ansdell (Eds.), *Community music therapy* (pp. 114–128). Jessica Kingsley.

Davis, C. M. (1990). What is empathy, and can empathy be taught? *Physical Therapy, 70*(11), 707–711.

Davis, L. (2010). Constructing normalcy. In L. Davis (Ed.), *The disability studies reader* (3rd ed., pp. 3–19). Routledge.
Davis, M. (1994). *Empathy: A social psychological approach.* Westview Press.
Davis, M. H., Conklin, L., Smith, A., & Luce, C. (1996). Effect of perspective taking on the cognitive representation of persons: A merging of self and other. *Journal of Personality and Social Psychology, 70*(4), 713.
De Backer, J. (2008). Music and psychosis: A research report detailing the transition from sensorial play to musical form by psychotic patients. *Nordic Journal of Music Therapy, 17*(2), 89–104.
De Backer, J., & Sutton, J. (2014). Therapeutic interventions in psychodynamic music therapy. In J. De Backer & J. Sutton (Eds.), *The music in music therapy: Psychodynamic music therapy in Europe: Clinical, theoretical and research approaches* (pp. 338–350). Jessica Kingsley.
De Kock, K. (2003). *Experiencing time and repetition: Finding common ground between traditional and modern music therapy practices* [Master's dissertation, University of Pretoria].
De Landa, M. (2006). *A new philosophy of society: Assemblage theory and social complexity.* Continuum.
De Leersnyder, J., Boiger, M., & Mesquita, B. (2013). Cultural regulation of emotion: Individual, relational, and structural sources. *Frontiers in Psychology, 4*, 55.
De Quincey, C. (2005). *Radical knowing: Understanding consciousness through relationship.* Park Street Press.
De Vaus, J., Hornsey, M. J., Kuppens, P., & Bastian, B. (2018). Exploring the East-West divide in prevalence of affective disorder: A case for cultural differences in coping with negative emotion. *Personality and Social Psychology Review, 22*(3), 285–304.
de Vignemont, F. (2006). When do we empathize? In G. Bock & J. Goode (Eds.), *Empathy and fairness* (pp. 181–196). Wiley.
de Vignemont, F., & Singer, T. (2006). The emphatic brain: How, when, and why? *Trends in the Cognitive Sciences, 10*, 435–441.
de Waal, F. (2008). Putting the altruism back into altruism: The evolution of empathy. *Annual Review of Psychology, 59*, 279–300.
de Waal, F. (2009). *The age of empathy.* Harmony Books.
Decety, J., & Jackson, P. L. (2004). The functional architecture of human empathy. *Behavioral and Cognitive Neuroscience Review, 3*(2), 406–412.
Decety, J., & Lamm, C. (2006). Human empathy through the lens of social neuroscience. *The Scientific World Journal, 6*, 1146–1163.

Decety, J., & Lamm, C. (2011). Empathy vs. personal distress. In J. Decety, & W. Ickes (Eds.), *The social neuroscience of empathy* (pp. 199–214). MIT Press.

Dekeyser, M., Elliott, R., & Leijssen, M. (2009). Empathy in psychotherapy: Dialogue and embodied understanding. In J. Decety & W. Ickes (Eds.), *The social neuroscience of empathy* (pp. 113–124). The MIT Press.

Deleuze, G. (1991). *Empiricism and subjectivity: An essay on Hume's theory of human nature* (C. Boundas, Trans.). Columbia University Press.

Deleuze, G. (1992). *Expressionism in philosophy: Spinoza* (M. Joughin, Trans.). Zone Books.

Deleuze, G. (1994). *Difference and repetition.* The Athlone Press.

Deleuze, G. (1995). *Negotiations: 1972–1990* (M. Joughin, Trans.). Columbia University Press.

Deleuze, G. (2001). *Pure immanence: Essays on a life* (A. Boyman, Trans.). Zone Books.

Deleuze, G. (2003). *Francis Bacon: The logic of sensation* (D. W. Smith, Trans.). Continuum.

Deleuze, G., & Guattari, F. (1987). *A thousand plateaus: Capitalism and schizophrenia.* Athlone.

Deleuze, G., & Parnet, C. (1987). *Dialogues.* Athlone.

Demichelis, O. P., Coundouris, S. P., Grainger, S. A., & Henry, J. D. (2020). Empathy and theory of mind in Alzheimer's disease: A meta-analysis. *Journal of the International Neuropsychological Society, 26*(10), 963–977.

Demmrich, S. (2020). Music as a trigger of religious experience: What role does culture play? *Psychology of Music, 48*(1), 35–49.

DeNora, T. (2000). *Music in everyday life.* Cambridge University Press.

DeNora, T. (2003). *After Adorno: Rethinking music sociology.* Cambridge University Press.

DeNora, T. (2004). *Music in everyday life.* Cambridge University Press.

DeNora, T. (2006). Music and self-identity. In A. Bennett, D. Shank, & J. Toynbee (Eds.), *The popular music studies reader* (pp. 141–147). Abingdon.

DeNora, T. (2011a). Practical consciousness and social relation in MusEcological perspective. In D. Clarke & E. Clarke (Eds.), *Music and consciousness: Philosophical, psychological, and cultural perspectives* (pp. 309–326). Oxford University Press.

DeNora, T. (2011b). *Music in action: Selected essays in sonic ecology.* Ashgate.

DeNora, T., & Ansdell, G. (2017). Music in action: Tinkering, testing and tracing over time. *Qualitative Research, 17*(2), 231–245.

Denzin, N. (2009). *On understanding emotion.* Transaction Publishers.

Derrington, P. (2012). 'Yeah, I'll do music!' Working with secondary-aged students who have complex emotional and behavioural difficulties. In J. Tomlinson, P. Derrington, & A. Oldfield (Eds.), *Music therapy in schools* (pp. 195–211). Jessica Kingsley.

Dervin, F. (2011). A plea for change in research on intercultural discourses: A 'liquid' approach to the study of the acculturation of Chinese students. *Journal of Multicultural Discourses, 6*(1), 37–52.

Dewall, C., Baumeister, R., Gailliot, M., & Maner, J. (2008). Depletion makes the heart grow less helpful: Helping as a function of self-regulatory energy and genetic relatedness. *Personality & Social Psychology Bulletin, 34*, 1653–1662.

DeWall, C., Lambert, N., Pond, R., Kashdan, T., & Fincham, F. (2012). A grateful heart is a nonviolent heart: Cross-sectional, experience sampling, longitudinal, and experimental evidence. *Social Psychological and Personality Science, 3*(2), 232–240.

DeYoung, P. A. (2014). *Relational psychotherapy: A primer.* Routledge.

Dileo, C. (2021). *Ethical thinking in music therapy* (2nd ed.). Jeffrey Books.

Dimaio, L. (2010). Music therapy entrainment: A humanistic music therapist's perspective of using music therapy entrainment with hospice clients experiencing pain. *Music Therapy Perspectives, 28*(2), 106–115.

Dimitriadis, T., & Smeijsters, H. (2011). Autistic spectrum disorder and music therapy: Theory underpinning practice. *Nordic Journal of Music Therapy, 20*(2), 108–122.

Dindoyal, L. (2018). 'In the therapist's head and heart': An investigation into the profound impact that motherhood has on the work of a music therapist. *British Journal of Music Therapy, 32*(2), 105–110.

Dixon, T. (2003). *From passions to emotions.* Cambridge University Press.

Donati, P. (2011). *Relational sociology: A new paradigm for the social sciences.* Routledge.

Doré, B. P., Morris, R. R., Burr, D. A., Picard, R. W., & Ochsner, K. N. (2017). Helping others regulate emotion predicts increased regulation of one's own emotions and decreased symptoms of depression. *Personality and Social Psychology Bulletin, 43*(5), 729–739.

Dorlando, H. (2011). *Communal empathy in native American older adults* [Doctoral thesis, University of Montana].

dos Santos, A. (2018). *Empathy and aggression in group music therapy with adolescents: Comparing the affordances of two paradigms* [Doctoral dissertation, University of Pretoria].

dos Santos, A. (2019a). Empathy and aggression in group music therapy with teenagers: A descriptive phenomenological study. *Music Therapy Perspectives, 37*(1), 14–27.

dos Santos, A. (2019b). Group music therapy with adolescents referred for aggression. In K. McFerran, P. Derrington, & S. Saarikallio (Eds.), *Handbook of music, adolescents, and wellbeing* (pp. 15–24). Oxford University Press.

dos Santos, A. (2020). The usefulness of aggression as explored by becoming-teenagers in group music therapy. *Nordic Journal of Music Therapy, 29*(2), 150–173. https://doi.org/10.1080/08098131.2019.1649712

dos Santos, A., & Brown, T. (2021). Music therapists' empathic experiences of shared and differing orientations to religion and spirituality in the client-therapist relationship. *The Arts in Psychotherapy, 74*, 101786.

Dovidio, J., Johnson, J., Gaertner, S., Pearson, A., Saguy, T., & Ashburn-Nardo, L. (2010). Empathy and intergroup relations. In M. Mikulincer & P. Shaver (Eds.), *Prosocial motives, emotions, and behavior* (pp. 393–408). American Psychological Association.

Drakulic, C. (1993). *The Balkan Express: Fragments from the other side of the war.* Norton.

Duan, C., & Hill, C. E. (1996). The current state of empathy research. *Journal of Counseling Psychology, 43*(3), 261.

DuBose, J., MacAllister, L., Hadi, K., & Sakallaris, B. (2018). Exploring the concept of healing spaces. *HERD: Health Environments Research & Design Journal, 11*(1), 43–56.

Duff, C. (2012). Exploring the role of 'enabling places' in promoting recovery from mental illness: A qualitative test of a relational model. *Health & Place, 18*(6), 1388–1395.

Duff, C. (2014). *Assemblages of health: Deleuze's empiricism and the ethology of life*. Springer.

Dunford, R. (2017). Toward a decolonial global ethics. *Journal of Global Ethics, 13*(3), 380–397.

Dunlap, A. L. (2017). *Women with addictions' experience in music therapy* [Master's thesis, College of Fine Arts of Ohio University]. https://etd.ohiolink.edu/

Duranti, A. (2010). Husserl, intersubjectivity and anthropology. *Anthropological Theory, 10*(1–2), 16–35.

Durkheim, E. (1912/1964). *The elementary forms of religious life.* Free Press.

Eagle, M., & Wolitzky, D. L. (1997). Empathy: A psychoanalytic perspective. In A. C. Bohart & L. S. Greenberg (Eds.), *Empathy reconsidered:*

New directions in psychotherapy (pp. 217–244). American Psychological Association.
Eagleton, T. (2000). *The idea of culture.* Blackwell.
Eakin, P. J. (2019). *How our lives become stories.* Cornell University Press.
Ebersöhn, L. (2019). *Flocking together: An indigenous psychology theory of resilience in Southern Africa.* Springer.
Ebisch, S. J. H., Perrucci, M. G., Ferretti, A., Gratta, C. D., Romani, G. L., Gallese, V. (2008). The sense of touch: Embodied simulation in a visuotactile mirroring mechanism for observed animate or inanimate touch. *Journal of Cognitive Neuroscience, 20*(9), 1611–1623.
Ebisch, S. J., Aureli, T., Bafunno, D., Cardone, D., Romani, G. L., & Merla, A. (2012). Mother and child in synchrony: Thermal facial imprints of autonomic contagion. *Biological Psychology, 89*(1), 123–129.
Ebish, S. J. H., Perrucci, M. G., Ferretti, A., Del Gratta, C., Romani, G. L., & Gallese, V. (2008). The sense of touch: Embodied simulation in a visuotactile mirroring mechanism for the sight of any touch. *Journal of Cognitive Neuroscience, 20,* 1611–1623.
Echard, A. (2019). Making sense of self: An autoethnographic study of identity formation for adolescents in music therapy. *Music Therapy Perspectives, 37*(2), 141–150.
Eckroth-Bucher, M. (2010). Self-awareness: A review and analysis of a basic nursing concept. *Advances in Nursing Science, 33*(4), 297–309.
Edwards, J. (2012). We need to talk about epistemology: Orientations, meaning, and interpretation within music therapy research. *Journal of Music Therapy, 49*(4), 372–394.
Edwards, S. (2010). A Rogerian perspective on empathic patterns in Southern African healing. *Journal of Psychology in Africa, 20*(2), 321–326.
Egan, G. (2014). *The skilled helper: A problem-management and opportunity-development approach to helping* (10th ed.). Cenage Learning.
Eisenberg, N. (2000). Emotion, regulation, and moral development. *Annual Review of Psychology, 51,* 665–697.
Eisenberg, N., & Eggum, N. D. (2009). Empathic responding: Sympathy and personal distress. In J. Decety & W. Ickes (Eds.), *The social neuroscience of empathy* (pp. 71–83). MIT Press.
Eisenberg, N., Fabes, R. A., Murphy, B., Karbon, M., Maszk, P., Smith, M., O'Boyle, C., & Suh, K. (1994). The relations of emotionality and regulation to dispositional and situational empathy-related responding. *Journal of Personality and Social Psychology, 66*(4), 776.

Eisenberg, N., Michalik, N., Spinrad, T. L., Hofer, C., Kupfer, A., Valiente, C., Liew, J., Cumberland, A., & Reiser, M. (2007). The relations of effortful control and impulsivity to children's sympathy: A longitudinal study. *Cognitive Development, 22*(4), 544–567.

Eisenberg, N., Spinrad, T. L., & Knafo-Noam, A. (2015). Prosocial development. In L. Liben & U. Muller (Eds.), *Handbook of child psychology and developmental science* (pp. 1–47). Wiley.

Elefant, C. (2010). Musical inclusion, intergroup relations, and community development. In B. Stige, G. Ansdell, C. Elefant, & M. Pavlicevic (Eds.), *Where music helps: Community music therapy in action and reflection* (pp. 75–92). Ashgate.

Elefant, C. (2017). Reflections: Giving voice: Participatory action research with a marginalized group. In B. Stige, G. Ansdell, C. Elefant, & M. Pavlicvic (Eds.), *Where music helps: Community music therapy in action and reflection* (pp. 199–216). Routledge.

Elfenbein, H. (2014). The many faces of emotional contagion: An affective process theory of affective linkage. *Organizational Psychology Review, 4*(4), 326–362.

Ellingson, L. (2011). *Engaging crystallisation in qualitative research*. Sage.

Elliott, R., Bohart, A., Watson, J., & Greenberg, L. (2011). Empathy. In J. Norcross (Ed.), *Psychotherapy relationships that work: Evidence-based responsiveness* (2nd ed., pp. 132–152). Oxford University Press.

Ellis, T. E., Schwartz, J. A., & Rufino, K. A. (2018). Negative reactions of therapists working with suicidal patients: A CBT/Mindfulness perspective on "countertransference." *International Journal of Cognitive Therapy, 11*(1), 80–99.

Emanuel, C. (2016). The disabled: The most othered others. In D. Goodman & E. Severson (Eds.), *The ethical turn: Otherness and subjectivity in contemporary psychoanalysis* (pp. 270–285). Routledge.

Emirbayer, M. (1997). Manifesto for a relational sociology. *American Journal of Sociology, 103*(2), 281–317.

Enfield, N. (2017). Elements of agency. In N. Enfield & P. Kockelman (Eds.), *Distributed agency* (pp. 3–8). Oxford University Press.

Epley, N., Keysar, B., Van Boven, L., & Gilovich, T. (2004). Perspective taking as egocentric anchoring and adjustment. *Journal of Personality and Social Psychology, 87*(3), 327.

Epley, N., & Waytz, A. (2009). Mind perception. In S. T. Fiske, D. T. Gilbert, & G. Lindzey (Eds.), *The handbook of social psychology* (5th ed., pp. 498–541). Wiley.

Epp, E. (2007). Locating the autonomous voice: Self-expression in music-centered music therapy. *Voices: A World Forum for Music Therapy, 7*(1).
Esin, C. (2011). Narrative analysis approaches. In N. Frost (Ed.), *Qualitative research methods in psychology: Combining core approaches* (pp. 92–118). Open University Press.
Etherington, K. (2004). *Becoming a reflexive researcher.* Jessica Kingsley.
Eurich, T. (2017). *Insight*. Crown Business.
Evans, D. (2001). *Emotion: A very short introduction.* Oxford University Press.
Eyal, T., Steffel, M., & Epley, N. (2018). Perspective mistaking: Accurately understanding the mind of another requires getting perspective, not taking perspective. *Journal of Personality and Social Psychology, 114*(4), 547.
Eyre, L. (2007). The marriage of music and narrative: Explorations in art, therapy, and research. *Voices: A World Forum for Music Therapy, 7*(3).
Eyuboglu, D., Bolat, N., & Eyuboglu, M. (2018). Empathy and theory of mind abilities of children with specific learning disorder (SLD). *Psychiatry and Clinical Psychopharmacology, 28*(2), 136–141.
Fagiano, M. (2019). Relational empathy as an instrument of democratic hope in action. *The Journal of Speculative Philosophy, 33*(2), 200–219.
Fanon, F. (2004). *The wretched of the earth.* Grove Press.
Fansler, V., Reed, R., Bautista, E., Arnett, A., Perkins, F., & Hadley, S. (2019). Playing in the borderlands: The transformative possibilities of queering music therapy pedagogy. *Voices: A World Forum of Music Therapy, 19*(3),
Feldman, R. (2007). Parent-infant synchrony and the construction of shared timing; physiological precursors, developmental outcomes, and risk conditions. *Journal of Child Psychology and Psychiatry, 48*(3–4), 329–354.
FeldmanHall, O., Dalgleish, T., Evans, D., & Mobbs, D. (2015). Empathic concern drives costly altruism. *NeuroImage, 105*, 347–356.
Feller, C. P., & Cottone, R. R. (2003). The importance of empathy in the therapeutic alliance. *The Journal of Humanistic Counseling, Education and Development, 42*(1), 53–61.
Ferrando, F. (2019). *Philosophical posthumanism.* Bloomsbury Academic.
Ferrara, L. (1984). Phenomenology as a tool for musical analysis. *The Musical Quarterly, 70*(3), 355–373.
Ferri, G. (2018). *Intercultural communication: Critical approaches and future challenges.* Palgrave Macmillan.
Fields, J., Copp, M., & Kleinman, S. (2006). Symbolic interactionism, inequality, and emotions. In J. Stets & J. Turner (Eds.), *Handbook of the Sociology of Emotions* (pp. 155–178). Springer.

Fine, M. (1997). Witnessing whiteness. In M. Fine, L. Weis, L. Powell, & L. Wong (Eds.), *Off White: Readings on race, power, and society* (pp. 57–66). Routledge.

Finlay, L. (2009). Debating phenomenological research methods. *Phenomenology and Practice, 3*(1), 6–25.

Finlay, L. (2011). *Phenomenology for therapists.* Wiley-Blackwell.

Finlay, L. (2014). Engaging phenomenological analysis. *Qualitative Research in Psychology, 11*, 121–141.

Fishman, G. (1999). Knowing another from a dynamic systems point of view: The need for a multimodal concept of empathy. *The Psychoanalytic Quarterly, 68*(3), 376–400.

Fiske, S. T. (1993). Controlling other people: The impact of power on stereotyping. *American Psychologist, 48*(6), 621–628.

Fleischacker, S. (2019). *Being me being you: Adam Smith and empathy.* University of Chicago Press.

Flower, C. (2019). *Music therapy with children and parents: Toward an ecological attitude* [Doctoral dissertation, Goldsmiths, University of London].

Floyd, J. J. (2010). Listening: A dialogic perspective. In A. D. Wolvin (Ed.), *Listening and human communication in the 21st century* (pp. 127–140). Wiley-Blackwell.

Fogel, A., Nwokah, E., Dedo, J. Y., Messinger, D., Dickson, K. L., Matusov, E., & Holt, S. A. (1992). Social process theory of emotion: A dynamic systems approach. *Social Development, 1*, 122–142.

Foley, G. N., & Gentile, J. P. (2010). Nonverbal communication in psychotherapy. *Psychiatry, 7*(6), 38.

Fonagy, P. (1991). Thinking about thinking: Some clinical and theoretical consideration in the treatment of a borderline patient. *International Journal of Psychoanalysis, 72*, 639–656.

Fonagy, P. (2001). *Attachment theory and psychoanalysis.* Other Press.

Fonagy, P., & Allison, E. (2014). The role of mentalizing and epistemic trust in the therapeutic relationship. *Psychotherapy, 51*, 372–380.

Forinash, M., & Gonzalez, D. (1989). A phenomenological perspective of music therapy. *Music Therapy, 8*(1), 35–46.

Forster, D. A. (2010). A generous ontology: Identity as a process of intersubjective discovery—An African theological contribution. *HTS: Theological Studies, 66*(1), 1–12.

Fosha, D. (2001). The dyadic regulation of affect. *Journal of Clinical Psychology, 57*(2), 227–242.

Fouché, S., & Stevens, M. (2018). Co-creating spaces for resilience to flourish. *Voices: A World Forum for Music Therapy, 18*(4).
Fouché, S., & Torrance, K. (2005). Lose yourself in the music, the moment, Yo! Music therapy with an adolescent group involved in gangsterism. *Voices: A World Forum for Music Therapy, 5*(3).
Fox, E. I., & McKinney, C. H. (2016). The Bonny method of guided imagery and music for music therapy interns. *Music Therapy Perspectives, 34*(1), 90–98.
Fox, N. J., & Alldred, P. (2013). The sexuality-assemblage: Desire, affect, anti-humanism. *The Sociological Review, 61*(4), 769–789.
Frank, A. (2015). Practicing dialogical narrative analysis. In J. A. Holstein & J. F. Gubrium (Eds.), *Varieties of narrative analysis* (pp. 33–52). Sage.
Freire, P. (1970/2000). *Pedagogy of the oppressed* (30th Anniversary ed.). Bloomsbury Press.
Freud, S. (1922/1949). Group psychology and the analysis of the ego. In E. Jones (Eds), *International Psychoanalytic Library, No. 6.* (J. Strachey, Trans.). Hogarth Press.
Frid, E. (2019). Accessible digital musical instruments: A review of musical interfaces in inclusive music practice. *Multimodal Technologies and Interaction, 3*(3), 57.
Frie, R. (2010). Compassion, dialogue, and context: On understanding the other. *International Journal of Psychoanalytic Self Psychology, 5*, 451–466.
Frith, C. D. (1992). *The cognitive neuropsychology of schizophrenia.* Lawrence Erlbaum.
Frith, C. D. (2007). The social brain? *Philosophical Transactions of the Royal Society, Series b, Biological Sciences, 362*, 671–678.
Froese, T., & Di Paolo, E. (2011). The enactive approach: Theoretical sketches from cell to society. *Pragmatics & Cognition, 19*(1), 1–36.
Fuchs, T. (2013). Depression, intercorporeality, and interaffectivity. *Journal of Consciousness Studies, 20*(7–8), 219–238.
Funk, M., & Coeckelbergh, M. (2013). Is gesture knowledge? A philosophical approach to epistemology of musical gestures. In H. De Preester (Ed.), *Moving imagination: Explorations of gesture and inner movement* (pp. 113–132). John Benjamins.
Gadamer, H. (1976). *Philosophical hermeneutics* (D. E. Linge, Trans.). University of California.
Gadamer, H. (2007). Letter exchange with Karl Löwith on being and time. In T. Kisiel & T. Sheehan (Eds.), *Becoming Heidegger: On the trail of his*

occasional writings, 1910–1927 (pp. 289–303). Northwestern University Press.

Galati, A., & Brennan, S. E. (2010). Attenuating information in spoken communication: For the speaker, or for the addressee? *Journal of Memory and Language, 62*, 35–51.

Galinsky, A., Ku, G., & Wang, C. (2005). Perspective-taking and self-other overlap. *Group Processes & Intergroup Relations, 8*(2), 109–124.

Galinsky, A., Maddux, W., Gilin, D., & White, J. (2008). Why it pays to get inside the head of your opponent. *Psychological Science, 19*(4), 378–384.

Gallagher, H., & Frith, C. (2003). Functioning imaging of 'theory of mind.' *Trends in Cognitive Sciences, 7*(2), 77–83.

Gallagher, S. (2017). *Enactivist interventions: Rethinking the mind.* Oxford University Press.

Gallese, V. (2003). The roots of empathy: The shared manifold hypothesis and the neural basis of intersubjectivity. *Psychopathology, 36*(4), 171–180.

Gallese, V. (2009). Mirror neurons, embodied simulation, and the neural basis of social identification. *Psychoanalytic Dialogues, 19*(5), 519–536.

Gallese, V., & Goldman, A. (1998). Mirror neurons and the simulation theory of mind-reading. *Trends in Cognitive Sciences, 2*(12), 493–501.

Gallese, V., & Guerra, M. (2015). *The empathic screen: Cinema and neuroscience.* Oxford University Press.

Ganeri, J. (2017). *Attention, not self*. Oxford University Press.

Gardstrom, S., & Hiller, J. (2010). Song discussion as music psychotherapy. *Music Therapy Perspectives, 28*(2), 147–156.

Gardstrom, S., & Hiller, J. (2016). Resistances in group music therapy with women and men with substance use disorders. *Voices: A World Forum for Music Therapy, 16*(3).

Garred, R. (2001, November). The ontology of music in music therapy. *Voices: A World Forum for Music Therapy, 1*(3). https://voices.no/index.php/voices/article/view/1604/1363

Garred, R. (2006). *Music as therapy: A dialogical perspective.* Barcelona Publishers.

Gawerc, M. I. (2006). Peace-building: Theoretical and concrete perspectives. *Peace & Change, 31*(4), 435–478.

Gaztambide-Fernandez, R. (2012). Decolonization and the pedagogy of solidarity. *Decolonization: Indigeneity, Education and Society, 1*(1), 41–67.

Genat, B. with Bushby, S., McGuire, M., taylor, E., Walley, Y., & Weston, T. (2006). *Aboriginal healthworkers: Primary health care at the margins.* University of Western Australia Press.

Gerdes, K. E., & Segal, E. (2011). Importance of empathy for social work practice: Integrating new science. *Social Work, 56*(2), 141–148.
Gergen, K. (2000). The coming of creative confluence in therapeutic practice. *Psychotherapy: Theory, Research, Practice, Training, 37*(4), 364.
Gergen, K. (2009). *Relational being: Beyond self and community*. Oxford University Press.
Gergen, K. (2018). Human essence: Toward a relational reconstruction. In M. van Zomerpen & J. Dovidio (Eds.), *The Oxford handbook of Human essence* (pp. 247–260). Oxford University Press.
Gergen, K., & Gergen, M. (2010). Scanning the landscape of narrative inquiry. *Social and Personality Psychology Compass, 4*(9), 728–735.
Germer, C., Siegel, R., & Fulton, P. (2013). *Mindfulness and psychotherapy*. Guilford press.
Ghetti, C. (2012). Music therapy as procedural support for invasive medical procedures: Toward the development of music therapy theory. *Nordic Journal of Music Therapy, 21*(1), 3–35.
Gibson, J. (2017). A relational approach to suffering: A reappraisal of suffering in the helping relationship. *Journal of Humanistic Psychology, 57*(3), 281–300.
Gilbert, J. (2004). Becoming-music: The rhizomatic moment of improvisation. In I. Buchanan & M. Swiboda (Eds.), *Deleuze and music* (pp. 118–139). Edinburgh University Press.
Gilbert, P. (2009). *The compassionate mind: A new approach to life's challenges*. Constable and Robinson.
Gilbert, P. (2014). The origins and nature of compassion focused therapy. *British Journal of Clinical Psychology, 53*(1), 6–41.
Gilbert, P., & Woodyatt, L. (2017). An evolutionary approach to shame-based self-criticism, self-forgiveness, and compassion. In L. Woodyatt, E. Worthington, M. Wenzel, & B. Griffin (Eds.), *Handbook of the psychology of self-forgiveness* (pp. 29–41). Springer.
Gillespie, A., & Cornish, F. (2010). Intersubjectivity: Towards a dialogical analysis. *Journal for the Theory of Social Behaviour, 40*(1), 19–46.
Gillespie, A., Howarth, C. S., & Cornish, F. (2012). Four problems for researchers using social categories. *Culture Psychology, 18*(3), 391–402.
Gilewski, M. (1993). The use of transpersonal empathy with child abuse survivors. In *Proceedings from The American Psychological Association* (Toronto, Canada). https://files.eric.ed.gov/fulltext/ED371247.pdf

Gillberg, C. (2007). Non-autism childhood empathy disorders. In T. Farrow & P. Woodruff (Eds.), *Empathy in mental illness* (pp. 111–125). Cambridge University Press.

Gleichgerrcht, E., & Decety, J. (2013). Empathy in clinical practice: How individual dispositions, gender, and experience moderate empathic concern, burnout, and emotional distress in physicians. *PLoS ONE, 8*(4), e61526.

Goebl, W., & Palmer, C. (2009). Synchronization of timing and motion among performing musicians. *Music Perception, 26*(5), 427–438.

Goetz, J., Keltner, D., & Simon-Thomas, E. (2010). Compassion: An evolutionary analysis and empirical review. *Psychological Bulletin, 136*(3), 351–374.

Goldie, P. (2000). *The emotions: A philosophical exploration.* Oxford University Press.

Goldie, P. (2011). Self-forgiveness and the narrative sense of self. In C. Fricke (Ed.), *The ethics of forgiveness: A collection of essays* (pp. 81–94). Routledge.

Goldman, A. (2012). Theory of mind. In E. Margolis, R. Samuels, & S. Stich (Eds.), *The Oxford handbook of philosophy and cognitive science* (pp. 1–25). Oxford University Press.

Gonzalez, P. J. (2011). The impact of music therapists' music cultures on the development of their professional frameworks. *Qualitative Inquiries in Music Therapy, 6*, 1–13.

Goodwin, S. A., Gubin, A., Fiske, S. T., & Yzerbyt, V. Y. (2000). Power can bias impression processes: Stereotyping subordinates by default and by design. *Group Processes & Intergroup Relations, 3*(3), 227–256.

Gordon, M. (2011). Listening as embracing the other. *Educational Theory, 61*(2), 207–219.

Gordon, R. M. (1986). Folk psychology as simulation. *Mind and Language, 1*, 158–171.

Gottesman, S. (2017). Hear and be heard: Learning with and through music as a dialogical space for co-creating youth led conflict transformation. *Voices: A World Forum for Music Therapy, 17*(1).

Gray, C. D., Hosie, J. A., Russell, P. A., & Ormel, E. A. (2001). Emotional development in deaf children: Facial expressions, display rules, and theory of mind. In D. Clark, M. Marschark, & M. Karchmer (Eds.), *Context, cognition, and deafness* (pp. 135–160). Gallaudet University Press.

Graziano, M. (2013). *Consciousness and the social brain.* Oxford University Press.

Greason, P., & Cashwell, C. (2009). Mindfulness and counseling self-efficacy: The mediating role of attention and empathy. *Counselor Education and Supervision, 49*(1), 2–19.

Green, B. (2003). *The mastery of music: Ten pathways to true artistry.* Broadway Books.

Green, M., & Brock, T. (2000). The role of transportation in the persuasiveness of public narratives. *Journal of Personality and Social Psychology, 79*(5), 701–721.

Greenberg, D., Baron-Cohen, S., Rosenberg, N., Fonagy, P., & Rentfrow, P. (2018). Elevated empathy in adults following childhood trauma. *PLoS ONE, 13*(10), e0203886.

Greene, J., & Haidt, J. (2002). How (and where) does moral judgment work? *Trends in Cognitive Sciences, 6*, 517–523.

Greenhough, B., & Roe, E. (2010). From ethical principles to response-able practice. *Environment and Planning d: Society and Space, 28*(1), 43–45.

Gregg, M., & Seigworth, G. (2010). An inventory of shimmers. In M. Gregg & G. Seigworth (Eds.), *The affect theory reader* (pp. 1–28). Duke University Press.

Grey, F. (2016). Benevolent othering: Speaking positively about mental health service users. *Philosophy, Psychiatry, & Psychology, 23*(3), 241–251.

Grimmer, M. S., & Schwantes, M. (2018). Cross-cultural music therapy: Reflections of American music therapists working internationally. *The Arts in Psychotherapy, 61*, 21–32.

Groark, K. (2008). Social opacity and the dynamics of empathic in-sight among the Tzotzil Maya of Chiapas, Mexico. *Ethos, 36*(4), 427–448.

Grocke, D., & Wigram, T. (2006). *Receptive methods in music therapy.* Jessica Kingsley.

Gross, J. (1998). The emerging field of emotion regulation: An integrative review. *Review of General Psychology, 2*, 271–299.

Gross, J., & Cassidy, J. (2019). Expressive suppression of negative emotions in children and adolescents: Theory, data, and a guide for future research. *Developmental Psychology, 55*(9), 1938.

Gross, J., & John, O. (2003). Individual differences in two emotion regulation processes: Implications for affect, relationships, and well-being. *Journal of Personality and Social Psychology, 85*(2), 348–362.

Gruenfeld, D., Inesi, M., Magee, J., & Galinsky, A. (2008). Power and the objectification of social targets. *Journal of Personality and Social Psychology, 95*(1), 111–127.

Grynberg, D., & Pollatos, O. (2015). Perceiving one's body shapes empathy. *Physiology & Behavior, 140*, 54–60.

Gu, X., Eilam-Stock, T., Zhou, T., Anagnostou, E., Kolevzon, A., Soorya, L., Hof, P., Friston, K., & Fan, J. (2015). Autonomic and brain responses associated with empathy deficits in Autism Spectrum Disorder. *Human Brain Mapping, 36*, 3323–3338.

Guilfoyle, M. (2015). Listening in narrative therapy: Double listening and empathic positioning. *South African Journal of Psychology, 45*(1), 36–49.

Gureje, O., Simon, G., & Üstün, T. (1997). Somatization in cross-cultural perspective: A world health organization study in primary care. *American Journal of Psychiatry, 154*, 989–995.

Gutsell, J., & Inzlicht, M. (2010). Empathy constrained: Prejudice predicts reduced mental simulation of actions during observation of outgroups. *Journal of Experimental Social Psychology, 46*(5), 841–845.

Guyotte, K., & Kuntz, A. (2018). Becoming openly faithful: Qualitative pedagogy and paradigmatic slippage. *International Review of Qualitative Research, 11*(3), 256–270.

Habermas, J. (2008). *Between naturalism and religion*. Polity Press.

Hackel, L., Zaki, J., & Van Bavel, J. (2017). Social identity shapes social valuation: Evidence from prosocial behavior and vicarious reward. *Social Cognitive and Affective Neuroscience, 12*(8), 1219–1228.

Hadley, S. (2013). Dominant narratives: Complicity and the need for vigilance in the creative arts therapies. *The Arts in Psychotherapy, 4*, 373–381.

Hadley, S., & Norris, M. S. (2016). Musical multicultural competency in music therapy: The first step. *Music Therapy Perspectives, 34*(2), 129–137.

Hald, S. V., Baker, F. A., & Ridder, H. M. (2017). A preliminary evaluation of the interpersonal music-communication competence scales. *Nordic Journal of Music Therapy, 26*(1), 40–61.

Haley, B., Heo, S., Wright, P., Barone, C., Rao Rettiganti, M., & Anders, M. (2017). Relationships among active listening, self-awareness, empathy, and patient-centered care in associate and baccalaureate degree nursing students. *NursingPlus Open, 3*, 11–16.

Hall, J., & Schwartz, R. (2019). Empathy present and future. *The Journal of Social Psychology, 159*(3), 225–243.

Hall, S. (2009). *Anger, rage and relationship*. Routledge.

Hallward, P. (2006). *Out of the world: Deleuze and the philosophy of creation*. Verso.

Halperin, E. (2015). *Emotions in conflict: Inhibitors and facilitators of peacemaking*. Routledge.

Halpern, J. (2001). *From detached concern to empathy*. Oxford University Press.
Ham, J., & Tronick, E. (2009). Relational psychophysiology: Lessons from mother–infant physiology research on dyadically expanded states of consciousness. *Psychotherapy Research, 19*(6), 619–632.
Hamilton, S., Pinfold, V., Cotney, J., Couperthwaite, L., Matthews, J., Barret, K., Warren, S., Corker, E., Rose, D., Thornicroft, G., & Henderson, C. (2016). Qualitative analysis of mental health service users' reported experiences of discrimination. *Acta Psychiatrica Scandinavica, 134*, 14–22.
Han, S. (2018). Neurocognitive basis of racial ingroup bias in empathy. *Trends in Cognitive Sciences, 22*(5), 400–421.
Haraway, D. (1992). The promises of monsters: A regenerative politics for inappropriate/d others. In L. Grossberg, C. Nelson, & P. Treichler (Eds.), *Cultural studies* (pp. 295–337). Routledge.
Haraway, D. (2016). *Staying with the trouble: Making kin in the Chthulucene.* Duke University Press.
Hare, R. D. (1991). *The Hare psychopathy checklist—Revised.* Multi-Health Systems.
Harris, C., Hemer, S., & Chur-Hansen, A. (2021). Emotion as motivator: Parents, professionals and diagnosing childhood deafness. *Medical Anthropology, 40*(3), 254–266.
Hart, S., Cox, D., & Hare, R. (1995). *The Hare psychopathy checklist: Screening version.* Multi-Health Systems.
Hart, T. (1997). Transcendental empathy in the therapeutic encounter. *The Humanistic Psychologist, 25*(3), 245–270.
Harvey, G. (2013). Introduction. In G. Harvey (Ed.), *The handbook of contemporary animism* (pp. 15–16). Routledge.
Haslbeck, F. (2016). Three little wonders: Music therapy with families in neonatal care. In S. Lindahl Jacobsen & G. Thompson (Eds.), *Music therapy with families: Therapeutic approaches and theoretical perspectives* (pp. 19–44). Jessica Kingsley.
Hasler, J. (2017). Healing rhythms: Music therapy for attachment. In A. Hendry & J. Hasler (Eds.), *Creative therapies for complex trauma* (pp. 135–153). Jessica Kingsley.
Hasson, Y., Tamir, M., Brahms, K. S., Cohrs, J. C., & Halperin, E. (2018). Are liberals and conservatives equally motivated to feel empathy toward others? *Personality and Social Psychology Bulletin, 44*(10), 1449–1459.
Hatfield, E., Cacioppo, J., & Rapson, R. (1994). *Emotional contagion.* Cambridge University Press.

Hausman, C. R. (1991). Language and metaphysics: The ontology of metaphor. *Philosophy & Rhetoric, 24*(1), 25–42.
Haviland, J. (2014). *Exploring empathy in music therapy* [Masters Dissertation, Molloy College].
Hawkesworth, M. (2016). *Embodied power: Demystifying disembodied politics.* Routledge.
Hayes, J. A., Gelso, C. J., & Hummel, A. M. (2011). Managing countertransference. *Psychotherapy, 48*, 88–97.
Hays, P. (1996). Addressing the complexities of a culture and gender in counseling. *Journal of Counseling and Development, 74*, 332–337.
Hays, T., & Minichiello, V. (2005). The meaning of music in the lives of older people: A qualitative study. *Psychology of Music, 33*(4), 437–451.
Heidegger, M. (1962). *Being and time.* Harper & Row.
Hein, G., & Singer, T. (2008). I feel how you feel but not always: The empathic brain and its modulation. *Current Opinion Neurobiology, 18*(2), 153–158.
Henderson, C., Noblett, J., Parke, H., Clement, S., Caffrey, A., Gale-Grant, O., Schulze, B., Druss, B., & Thornicroft, G. (2014). Mental health-related stigma in health care and mental health-care settings. *The Lancet Psychiatry, 1*(6), 467–482.
Herman, J. (1997). *Trauma and recovery: The aftermath of violence: from domestic abuse to political terror.* Basic Books.
Hermansson, G. (1997). Boundaries and boundary management in counselling: The never-ending story. *British Journal of Guidance and Counselling, 25*(2), 133–146.
Hermberg, K. (2006). *Husserl's phenomenology: Knowledge, objectivity and others.* Continuum International Publishing Group.
Hesse-Biber, S. N. (2016). *The Oxford handbook of multimethod and mixed methods inquiry.* Oxford University Press.
Hickey-Moody, A., & Malins, P. (2007). Introduction: Gilles Deleuze and four movements in social thought. In A. Hickey-Moody & P. Malins (Eds.), *Deleuzian encounters: Studies in contemporary social issues* (pp. 1–26). Palgrave Macmillan.
Hiller, J. (2015). Aesthetic foundations of music therapy: Music and emotion. In B. Wheeler (Ed.), *Music therapy handbook* (pp. 29–39). The Guilford Press.
Hinojosa, K. (2021). *Building community and finding identity: Queer sounds, an inclusive school based music therapy group* [Doctoral dissertation, Saint Mary's College of California].

Hirschfeld, R., Montgomery, S. A., Keller, M. B., Kasper, S., Schatzberg, A. F., Möller, H.-J., et al. (2000). Social functioning in depression: A review. *Journal of Clinical Psychiatry, 61*, 268–275.
Hobson, P. (2007). Empathy and autism. In T. Farrow & P. Woodruff (Eds.), *Empathy in mental illness* (pp. 126–141). Cambridge University Press.
Hodges, S. D. (2005). Is how much you understand me in your head or mine? In B. F. Malle & S. D. Hodges (Eds.), *Other minds* (pp. 298–309). The Guilford Press.
Hoffer, E. (1955). *The passionate state of mind.* Harper and Brothers.
Hoffman, M. (2000). *Empathy and moral development*. Cambridge University Press.
Hoffman, M. L. (1991). Is empathy altruistic? *Psychological Inquiry, 2*(2), 131–133.
Hogeveen, J., Inzlicht, M., & Obhi, S. S. (2013). Power changes how the brain responds to others. *Journal of Experimental Psychology. General, 142*, 1–9.
Hojat, M. (2016). *Empathy in health professions education and patient care.* Springer.
Holdstock, T. L. (2000). *Re-examining psychology: Critical perspectives and African insights*. Routledge.
Hollan, D. (2008). Being there: On the imaginative aspects of understanding others and being understood. *Ethos, 36*(4), 475–489.
Hollan, D., & Throop, J. (2011). The anthropology of empathy: Introduction. In D. Hollan & J. Throop (Eds.), *The anthropology of empathy: Experiencing the lives of others in Pacific societies* (pp. 1–24). Berghahn Books.
Hollenstein, T., Allen, N. B., & Sheeber, L. (2016). Affective patterns in triadic family interactions: Associations with adolescent depression. *Development and Psychopathology, 28*(1), 85–96.
Holmes, M. (2004). Introduction: The importance of being angry: Anger in political life. *European Journal of Social Theory, 7*, 123–132.
Honan, E., & Sellers, M. (2006). *So how does it work?—Rhizomatic methodologies.* Paper presented at the AARE Annual Conference, Adelaide. http://www.aare.edu.au/publications-database.php/5086/so-how-does-it-work-rhizomatic-methodologies
hooks, b. (1991). Theory as liberatory practice. *Yale Journal of Law & Feminism, 4* (1).
hooks, b. (2014). *Teaching to transgress*. Routledge.
Hornstein, H. A. (1991). Empathic distress and altruism: Still inseparable. *Psychological Inquiry, 2*(2), 133–135.

Horwitz, E. B. (2018). Humanizing the working environment in health care through music and movement. In L. O. Bonde & T. Theorell (Eds.), *Music and public health: A Nordic perspective* (pp. 187–199). Springer.

Hosie, J. A., Gray, C. D., Russell, P. A., Scott, C., & Hunter, N. (1998). The matching of facial expressions by deaf and hearing children and their production and comprehension of emotion labels. *Motivation and Emotion, 22*(4), 293–313.

Howell, G. (2018). Harmony. *Music & Arts in Action, 6*(2), 45–58.

Hunt, M. (2005). Action research and music therapy: Group music therapy with young refugees in a school community. *Voices: A World Forum for Music Therapy, 5*(2).

Hunt, P., Denieffe, S., & Gooney, M. (2017). Burnout and its relationship to empathy in nursing: A review of the literature. *Journal of Research in Nursing, 22*(1–2), 7–22.

Hunter, C. (2020). I am with you in your pain: Privilege, humanity, and cultural humility in social work. *Reflections: Narratives of Professional Helping, 26*(2), 89–100.

Husserl, E. (1969). *Formal and transcendental logic*. Martinus Nijhoff.

Husserl, E. (1977). *Cartesian meditations* (D. Cairns, Trans.). Kluwer Academic.

Husserl, E. (1989). *Ideas pertaining to a pure phenomenology and to a phenomenological philosophy, second book: Studies in the phenomenology of constitution* (R. Rojcewicz & A. Schuwer, Trans.). Kluwer.

Hussey, D. L., Reed, A. M., Layman, D. L., & Pasiali, V. (2008). Music therapy and complex trauma: A protocol for developing social reciprocity. *Residential Treatment for Children & Youth, 24*(1–2), 111–129.

Hycner, R. (1993). *Between person and person: Toward a dialogical psychotherapy.* Gestalt Journal Press.

Iacoboni, M. (2009). Imitation, empathy, and mirror neurons. *Annual Review of Psychology, 60*, 653–670.

Ickes, W. (2009). Empathic accuracy: Its links to clinical, cognitive, developmental, social, and physiological psychology. In J. Decety & W. Ickes (Eds.), *The social neuroscience of empathy* (pp. 57–70). The MIT Press.

Illouz, E., Gilon, D., & Shachak, M. (2014). Emotions and cultural theory. In J. Stets & J. Turner (Eds.), *Handbook of the sociology of emotions* (Vol. II, pp. 221–244). Springer.

Impey, A. (2013). The poetics of transitional justice in Dinka songs in South Sudan. *UNISCI Discussion Papers, 33*, 57–77.

Indigenous Action. (2014). *Accomplices not allies: Abolishing the ally industrial network*. http://www.indigenousaction.org/accomplices-not-allies-abolishing-the-ally-industrial-complex/
Ingold, T. (2000). *The perception of the environment: Essays on livelihood, dwelling and skill*. Routledge.
Ingold, T. (2007). *Lines: A brief history*. Routledge.
Ingold, T. (2008). Bindings against boundaries: Entanglements of life in an open world. *Environment and Planning A, 40*, 1796–1810.
Ingold, T. (2010). The textility of making. *Cambridge Journal of Economics, 34*(1), 91–102.
Ip, K. I., Miller, A. L., Karasawa, M., Hirabayashi, H., Kazama, M., Wang, L., Olson, S. L., Kessler, D., & Tardif, T. (2021). Emotion expression and regulation in three cultures: Chinese, Japanese, and American preschoolers' reactions to disappointment. *Journal of Experimental Child Psychology, 201*, 104972.
Ip-Winfield, V., & Grocke, D. (2011). Group music therapy methods in cross-cultural aged care practice in Australia. *Australian Journal of Music Therapy, 22*, 55–80.
Ivey, A., Daniels, T., Zalaquett, C., & Ivey, M. (2017). Neuroscience of attention: Empathy and counselling skills. In T. Field, L. Jones, & L. Russell-Chapin (Eds.), *Neurocounseling: Brian-based clinical approaches* (pp. 83–100). American Counseling Association.
Jack, K., & Smith, A. (2007). Promoting self-awareness in nurses to improve nursing practice. *Nursing Standard, 21*(32), 47–52.
Jackendoff, R., & Lerdahl, F. (2006). The capacity for music: What is it, and what's special about it? *Cognition, 100*(1), 33–72.
Jackson, A. (2003). Rhizovocality. *Qualitative Studies in Education, 16*(5), 693–710.
Jackson, B. A., & Harvey Wingfield, A. (2013). Getting angry to get ahead: Black college men, emotional performance, and encouraging respectable masculinity. *Symbolic Interaction, 36*(3), 275–292.
Jackson, N., & Gardstrom, S. (2012). Undergraduate music therapy students' experiences as clients in short-term group music therapy. *Music Therapy Perspectives, 30*(1), 65–82.
Jackson, N. A. (2010). Models of response to client anger in music therapy. *The Arts in Psychotherapy, 37*(1), 46–55.
Jacobowitz, R. M. (1992). Music therapy in the short-term pediatric setting: Practical guidelines for the limited time frame. *Music Therapy, 11*(1), 45–64.

Janata, P., Tomic, S. T., & Haberman, J. M. (2012). Sensorimotor coupling in music and the psychology of the groove. *Journal of Experimental Psychology: General, 141*(1), 54.

Jankowiak-Siuda, K., & Zajkowski, W. (2013). A neural model of mechanisms of empathy deficits in narcissism. *Medical Science Monitor, 19*, 934–941.

Janus, S. I., Vink, A. C., Ridder, H. M., Geretsegger, M., Stige, B., Gold, C., & Zuidema, S. U. (2020). Developing consensus description of group music therapy characteristics for persons with dementia. *Nordic Journal of Music Therapy*, 1–17.

Jardine, J. (2014). Husserl and Stein on the phenomenology of empathy: Perception and explication. *Synthesis Philosophica, 58*, 273–288.

Jaspers, K. (1997). *General psychopathology* (J. Hoenig & M. Hamilton, Trans.). Johns Hopkins University Press.

Jeffrey, D., & Downie, R. (2016). Empathy-can it be taught? *Journal of the Royal College of Physicians of Edinburgh, 46*(2), 107–112.

Johansson, L. (2016). Post-qualitative line of flight and the confabulative conversation: A methodological ethnography. *International Journal of Qualitative Studies in Education, 29*(4), 445–466.

Johns, U. T. (2018). Exploring musical dynamics in therapeutic interplay with children: A multilayered method of microanalysis. *Nordic Journal of Music Therapy, 27*(3), 197–217.

Johnson, D. R., Cushman, G. K., Borden, L. A., & McCune, M. S. (2013). Potentiating empathic growth: Generating imagery while reading fiction increases empathy and prosocial behavior. *Psychology of Aesthetics, Creativity, and the Arts, 7*(3), 306.

Johnson, G. (2009). Theories of emotion. In J. Fieser & B. Dowden (Eds.), *The internet encyclopedia of philosophy*. https://iep.utm.edu/emotion/

Johnson, J. L., Bottorff, J. L., Browne, A. J., Grewal, S., Hilton, B. A., & Clarke, H. (2004). Othering and being othered in the context of health care services. *Health Communication, 16*(2), 255–271.

Johnston, K. (2018). Great reckonings in more accessible rooms: The provocative reimaginings of disability theatre. In B. Hadley & D. McDonald (Eds.), *The Routledge handbook of disability arts, culture, and media* (pp. 19–35). Routledge.

Jolliffe, D., & Farrington, D. P. (2006). Examining the relationship between low empathy and bullying. *Aggressive Behavior: Official Journal of the International Society for Research on Aggression, 32*(6), 540–550.

Jones, A., Gutierrez, R., & Ludlow, A. (2021). Emotion production of facial expressions: A comparison of deaf and hearing children. *Journal of Communication Disorders,* 106113.

Jones, A., & Oldfield, A. (1999). Sharing sessions with John. In J. Hibben (Ed.), *Inside music therapy: Client experiences* (pp. 168–173). Barcelona Publishers.

Jones, S. (2011). Supportive listening. *The International Journal of Listening, 25*(1–2), 85–103.

Jones, S. (2018). *Effect of vocal style on perceived empathy, rapport, patient engagement, and competency of music therapists* [Masters Thesis, Florida State University].

Jordan, J. (2000). The role of mutual empathy in relation/cultural therapy. *Psychotherapy in Practice, 56*(5), 1005–1016.

Jurist, E. L. (2010). Mentalizing minds. *Psychoanalytic Inquiry, 30*(4), 289–300.

Juslin, P. (2013). From everyday emotions to aesthetic emotions: Towards a unified theory of musical emotions. *Physics of Life Reviews, 10*(3), 235–266.

Juslin, P. (2016). Emotional reactions to music. In S. Hallam, I. Cross, & M. Thaut (Eds.), *The Oxford handbook of music psychology* (pp. 197–213). Oxford University Press.

Juslin, P., & Sloboda, J. (2010). *Handbook of music and emotion: Theory, research, and applications.* Oxford University Press.

Juslin, P., & Västfjäll, D. (2008). Emotional responses to music: The need to consider underlying mechanisms. *Behavioral and Brain Sciences, 31*(6), 751.

Kagumba, A. K. (2021). *Orchestrating social competence: On the transformative work of musicking in two Ugandan NGOs (M-LISADA and Brass for Africa)* [Doctoral dissertation, Texas Tech University].

Kamens, S. (2019). De-othering "schizophrenia." *Theory & Psychology, 29*(2), 200–218.

Kaminer, D. (2006). Healing processes in trauma narratives: A review. *South African Journal of Psychology, 36*(3), 481–499.

Kan, Y., Mimura, M., Kamijima, K., & Kawamura, M. (2004). Recognition of emotion from moving facial and prosodic stimuli in depressed patients. *Journal of Neurology, Neurosurgery & Psychiatry, 75*(12), 1667–1671.

Kaneh-Shalit, T. (2017). The goal is not to cheer you up: Empathetic care in Israeli life coaching. *Ethos, 45*(1), 98–115.

Kantor, J. (2020, June). "How well do I know you?": Intersubjective perspectives in music therapy when working with persons with profound intellectual and multiple disability. *Voices: A World Forum for Music Therapy, 20*(2).

Kappas, A. (2013). Social regulation of emotion: Messy layers. *Frontiers in Psychology, 4*, 51.

Karasu, T. B. (1992). *Wisdom in the practice of psychotherapy*. Basic Books.

Kasari, C., Freeman, S. F., & Bass, W. (2003). Empathy and response to distress in children with Down syndrome. *Journal of Child Psychology and Psychiatry, 44*(3), 424–431.

Kavaliova-Moussi, A. (2017). Discovering Arab/Middle Eastern culture. In A. Whitehead-Pleaux & X. Tan (Eds.), *Cultural intersections in music therapy: Music, health, and the person* (pp. 91–104). Barcelona Publishers.

Keil, C. (1994). Participatory discrepancies and the power of music. In C. Keil & S. Feld (Eds.), *Music grooves* (pp. 275–283). University of Chicago Press.

Keller, M. S. (2012). Zoomusicology and ethnomusicology: A marriage to celebrate in heaven. *Yearbook for Traditional Music, 44*, 166–183.

Kenny, C. (1999). Beyond this point there be dragons: Developing general theory in music therapy. *Nordic Journal of Music Therapy, 8*(2), 127–136.

Kenny, C. (2016). The field of play: A focus on energy and the ecology of being and playing. In J. Edwards (Ed.), *The Oxford handbook of music therapy* (pp. 472–482). Oxford University Press.

Kerig, P. (2020). Emotion dysregulation and childhood trauma. In T. Beauchaine & S. Crowell (Eds.), *The Oxford handbook of emotion dysregulation* (pp. 265–282). Oxford University Press.

Kerr-Gaffney, J., Harrison, A., & Tchanturia, K. (2019). Cognitive and affective empathy in eating disorders: A systematic review and meta-analysis. *Frontiers in Psychiatry, 10*, 102.

Kesner, L., & Horáček, J. (2017). Empathy-related responses to depicted people in art works. *Frontiers in Psychology, 8*, 228.

Kessler, N. (2019). *Ontology and closeness in human-nature relationships.* Springer.

Keysers, C., Kohler, E., Umiltà, M. A., Nanetti, L., Fogassi, L., & Gallese, V. (2003). Audiovisual mirror neurons and action recognition. *Experimental Brain Research, 153*(4), 628–636.

Kidwell, M. (2014). Music therapy and spirituality: How can I keep from singing? *Music Therapy Perspectives, 32*, 129–134.

Kim, J. (2014). The trauma of parting: Endings of music therapy with children with autism spectrum disorders. *Nordic Journal of Music Therapy, 23*(3), 263–281.
Kim, S. (2013). *Multimodal quantification of interpersonal physiological synchrony between non-verbal individuals with severe disabilities and their caregivers during music therapy* [Doctoral dissertation, University of Toronto].
Kim, S., & Whitehead-Pleaux, A. (2015). Music therapy and cultural diversity. In B. Wheeler (Ed.), *Music therapy handbook* (pp. 51–63). Guilford Press.
Kindler, R. C., & Gray, A. A. (2010). Theater and therapy: How improvisation informs the analytic hour. *Psychoanalytic Inquiry, 30*, 254–266.
King, K. (2021). Musical and cultural considerations for building rapport in music therapy practice. In M. Belgrave & S. Kim (Eds.), *Music therapy in a multicultural context: A handbook for music therapy students and professionals* (pp. 43–74). Barcelona.
King, R. (2018). *Mindfulness of race: Transforming racism from the inside out.* Sounds True.
Kinman, C. J., Finck, P., & Hoffman, L. (2004). Response-able practice. In T. Strong & D. Pare (Eds.), *Furthering talk* (pp. 233–251). Springer.
Kirmayer, L. J. (2008). Empathy and alterity in cultural psychiatry. *Ethos, 36*(4), 457–474.
Kitanaka, J. (2012) *Depression in Japan: Psychiatric cures for a society in distress.* Princeton University Press.
Kitayama, S., Mesquita, B., & Karasawa, M. (2006). The emotional basis of independent and interdependent selves: Socially disengaging and engaging emotions in the U.S. and Japan. *Journal of Personality and Social Psychology, 91*(5), 890–903.
Kittay, J. (2008). The sound surround. *Nordic Journal of Music Therapy, 17*(1), 41–54.
Klein, M. (2004). Chopin's Fourth Ballade as musical narrative. *Music Theory Spectrum—The Journal of the Society for Music Theory, 26*(1), 23–55.
Kleres, J. (2011). Emotions and narrative analysis: A methodological approach. *Journal for the Theory of Social Behaviour, 41*(2), 182–202.
Klyve, G. P. (2019). Whose knowledge? Epistemic injustice and challenges in hearing childrens' voices. *Voices: A World Forum for Music Therapy, 19*(3).
Knaak, S., Mantler, E., & Szeto, A. (2017). Mental illness-related stigma in healthcare: Barriers to access and care and evidence-based solutions. *Healthcare Management Forum, 30*(2), 111–116.

Knafo, A., Zahn-Waxler, C., van Hulle, C., Robinson, J. L., & Rhee, S. H. (2008). The developmental origins of a disposition toward empathy: Genetic and environmental contributions. *Emotion, 8*, 737–752.

Knapp, M., & Hall, J. A. (2010). *Nonverbal communication in human interaction* (7th ed.). Cengage Learning.

Knapp, S., Gottlieb, M. C., & Handelsman, M. M. (2017). Self-awareness questions for effective psychotherapists: Helping good psychotherapists become even better. *Practice Innovations, 2*(4), 163.

Knight, A. (2013). Uses of iPad® applications in music therapy. *Music Therapy Perspectives, 31*(2), 189–196.

Knott, D., & Block, S. (2020). Virtual music therapy: Developing new approaches to service delivery. *Music Therapy Perspectives, 38*(2), 151–156.

Kobin, C., & Tyson, E. (2006). Thematic analysis of hip-hop music: Can hip-hop in therapy facilitate empathic connections when working with clients in urban settings? *The Arts in Psychotherapy, 33*(4), 343–356.

Koch, S., & Fuchs, T. (2011). Embodied arts therapies. *The Arts in Psychotherapy, 38*, 278–280.

Koelsch, S., & Skouras, S. (2014). Functional centrality of amygdala, striatum and hypothalamus in a "small-world" network underlying joy: An fMRI study with music. *Human Brain Mapping, 35*(7), 3485–3498.

Koenig, J. (2021). *The dictionary of obscure sorrows*. Simon and Schuster.

Kofoed, J., & Ringrose, J. (2012). Travelling and sticky affects: Exploring teens and sexualized cyberbullying through a Butlerian-Deleuzian-Guattarian lens. *Discourse: Studies in the Cultural Politics of Education, 33*(1), 5–20.

Kogut, T., & Ritov, I. (2005). The "identified victim" effect: An identified group, or just a single individual? *Journal of Behavioral Decision Making, 18*(3), 157–167.

Kohler, E., Keysers, C., Umilta, M. A., Fogassi, L., Gallese, V., & Rizzolatti, G. (2002). Hearing sounds, understanding actions: Action representation in mirror neurons. *Science, 297*(5582), 846–848.

Kohut, H. (1968). The psychoanalytic treatment of narcissistic personality disorders, outline of a systematic approach. *The Psychoanalytic Study of the Child, 23*, 86–113.

Kohut, H. (1971). *The analysis of the self.* International Universities Press.

Kohut, H. (1977). *The restoration of the self.* International Universities Press.

Kohut, H. (1984). *How does analysis cure?* University of Chicago Press.

Kopf, M. (2010). Trauma, narrative and the art of witnessing. In B. Haehnel & M. Ulz (Eds.), *Slavery in art and literature* (pp. 41–58). Frank & Timme.

Korpela, K. M., Ylén, M., Tyrväinen, L., & Silvennoinen, H. (2008). Determinants of restorative experiences in everyday favorite places. *Health & Place, 14*(4), 636–652.

Kottler, J., & Carlson, J. (2014). *On being a master therapist*. Wiley.

Kovach, M. (2012). *Indigenous methods: Characteristics, conversation, and contexts*. University of Toronto Press.

Kraus, M. W., Cote, S., & Keltner, D. (2010). Social class, contextualism, and empathic accuracy. *Psychological Science, 21*(11), 1716–1723.

Krol, S., & Bartz, J. (2021). The self and empathy: Lacking a clear and stable sense of self undermines empathy and helping behavior. *Emotion, Advance Online Publication*. https://doi.org/10.1037/emo0000943

Kroll, S. L., Wunderli, M. D., Vonmoos, M., Hulka, L. M., Preller, K. H., Bosch, O. G., Baumgartner, M. R., & Quednow, B. B. (2018). Sociocognitive functioning in stimulant polysubstance users. *Drug and Alcohol Dependence, 190*, 94–103.

Krueger, J. (2013). Empathy, enaction, and shared musical experience: Evidence from infant cognition. In T. Cochrane, B. Fantini, & K. Scherer (Eds.), *The emotional power of music* (pp. 177–196). Oxford University Press.

Krueger, J. (2014). Affordances and the musically extended mind. *Frontiers in Psychology, 4*, 1003.

Kwan, M. (2010). Music therapists' experiences with adults in pain: Implications for clinical practice. *Qualitative Inquiries in Music Therapy, 5*, 43.

Kwuon, S. (2009). An examination of cue-redundancy theory in cross-cultural decoding of emotions in music. *Journal of Music Therapy, 46*(3), 217–237.

Labanyi, J. (2010). Doing things: Emotion, affect, and materiality. *Journal of Spanish Cultural Studies,* 11(3–4). Special Issue: Cultural/political reflection—Lines, routes, spaces

Labov, W. (1972). *Language in the inner city*. Blackwell.

LaCapra, D. (2001). *Writing history, writing trauma*. Johns Hopkins University Press.

Laiho, A. (2009). The psychological functions of music in adolescence. *Nordic Journal of Music Therapy, 13*(1), 47–63.

Lakeman, R. (2020). Advanced empathy: A key to supporting people experiencing psychosis or other extreme states. *Psychotherapy and Counselling Journal of Australia, 8*(1).

Lambert, M. J., & Barley, D. E. (2001). Research summary on the therapeutic relationship and psychotherapy outcome. *Psychotherapy: Theory, Research, Practice, Training, 38*(4), 357.

Lamm, C., Batson, C. D., & Decety, J. (2007). The neural substrate of human empathy: Effects of perspective-taking and cognitive appraisal. *Journal of Cognitive Neuroscience, 19*(1), 42–58.

Lamm, C., Decety, J., & Singer, T. (2011). Meta-analytic evidence for common and distinct neural networks associated with directly experienced pain and empathy for pain. *NeuroImage, 54*, 2492–2502.

Lancione, M. (2013). Homeless people and the city of abstract machines: Assemblage thinking and the performative approach to homelessness. *Area, 45*(3), 358–364.

Landau, M. J., Vess, M., Arndt, J., Rothschild, Z. K., Sullivan, D., & Atchley, R. A. (2011). Embodied metaphor and the "true" self: Priming entity expansion and protection influences intrinsic self-expressions in self-perceptions and interpersonal behavior. *Journal of Experimental Social Psychology, 47*(1), 79–87.

Langdon, P. E., Murphy, G. H., Clare, I. C. H., Stevenson, T., & Palmer, E. J. (2011). Relationships among moral reasoning, empathy and distorted cognitions in men with intellectual disabilities and a history of criminal offending. *American Association on Intellectual and Developmental Disabilities, 116*(6), 438–456.

Langer, S. K. (1953). *Feeling and form: A theory of art.* Routledge.

LaPierre, L. L. (1994). A model for describing spirituality. *Journal of Religion and Health, 33*(2), 153–161.

Latour, B. (1996). On interobjectivity. *Mind, Culture and Activity, 3*(4), 228–245.

Laub, D. (2013). Bearing witness, or the vicissitudes of listening. In S. Felman & D. Laub (Eds.), *Testimony* (pp. 77–94). Routledge.

Laukka, P., Eerola, T., Thingujam, N. S., Yamasaki, T., & Beller, G. (2013). Universal and culture-specific factors in the recognition and performance of musical affect expressions. *Emotion, 13*(3), 434–449.

Laurence, F. (2008). Music and empathy. In O. Urbain (Ed.), *Music and conflict: Harmonies and dissonances in geopolitics* (pp. 85–98). I. B. Tauris.

Lavallée, L. F. (2009). Practical application of an indigenous research framework and two qualitative indigenous research methods: Sharing Circles and Anishnaabe symbol-based reflection. *International Journal of Qualitative Methods, 8*(1), 21–40.

Law J. (2004). *After method: Mess in social science research.* Routledge.

Law, J., & Urry, J. (2004). Enacting the social. *Economy and Society, 33*(3), 390–410.

Layman, D. L., Hussey, D. L., & Reed, A. M. (2013). The beech brook group therapy assessment tool: A pilot study. *Journal of Music Therapy, 50*(3), 155–175.

Lazard, L., & McAvoy, J. (2020). Doing reflexivity in psychological research: What's the point? What's the practice? *Qualitative Research in Psychology, 17*(2), 159–177.

Le Poidevin, R. (2005). Narrative. In T. Honderich (Ed.), *The Oxford companion to philosophy* (pp. 638–639). Oxford University Press.

LeBaron, M., & Alexander, N. M. (2012). Dancing to the rhythm of the role-play, applying dance intelligence to conflict resolution. *Applying Dance Intelligence to Conflict Resolution, 33*, 2.

Lederach, J. (2015). *Little book of conflict transformation.* Simon and Schuster.

Lederach, J. P. (2005). *The moral imagination.* Oxford University Press.

Lee, C. (1996). *Music at the edge: The music therapy experience of a musician with AIDS.* Routledge.

Lee, J. H. (2016). A qualitative inquiry of the lived experiences of music therapists who have survived cancer who are working with medical and hospice patients. *Frontiers in Psychology, 7*, 1840.

Lee, K. (2007). Empathy deficits in schizophrenia. In T. Farrow & P. Woodruff (Eds.), *Empathy in mental illness* (pp. 17–32). Cambridge University Press.

Lee Soon, R. (2016). Nohana i waena i na mo'olelo/living between the stories: Contextualising drama therapy within an indigenous Hawaiian epistemology. *Drama Therapy Review, 2*(2), 257–271.

Lehne, M., Rohrmeier, M., & Koelsch, S. (2014). Tension-related activity in the orbitofrontal cortex and amygdala: An fMRI study with music. *Social Cognitive and Affective Neuroscience, 9*(10), 1515–1523.

Leiker, E. K., Meffert, H., Thornton, L. C., Taylor, B. K., Aloi, J., Abdel-Rahim, H., Shah N., Tyler P. M., White S. F., Filbey F., Pope K., Do M. D., & Blair, R. J. R. (2019). Alcohol use disorder and cannabis use disorder symptomatology in adolescents are differentially related to dysfunction in brain regions supporting face processing. *Psychiatry Research: Neuroimaging, 292*, 62–71.

Leite, T. (2003). Music, metaphor and "being with the other". *Voices: A World Forum for Music Therapy, 3*(2).

Leman, M. (2007). *Embodied music cognition and mediation technology.* The MIT Press.

Lench, H., & Carpenter, Z. (2018). What do emotions do for us? In H. Lench (Ed.), *The functions of emotions: When and why emotions help us* (pp. 1–8). Springer.

Leonard, H. (2020). The arts are for freedom: Centering Black embodied music to make music free. *Journal of Performing Art Leadership in Higher Education, 11*, 4–25.

Lerner M. (1980). *The belief in a just world: A fundamental delusion.* Plenum.

Levenson, R. W., Ekman, P., Heider, K., & Friesen, W. V. (1992). Emotion and autonomic nervous system activity in the Minangkabau of West Sumatra. *Journal of Personality and Social Psychology, 62*(6), 972–988.

Levinas, E. (1948). *Le Temps et l'Autre*. Presses Universitaires de France.

Levine, P. (2010). *In an unspoken voice: How the body releases trauma and restores goodness.* North Atlantic Books.

Levine, L. J., Lench, H. C., Kaplan, R. L., & Safer, M. A. (2012). Accuracy and artifact: Re-examining the intensity bias in affective forecasting. *Journal of Personality and Social Psychology, 103*, 584–605.

Levinson, J. (2006). *Contemplating art: Essays in aesthetics.* Clarendon Press.

Levitin, D. (2006). *This is your brain on music.* Dutton.

Lewis, C. S. (1990). *Studies in words*. Cambridge University Press.

Liebmann, M. (2007). *Restorative justice: How it works.* Jessica Kingsley.

Lim, M. (2007). The ethics of alterity and the teaching of otherness. *Business Ethics: A European Review, 16*(3), 251–263.

Lincoln, Y. S., & Guba, E. G. (1985). Establishing trustworthiness. *Naturalistic inquiry, 289*(331), 289–327.

Lindvang, C. (2013). Resonant learning: A qualitative inquiry into music therapy students' self-experiential learning processes. *Qualitative Inquiries in Music Therapy, 8*, 1–31.

Lindvang, C., & Frederiksen, B. (1999). Suitability for music therapy: Evaluating music therapy as an indicated treatment in psychiatry. *Nordic Journal of Music Therapy, 8*(1), 47–57.

Lipari, L. (2010). Listening, thinking, being. *Communication Theory, 20*, 348–362.

Lipps, T. (1903/1931). Empathy, inward imitation, and sense feelings. In E. F. Carritt (Ed.), *Philosophies of beauty: From socrates to Robert Bridges being the sources of aesthetic theory* (pp. 252–258). Clarendon Press.

Lipson, J. (2019). Seeking the uncensored self. In B. MacWilliam, B. Harris, D. Trottier, & K. Long (Eds.), *Creative arts therapies and the LGBTQ community: Theory and practice* (pp. 171–184). Jessica Kingsley.

Lishner, D. A., Batson, C. D., & Huss, E. (2011). Tenderness and sympathy: Distinct empathic emotions elicited by different forms of need. *Personality and Social Psychology Bulletin, 37*, 614–625.

Lively, K. J., & Weed, E. A. (2014). Emotions management: Sociological insight into what, how, why, and to what end? *Emotion Review, 6*(3), 202–207.

Loaiza, J. (2016). Musicking, embodiment and participatory enaction of music: Outline and key points. *Connection Science, 28*(4), 410–422.

Lockwood, P. L. (2016). The anatomy of empathy: Vicarious experience and disorders of social cognition. *Behavioural Brain Research, 311*, 255–266.

Loewenstein, G. (2005). Hot-cold empathy gaps and medical decision making. *Health Psychology, 24*, S49.

Long Soldier, L. (2017). *Whereas*. Graywolf Press.

López-Pérez, B., Carrera, P., Ambrona, T., & Oceja, L. (2014). Testing the qualitative differences between empathy and personal distress: Measuring core affect and self-orientation. *The Social Science Journal, 51*, 676–680.

Lotter, C. (2017). *The qualitative affordances of active and receptive music therapy techniques in major depressive disorder and schizophrenia-spectrum psychotic disorders* [Doctoral thesis, University of Pretoria].

Lougheed, J. P., Koval, P., & Hollenstein, T. (2016). Sharing the burden: The interpersonal regulation of emotional arousal in mother–daughter dyads. *Emotion, 16*(1), 83.

Louw, W. (2013). *Community-based educational programmes as support structures for adolescents within the context of HIV and AIDS* [Doctoral dissertation, University of Pretoria].

Low, M. Y., Kuek Ser, S. T., & Kalsi, G. K. (2020). Rojak: An ethnographic exploration of pluralism and music therapy in post-British-colonial Malaysia. *Music Therapy Perspectives, 38*(2), 119–125.

Lugones, M. (1987). Playfulness, "world"—Travelling, and loving perception. *Hypatia, 2*(2), 3–19.

Lutz, C., & White, G. M. (1986). The anthropology of emotions. *Annual Review of Anthropology, 15*, 405–436.

Ly, R. (2014). *Beyond strategies: Infusing empathy and indigenous approaches in the elementary classroom* [Masters dissertation, University of Toronto].

Lyman, P. (2004). The domestication of anger: The use and abuse of anger in politics. *European Journal of Social Theory, 7*(2), 133–147.

Ma-Kellams, C., & Lerner, J. (2016). Trust your gut or think carefully? Examining whether an intuitive, versus a systematic, mode of thought produces greater empathic accuracy. *Journal of Personality and Social Psychology, 111*(5), 674.

MacLure, M. (2013). Researching without representation? Language and materiality in post-qualitative methodology. *International Journal of Qualitative Studies in Education, 26*(6), 658–667.

Magnusson, T. (2009). Of epistemic tools: Musical instruments as cognitive extensions. *Organised Sound, 14*, 168–176.

Maibom, H. (2017). Affective empathy. In H. Maibom (Ed.), *The Routledge handbook of philosophy of empathy* (pp. 22–32). Routledge.

Malloch, S., & Trevarthen, C. (2009). Musicality: Communicating the vitality and interests of life. In S. Malloch & C. Trevarthen (Eds.), *Communicative musicality* (pp. 1–10). Oxford University Press.

Maltsberger, J. (2011). Empathy and the historical context, or how we learned to listen to patients. In K. Michel & D. Jobes (Eds.), *Building a therapeutic alliance with the suicidal patient* (pp. 29–48). American Psychological Association.

Mann, D. (2020). *Gestalt therapy: 100 key points and techniques*. Routledge.

Mar, R. A., & Oatley, K. (2008). The function of fiction is the abstraction and simulation of social experience. *Perspectives on Psychological Science, 3*, 173–192.

Marangoni, G., Garcia, S., Ickes, W., & Teng, G. (1995). Empathic accuracy in a clinically relevant setting. *Journal of Personality and Social Psychology, 68*, 854–886.

Marci, C. D., Ham, J., Moran, E., & Orr, S. P. (2007). Physiologic correlates of perceived therapist empathy and social-emotional process during psychotherapy. *The Journal of Nervous and Mental Disease, 195*(2), 103–111.

Markovitch, N., Netzer, L., & Tamir, M. (2017). What you like is what you try to get: Attitudes toward emotions and situation selection. *Emotion, 17*, 728–739.

Markus, H. R., & Kitayama, S. (1994). The cultural shaping of emotion: A conceptual framework. In S. Kitayama & H. R. Markus (Eds.), *Emotion and culture: Empirical studies of mutual influence* (pp. 339–351). American Psychological Association.

Masny, D. (2013). Rhizoanalytic pathways in qualitative research. *Qualitative Inquiry, 19*(5), 339–348.

Masny, D. (2016). Problematizing qualitative research: Reading a data assemblage with rhizoanalysis. *Qualitative Inquiry, 22*(8), 666–675.

Massumi, B. (2002). *Parables for the virtual: Movement, affect, sensation.* Duke University Press.

Masuda, T., Ellsworth, P., Mesquita, B., Leu, J., Tanida, S., & Van de Veerdonk, E. (2008). Placing the face in context: Cultural differences in the perception of facial emotion. *Journal of Personality and Social Psychology, 94*(3), 365–381.

Masuda, T., Wang, H., Ishii, K., & Ito, K. (2012). Do surrounding figures' emotions affect judgment of the target figure's emotion? Comparing the eye-movement patterns of European Canadians, Asian Canadians, Asian international students, and Japanese. *Frontiers in Integrative Neuroscience, 6*, 72.

Matney, W. (2021). Music therapy as multiplicity: Implications for music therapy philosophy and theory. *Nordic Journal of Music Therapy, 30*(1), 3–23.

Matusov, E. (1996). Intersubjectivity without agreement. *Mind, Culture, and Activity, 3*(1), 25–45.

Maus, F. E. (1997). Narrative, drama, and emotion in instrumental music. *The Journal of Aesthetics and Art Criticism, 55*(3), 293–303.

Mawere, M., & Mubaya, T. (2016). *African philosophy and thought systems: A search for a culture and philosophy of belonging.* Langaa Research and Publishing.

Mazza, M., Tempesta, D., Pino, M. C., Nigri, A., Catalucci, A., Guadagni, V., Iaria, G., & Ferrara, M. (2015). Neural activity related to cognitive and emotional empathy in post-traumatic stress disorder. *Behavioural Brain Research, 282,* 37–45.

Mazzei, L., & Jackson, A. (2017). Voice in the agentic assemblage. *Educational Philosophy and Theory, 49*(11), 1090–1098.

Mbeki, T. (1996). *Statement on behalf of the African National Congress, on the occasion of the adoption by the Constitutional Assembly of 'The Republic of South Africa Constitutional Bill 1996.* Cape Town, 8 May—issued by the Office of the Deputy President. http://www.info.gov.za/speeches/1996/960819_23196.htm

Mbembe, A. (2015, April 23). *Decolonizing knowledge and the question of the archive.* Presentation at the University of the Witwatersrand, Johannesburg. https://wiser.wits.ac.za/system/files/Achille%20Mbembe%20-%20Decolonizing%20Knowledge%20and%2020the%20Question%20of%20the%20Archive.pdf

Mbiti, J. (1969). *African religion and philosophy*. Heinemann.

McCaffrey, T., Carr, C., Solli, H. P., & Hense, C. (2018). Music therapy and recovery in mental health: Seeking a way forward. *Voices: A World Forum for Music Therapy, 18*(1). https://doi.org/10.15845/voices.v18i1.918

McCaffrey, T., Higgins, P., Monahan, C., Moloney, S., Nelligan, S., Clancy, A., & Cheung, P. S. (2021). Exploring the role and impact of group songwriting with multiple stakeholders in recovery-oriented mental health services. *Nordic Journal of Music Therapy, 30*(1), 41–60.

McCormack, B., & McCance, T. (2016). *Person-centred practice in nursing and health care: Theory and practice.* Wiley.

McFerran, K. S., Garrido, S., & Saarikallio, S. (2016). A critical interpretive synthesis of the literature linking music and adolescent mental health. *Youth & Society, 48*, 521–538.

McFerran, K., & Teggelove, K. (2011). Music therapy with young people in schools: After the Black Saturday fires. *Voices: A World Forum for Music Therapy, 11*(1).

McGann, H. S. (2012). *Finding a place for music therapy practice in a hospital child development service* [Masters thesis, Victoria University of Wellington].

McGarry, L. M., & Russo, F. A. (2011). Mirroring in dance/movement therapy: Potential mechanisms behind empathy enhancement. *The Arts in Psychotherapy, 38*(3), 178–184.

McGrath, M., & Oakley, B. (2012). Codependency and pathological altruism. In B. Oakley, A. Knafo, G. Madhavan, & D. Sloan Wilson (Eds.), *Pathological altruism* (pp. 49–74). Oxford University Press.

McIntosh, D. N., Druckman, D., & Zajonc, R. B. (1994). Socially induced affect. In D. Druckman & R. A. Bjork (Eds.), *Learning, remembering, believing: Enhancing human performance* (pp. 251–276). National Academy Press.

McKearney, P. (2021). The limits of knowing other minds: Intellectual disability and the challenge of opacity. *Social Analysis, 65*(1), 1–22.

Mckeown, M., & Spandler, H. (2017). Exploring the case for truth and reconciliation in mental health services. *Mental Health Review Journal, 22*(2), 83–94.

McLeod, J. (2013). *An introduction to counselling*. Open University Press.

McLeod, K. (2017). *Wellbeing machine: How health emerges from the assemblages of everyday life.* Carolina Academic Press.

McPhie, J. (2019). *Mental health and wellbeing in the anthropocene: A posthuman inquiry.* Palgrave Macmillan.

Mcphie, J. (2018). I knock at the stone's front door: Performative pedagogies beyond the human story. *Parallax, 24*(3), 306–323.

Mead, G. H. (1934). *Mind, self and society* (Vol. 111). University of Chicago Press.

Meekums, B. (2012). Kinesthetic empathy and movement metaphor in dance movement psychotherapy. In D. Reynolds & M. Reason (Eds.), *Kinesthetic empathy in creative and cultural practices* (pp. 51–66). Intellect.

Meelberg, V. (2009). *Sonic strokes and musical gestures: The difference between musical affect and musical emotion.* In ESCOM 2009: 7th Triennial Conference of European Society for the Cognitive Sciences of Music.

Meffert, H., Gazzola, V., Den Boer, J. A., Bartels, A. A., & Keysers, C. (2013). Reduced spontaneous but relatively normal deliberate vicarious representations in psychopathy. *Brain, 136*(8), 2550–2562.

Meier, B. P., Hauser, D. J., Robinson, M. D., Friesen, C. K., & Schjeldahl, K. (2007). What's "up" with God? Vertical space as a representation of the divine. *Journal of Personality and Social Psychology, 93*, 699–710.

Mendenhall, E., Musau, A., Bosire, E., Mutiso, V., Ndetei, D., & Rock, M. (2020). What drives distress? Rethinking the roles of emotion and diagnosis among people with diabetes in Nairobi, Kenya. *Anthropology & Medicine, 27*(3), 252–267.

Menges, J. I., & Kilduff, M. (2015). Group emotions: Cutting the Gordian knots concerning terms, levels of analysis, and processes. *Academy of Management Annals, 9*(1), 845–928.

Menkiti, I. (2004). On the normative conception of person. In K. Wiredu (Ed.), *A companion to African philosophy* (pp. 324–331). Blackwell Publishing.

Menon, V., & Levitin, D. J. (2005). The rewards of music listening: Response and physiological connectivity of the mesolimbic system. *NeuroImage, 28*, 175–184.

Merleau-Ponty, M. (1962). *Phenomenology of perception* (C. Smith, Trans.). Routledge & Kegan Paul.

Merriam Webster. (2014). *Merriam Webster Online*. https://www.merriam-webster.com

Mesquita, B., de leersnyder, J., & Boiger, M. (2016). The cultural psychology of emotions. In L. Feldman Barrett, M. Lewis, J. Haviland-Jones (Eds.), *Handbook of emotions* (pp. 393–411).The Guilford Press.

Mesquita, B., Karasawa, M., Banjeri, I., Haire, A., & Kashiwagi, K. (2010). *Emotion as relationship acts: A study of cultural differences* [Unpublished manuscript, University of Leuven, Belgium].

Metell, M. (2014). Dis/Abling musicking: Reflections on a disability studies perspective in music therapy. *Voices: A World Forum for Music Therapy, 14*(3).

Metell, M. (2019). How we talk when we talk about disabled children and their families: An invitation to queer the discourse. *Voices: A World Forum for Music Therapy, 19*(3).

Meth, P. (2009). Marginalised men's emotions: Politics and place. *Geoforum, 40*(5), 853–863.

Metz, T. (2007). Toward an African moral theory. *Journal of Political Philosophy, 15*(3), 321–341.

Metzner, S. (2004). Some thoughts on receptive music therapy from a psychoanalytic viewpoint. *Nordic Journal of Music Therapy, 13*(2), 143–150.

Miall, H. (2004). Conflict transformation: A multi-dimensional task. In A. Austin, M. Fischer, & N. Ropers (Eds.), *Transforming ethnopolitical conflict* (pp. 67–89). Springer Fachmedien Wiesbaden.

Middleton, I. (2018). Trust. *Music and Arts in Action, 6*(2). https://musicandartsinaction.net/index.php/maia/article/view/187

Miga, E., Gdula, J. A., & Allen, J. (2012). Fighting fair: Adaptive marital conflict strategies as predictors of future adolescent peer and romantic relationship quality. *Social Development, 21*, 443–460.

Miller, W. (2018). *Listening well: The art of empathic understanding.* Wipf & Stock.

Milligan, C., & Bingley, A. (2007). Restorative places or scary places? The impact of woodland on the mental well-being of young adults. *Health and Place, 13*, 799–811.

Mitchell, C. (2002). Beyond resolution: What does conflict transformation actually transform? *Peace and Conflict Studies, 9*(1), 1–23.

Mitchell, S. (1991). Contemporary perspectives on self: Toward an integration. *Psychoanalytic Dialogues, 1*(2), 121–147.

Mol, A. (2006). *The multiple body: Ontology in medical practice.* Duke University Press.

Molnar-Szakacs, I. (2017). Music: The language of empathy. In E. King & C. Waddington (Eds.), *Music and empathy* (pp. 97–123). Routledge.

Molyneux, C., Hardy, T., Lin, Y., McKinnon, K., & Odell-Miller, H. (2020). Together in sound: Music therapy groups for people with dementia and their companions–moving online in response to a pandemic. *Approaches: An Interdisciplinary Journal of Music Therapy,* Advance Online Publication.

Montgomery, M. (1989). *Saying goodbye: A memoir for two fathers.* Alfred A. Knopf.

Moonga, N. U. (2019). *Exploring music therapy in the life of the batonga of Mazabuka Southern Zambia* [Master's dissertation, University of Pretoria].

Moors, A., & Kuppens, P. (2008). Distinguishing between two types of musical emotions and reconsidering the role of appraisal. *Behavioral and Brain Sciences, 31*, 588–589.

Moran, D. (2017). *Mindfulness and the music therapist: An approach to self-care* [Doctoral dissertation, Concordia University].

Mori, J., & Hayashi, M. (2006). The achievement of intersubjectivity through embodied completions: A study of interactions between first and second language speakers. *Applied Linguistics, 27*(2), 195–219.

Morris, S. (2019). Empathy on trial: A response to its critics. *Philosophical Psychology, 32*(4), 508–531.

Moskowitz, G. (2005). *Social cognition: Understanding self and others.* Guilford Press.

Most, T., & Aviner, C. (2009). Auditory, visual, and auditory–visual perception of emotions by individuals with cochlear implants, hearing aids, and normal hearing. *Journal of Deaf Studies and Deaf Education, 14*(4), 449–464.

Mucina, D. (2011). Story as research methodology. *AlterNative: An International Journal of Indigenous Peoples, 7*(1), 1–14.

Mulcahy, D. (2016). Policy matters: De/re/territorialising spaces of learning in Victorian government schools. *Journal of Education Policy, 31*(1), 81–97.

Mullen, P. (2018). Community music. *Music and Arts in Action, 6*(2), 4–15.

Mulligan, K., & Scherer, K. (2012). Toward a working definition of emotion. *Emotion Review, 4*(4), 345–357.

Muran, J. (2007). A relational turn on thick description. In J. C. Muran (Ed.), *Dialogues on difference: Studies of diversity in the therapeutic relationship* (pp. 257–274). American Psychological Association.

Murove, M. (2005). *The theory of self-interest in modern economic discourse: A critical study in the light of African humanism and process philosophical anthropology* [Doctoral thesis, University of South Africa].

Murphy, A. (2021). *The extended mind.* Houghton Mifflin Harcourt.

Murphy, K. (2007). Experiential learning in music therapy: Faculty and student perspectives. *Qualitative Inquiries in Music Therapy, 3*, 31–61.

Murray, M., Bowen, S., Verdugo, M., & Holtmannspötter, J. (2017). Care and relatedness among rural Mapuche women: Issues of carino and empathy. *Ethos, 45*(3), 367–385.

Næss, T., & Ruud, E. (2007). Audible gestures: From clinical improvisation to community music therapy. *Nordic Journal of Music Therapy, 16*(2), 160–171.

Naish, K., Reader, A., Houston-Price, C., Bremner, A., & Holmes, N. (2013). To eat or not to eat? Kinematics and muscle activity of reach-to-grasp movements are influenced by the action goal, but observers do not detect these differences. *Experimental Brain Research, 225*, 261–275.

Navne, L., Svendsen, M., & Gammeltoft, T. (2018). The attachment imperative: Parental experiences of relation-making in a Danish neonatal intensive care unit. *Medical Anthropology Quarterly, 32*(1), 120–137.

Nebelung, I., & Stensæth, K. (2018). Humanistic music therapy in child welfare. *Voices: A World Forum for Music Therapy, 18*(4).

Neff, K., & McGehee, P. (2010). Self-compassion and psychological resilience among adolescents and young adults. *Self and Identity, 9*(3), 225–240.

Neff, K., & Vonk, R. (2009). Self-compassion versus global self-esteem: Two different ways of relating to oneself. *Journal of Personality, 77*, 23–50.

Neff, K. D. (2003). Self-compassion: An alternative conceptualization of a healthy attitude toward oneself. *Self and Identity, 2*(2), 85–102.

Neisser, U. (1988). Five kinds of self-knowledge. *Philosophical Psychology, 1*(1), 35–59.

Nelligan, S., Hayes, T., & McCaffrey, T. (2020). A personal recovery narrative through Rap music in music therapy. In A. Hargreaves & A. Maguire (Eds.), *Schizophrenia: Triggers and treatments* (pp. 233–268). Nova Science.

Nelson, D. (2009). Feeling good and open-minded: The impact of positive affect on cross cultural empathic responding. *The Journal of Positive Psychology, 4*(1), 53–63.

Nelson, J., Hall, B., Anderson, J., Birtles, C., & Hemming, L. (2018). Self-Compassion as self-care: A simple and effective tool for counselor educators and counseling students. *Journal of Creativity in Mental Health, 13*(1), 121–133.

Nergaard, S. (2020). Living in translation. In F. Fernández & J. Evans (Eds.), *The Routledge handbook of translation and globalization* (pp. 147–160). Routledge.

Nesbitt, N. (2010). Critique and clinique: From sounding bodies to the musical event. In N. Nesbitt & B. Hulse (Eds.), *Sounding the virtual: Gilles Deleuze and the theory and philosophy of music* (pp. 159–179). Ashgate.

Niccolini, A. (2016). Terror (ism) in the classroom: Censorship, affect and uncivil bodies. *International Journal of Qualitative Studies in Education, 29*(7), 893–910.

Niedenthal, P., Rychlowska, M., & Wood, A. (2017). Feelings and contexts: Socioecological influences on the nonverbal expression of emotion. *Current Opinion in Psychology, 17*, 170–175.

Nietzsche, F. (2009). *Beyond good and evil* (I. Johnston, Trans.). Richer Resources Publications.
Niland, A. (2017). Singing and playing together: A community music group in an early intervention setting. *International Journal of Community Music, 10*(3), 273–288.
Nitsun, M. (1996). *The anti-group: Destructive forces in the group and their creative potential*. Routledge.
Noddings, N. (2003). *Happiness and education.* Cambridge University Press.
Noddings, N. (2010). Complexity in caring and empathy. *Abstracta, 6*(2), 6–12.
Noë, A. (2009). *Out of our heads.* Hill and Wang.
Nolan, P. (2005). Verbal processing within the music therapy relationship. *Music Therapy Perspectives, 23*(1), 18–28.
Nordgren, L., Banas, K., & MacDonald, G. (2011). Empathy gaps for social pain: Why people underestimate the pain of social suffering. *Journal of Personality and Social Psychology, 100*(1), 120.
Nordoff, P., & Robbins, C. (1971/2004). *Therapy in music for handicapped children.* Barcelona Publishers.
Nordoff, P., & Robbins, C. (2007). *Creative music therapy: A guide to fostering clinical musicianship* (2nd ed.). Barcelona Publishers.
Norris, M. (2019). *Between lines: A critical multimodal discourse analysis of Black aesthetics in a vocal music therapy group for chronic pain* [Doctoral dissertation, Drexel University].
North, F. (2014). Music, communication, relationship: A dual practitioner perspective from music therapy/speech and language therapy. *Psychology of Music, 42*(6), 776–790.
Nowak, A. (2011). *Introducing a pedagogy of empathic action as informed by social entrepreneurs.* McGill University.
Nussbaum, M. (1988). Narrative emotions: Beckett's genealogy of love. *Ethics, 98*(2), 225–254.
Nussbaum, M. (2003). *Upheavals of thought: The intelligence of emotions.* Cambridge University Press.
Nwoye, A. (2015). African psychology and the Africentric paradigm to clinical diagnosis and treatment. *South African Journal of Psychology, 45*(3), 305–317.
Nye, I. (2014). A response to Christine Atkinson's essay "'Dare we speak of love?' An exploration of love within the therapeutic relationship. *British Journal of Music Therapy, 28*(1), 36–39.

Nzewi, M. (2002). Backcloth to music and healing in traditional Africa society. *Voices, A World Forum for Music Therapy, 2*(2).

O'Connell, J. E. (2018). *Beyond prosocial motivations to empathize* [Doctoral dissertation, University of Ontario Institute of Technology].

O'Connor, L., Berry, J., Lewis, T., & Stiver, D. (2012). Empathy-based pathogenic guilt, pathological altruism, and psychopathology. In B. Oakley, A. Knafo, G. Madhavan, & D. Sloan Wilson (Eds.), *Pathological altruism* (pp. 10–30). Oxford University Press.

O'Hara, M. (1997). Relational empathy: Beyond modernist egocentricism to postmodern holistic contextualism. In A. C. Bohart & L. S. Greenberg (Eds.), *Empathy reconsidered* (pp. 295–319). American Psychological Association.

Oakley, B., Knafo, A., & McGrath, M. (2012). Pathological altruism—An introduction. In B. Oakley, A. Knafo, G. Madhavan, & D. Sloan Wilson (Eds.), *Pathological altruism* (pp. 3–9). Oxford University Press.

Odell-Miller, H. (2002). One man's journey and the importance of time: Music therapy in an NHS mental health day centre. In E. Richards & A. Davies (Eds.), *Music therapy and group work: Sound company* (pp. 63–76). Jessica Kingsley.

Ohrt, J. H., Foster, J. M., Hutchinson, T. S., & Ieva, K. P. (2009). Using music videos to enhance empathy in counselors-in-training. *Journal of Creativity in Mental Health, 4*(4), 320–333.

Olden, C. (1953). On adult empathy with children. *The Psychoanalytic Study of the Child, 8*(1), 111–126.

Oliver, L., Mitchell, D., Dziobek, I., MacKinley, J., Coleman, K., Rankin, K., & Finger, E. (2014). Parsing cognitive and emotional empathy deficits for negative and positive stimuli in frontotemporal dementia. *Neuropsychologia, 67*, 14–26.

Oliver, T. (2020). *The self delusion.* Weidenfeld & Nicolson.

Oliveros, P. (2005). *Deep listening: A composer's sound practice.* IUniverse.

Oosthuizen, H. (2019). The potential of paradox: Chaos and order as interdependent resources within short-term music therapy groups with young offenders in South Africa. *Qualitative Inquiries in Music Therapy, 14*(2).

Oosthuizen, H., & McFerran, K. (2021). Playing with chaos: Broadening possibilities for how music therapist's consider chaos in group work with young people. *Music Therapy Perspectives, 39*(1), 2–10.

Orange, D. (2003). *The post-Cartesian witness and the psychoanalytic profession.* Unpublished manuscript.

Orange, D. (2002). There is no outside: Empathy and authenticity in psychoanalytic process. *Psychoanalytic Psychology, 19*(4), 686.

Orange, D. M. (2008). Recognition as: Intersubjective vulnerability in the psychoanalytic dialogue. *International Journal of Psychoanalytic Self Psychology, 3*(2), 178–194.

Orange, D. (2009). Intersubjective systems theory: A fallibilist's journey. *Annals of the New York Academy of Sciences, 1159*(1), 237–248.

Orange, D., Atwood, G., & Stolorow, R. (2015). *Working intersubjectively: Contextualism in psychoanalytic practice.* Routledge.

Orth, J. (2005). Music therapy with traumatized refugees in a clinical setting. *Voices: A World Forum for Music Therapy, 5*(2).

Oswanski, L., & Donnenwerth, A. (2016). Allies for social justice. In A. Whitehead-Pleaux & X. Tan (Eds.), *Cultural intersections in music therapy: Music, health, and the person* (pp. 257–270). Barcelona Publishers.

Oudekerk, B. A., Allen, J. P., Hessel, E. T., & Molloy, L. E. (2015). The cascading development of autonomy and relatedness from adolescence to adulthood. *Child Development, 86*(2), 472–485.

Overy, K., & Molnar-Szakacs, I. (2009). Being together in time: Musical experience and the mirror neuron system. *Music Perception, 26*(5), 489–504.

Paden, C. E. (2018). *Adolescents with behavioral health concerns: Increasing empathy through lyric discussions of peer-chosen music.* Illinois State University.

Panfile, T. M., & Laible, D. J. (2012). Attachment security and child's empathy: The mediating role of emotion regulation. *Merrill-Palmer Quarterly, 58*(1), 1–21.

Papoušek, M., & Papoušek, H. (1981). Musical elements in the infant's vocalization: Their significance for communication, cognition, and creativity. In L. P. Lipsitt & C. K. Rovee-Collier (Eds.), *Advances in infancy research* (Vol. 1, pp. 163–224). Ablex.

Papp, L., Pendry, P., & Adam, E. (2009). Mother-adolescent physiological synchrony in naturalistic settings: Within-family cortisol associations and moderators. *Journal of Family Psychology, 23*(6), 882–894.

Pârlog, A. C. (2019). *Intersemiotic translation.* Springer.

Parr, H., &, Philo, C. (2003). Rural mental health and social geographies of caring. *Social & Cultural Geography, 4*(3), 471–488.

Parsons, C. E., Young, K. S., Jegindø, E. M. E., Vuust, P., Stein, A., & Kringelbach, M. L. (2014). Music training and empathy positively impact adults' sensitivity to infant distress. *Frontiers in Psychology, 5*, 1440.

Pavlicevic, M. (1997). *Music therapy in context.* Jessica Kingsley.

Pavlicevic, M. (1999a). Thoughts, words and deeds: Harmonies and counterpoints in music therapy theory: A response to Elaine Streeter's 'finding a balance between psychological thinking and musical awareness in music therapy theory: A psychoanalytic perspective. *British Journal of Music Therapy, 13*(2), 59–62.

Pavlicevic, M. (1999b). *Music therapy: Intimate notes.* Jessica Kingsley.

Pavlicevic, M. (2002). South Africa: Fragile rhythms and uncertain listenings: Perspectives from music therapy with South African children. In J. P. Sutton (Ed.), *Music, music therapy and trauma* (pp. 97–118). Jessica Kingsley Publishers.

Pavlicevic, M. (2004). Music therapy and the polyphony of near and far... *Voices: A World Forum for Music Therapy, 4*(1).

Pavlicevic, M. (2010). Action: Because it's cool. Community music therapy in Heideveld, South Africa. In B. Stige, G. Ansdell, C. Elefant, M. Pavlicevic (Eds.), *Where music helps: Community music therapy in action and reflection* (pp. 93–101). Ashgate.

Pavlicevic, M., & Ansdell, G. (2009). Between communicative musicality and collaborative musicing: A perspective from community music therapy. In S. Malloch & C. Trevarthen (Eds.), *Communicative musicality: Exploring the basis of human companionship* (pp. 357–376). Oxford University Press.

Pavlicevic, M., & Fouché, S. (2014). Reflections from the market place–community music therapy in context. *International Journal of Community Music, 7*(1), 57–74.

Pavlicevic, M., & Impey, A. (2013). Deep listening: Towards an imaginative reframing of health and well-being practices in international development. *Arts & Health, 5*(3), 238–252.

Pavlicevic, M., Tsiris, G., Wood, S., Powell, H., Graham, J., Sanderson, R., Millman, R., & Gibson, J. (2015). The 'ripple effect': Towards researching improvisational music therapy in dementia care homes. *Dementia, 14*(5), 659–679.

Pedersen, I. (1997). The music therapist's listening perspectives as source of information in improvised musical duets with grown-up: Psychiatric patients, suffering from Schizophrenia. *Nordic Journal of Music Therapy, 6*(2), 98–111.

Pedersen, P., Crethar, H., & Carlson, J. (2008). *Inclusive cultural empathy: Making relationships central in counseling and psychotherapy.* American Psychological Association.

Pedwell, C. (2012a). Affective (self-) transformations: Empathy, neoliberalism and international development. *Feminist Theory, 13*(2), 163–179.

Pedwell, C. (2012b). Economies of empathy: Obama, neoliberalism and social justice. *Environment and Planning D: Society and Space, 30*(2), 280–297.
Pedwell, C. (2014). *Affective relations: The transnational politics of empathy.* Palgrave Macmillan.
Pedwell, C. (2016). Decolonizing empathy: Thinking affect transnationally. *Samyukta: A Journal of Women's Studies, XVI*(1), 27–49.
Perec, G. (1973). *Species of spaces and other pieces.* Penguin.
Perel, E. (2018). In search of erotic intelligence. In A. R. Ben-Shahar, L. Lipkies, & N. Oster (Eds.), *Speaking of bodies* (pp. 49–55). Routledge.
Petitta, L., Jiang, L., & Härtel, C. (2017). Emotional contagion and burnout among nurses and doctors: Do joy and anger from different sources of stakeholders matter? *Stress and Health, 33*(4), 358–369.
Petitta, L., Probst, T., Ghezzi, V., & Barbaranelli, C. (2019). Cognitive failures in response to emotional contagion: Their effects on workplace accidents. *Accident Analysis & Prevention, 125*, 165–173.
Pfattheicher, S., Sassenrath, C., & Keller, J. (2019). Compassion magnifies third-party punishment. *Journal of Personality and Social Psychology, 117*(1), 124.
Phillips, A., & Taylor, B. (2009). *On kindness.* Penguin.
Phillips-Silver, J., Aktipis, C., & Bryant, G. (2010). The ecology of entrainment: Foundations of coordinated rhythmic movement. *Music Perception, 28*(1), 3–14.
Phillips-Silver, J., & Keller, P. (2012). Searching for roots of entrainment and joint action in early musical interactions. *Frontiers in Human Neuroscience, 6*(26), 1–11.
Pickering, A. (2010). *The mangle of practice: Time, agency and science.* University of Chicago Press.
Pieterse, A., Lee, M., Ritmeester, A., & Collins, N. (2013). Towards a model of self-awareness development for counselling and psychotherapy training. *Counselling Psychology Quarterly, 26*(2), 190–207.
Piff, P., Stancato, D., Cote, S., Mendoza-Denton, R., & Keltner, D. (2012). Higher social class predicts increased unethical behavior. *Proceedings of the National Academy of Sciences of the United States of America, 109*(11), 4086–4091.
Pinfold, V. (2000). Building up safe havens…all around the world: Users' experiences of living in the community with mental health problems. *Health and Place, 6*(2), 201–212.

Pizer, S. (2018). Core competency three: Deep listening/affective attunement. In R. Barsness (Ed.), *Core competencies of relational psychoanalysis* (pp. 104–120). Taylor & Francis.

Plutchik, R. (1994). *The psychology and biology of emotion.* Harper-Collins.

Poehlmann, J., Schwichtenberg, A. J., Bolt, D. M., Hane, A., Burnson, C., & Winters, J. (2011). Infant physiological regulation and maternal risks as predictors of dyadic interaction trajectories in families with a preterm infant. *Developmental Psychology, 47*, 91–105.

Poland, W. (2000). The analyst's witnessing and otherness. *Journal of American Psychoanalytic Association, 48*(1), 17–34.

Poquérusse, J., Pastore, L., Dellantonio, S., & Esposito, G. (2018). Alexithymia and autism spectrum disorder: A complex relationship. *Frontiers in Psychology, 9*, 1196.

Porat, R., Halperin, E., & Tamir, M. (2016). What we want is what we get: Group-based emotional preferences and conflict resolution. *Journal of Personality and Social Psychology, 110*(2), 167.

Potash, J. (2010). *Guided relational viewing: Art therapy for empathy and social change to increase understanding of people living with mental illness* [Doctoral dissertation, University of Hong Kong, Pokfulam, Hong Kong].

Potash, J., & Chen, J. (2014). Art-mediated peer-to-peer learning of empathy. *The Clinical Teacher, 11*, 327–331.

Potash, J., Chen, J., Lam, C., & Chau, V. (2014). Art-making in a family medicine clerkship: How does it affect medical student empathy? *BMC Medical Education, 14*, 247–256.

Potash, J., & Ho, R. (2011). Drawing involves caring: Fostering relationship building through art therapy for social change. *Art Therapy, 28*(2), 74–81.

Potash, J., Ho, R., Chick, J., & Au Yeung, F. (2013). Viewing and engaging in an art therapy exhibit by people living with mental illness: Implications for empathy and social change. *Public Health, 127*, 735–744.

Potter, N. (2006). What is manipulative behavior, anyway? *Journal of Personality Disorders, 20*(2), 139–156.

Potvin, N., & Argue, J. (2014). Theoretical considerations of spirit and spirituality in music therapy. *Music Therapy Perspectives, 32*(2), 118–128.

Potvin, N., Bradt, J., & Ghetti, C. (2018). A theoretical model of resource-oriented music therapy with informal hospice caregivers during pre-bereavement. *Journal of Music Therapy, 55*(1), 27–61.

Potvin, N., Bradt, J., & Kesslick, A. (2015). Expanding perspective on music therapy for symptom management in cancer care. *Journal of Music Therapy, 52*(1), 135–167.

Pound, E. (1963). *Translations.* New Directions.
Preckel, K., Kanske, P., & Singer, T. (2018). On the interaction of social affect and cognition: Empathy, compassion and theory of mind. *Current Opinion in Behavioral Sciences, 19*, 1–6.
Préfontaine, J. (2006). On becoming a music therapist. *Voices: A World Forum for Music Therapy, 6*(2).
Preston, S. D., & de Waal, F. B. M. (2002). Empathy: Its ultimate and proximate bases. *Behavioral and Brain Sciences, 25*, 1–72.
Preston-Roberts, P. (2011). An interview with Dr. Diane Austin. *Voices: A World Forum for Music Therapy, 11*(1).
Price, C., & Caouette, J. (2018). Introduction. In J. Caouette & C. Price (Eds.), *The moral psychology of compassion* (pp. ix–xviii). Rowman & Littlefield.
Price-Robertson, R., & Duff, C. (2015). Realism, materialism, and the assemblage: Thinking psychologically with Manuel DeLanda. *Theory & Psychology, 26*(1), 1–19.
Price-Robertson, R., Obradovic, A., & Morgan, B. (2017). Relational recovery: Beyond individualism in the recovery approach. *Advances in Mental Health, 15*(2), 108–120.
Priestley, M. (1975). *Music therapy in action.* Constable.
Priestley, M. (1983/1995). The meaning of music. *Nordic Journal of Music Therapy, 4*(1), 28–32.
Priestley, M. (1994) *Essays on analytical music therapy.* Barcelona Publishers
Pril, D. (2017). *Toward a better understanding of the interpersonal effects of power: Power decreases interpersonal sensitivity, but not toward people within the power relationship* [Doctoral dissertation, Universität zu Köln].
Procter, S. (2001). Empowering and enabling. *Voices: A World Forum for Music Therapy, 1*(2).
Procter, S. (2002a). Empowering and enabling—Music therapy in non-medical mental health provision. In Kenny, C. & Stige, B. (Eds.), *Contemporary voices in music therapy: Communication, culture and community* (pp. 95–107). Unipub.
Procter, S. (2002b). The therapeutic, musical relationship: A two-sided affair? *Voices: A World Forum for Music Therapy, 2*(3).
Procter, S. (2011). Reparative musicing: Thinking on the usefulness of social capital theory within music therapy. *Nordic Journal of Music Therapy, 20*(3), 242–262.
Prozesky, M. H. (2001). Well-fed animals and starving babies: Environmental and developmental challenges in process and African perspectives. In M.

Murove (Ed.), *African ethics: An anthology of comparative and applied ethics* (pp. 298–307). University of KwaZulu-Natal Press.

Pruitt, L. (2011). Creating a musical dialogue for peace. *International Journal of Peace Studies, 16*(1), 81–103.

Psychogiou, L., Daley, D., Thompson, M. J., & Sonuga-Barke, E. J. (2008). Parenting empathy: Associations with dimensions of parent and child psychopathology. *British Journal of Developmental Psychology, 26*(2), 221–232.

Purdon, C. (2002). The role of music in analytical music therapy: Music as a carrier of stories. In J. Eschen (Ed.), *Analytical music therapy* (pp. 104–114). Jessica Kingsley.

Quednow, B. B. (2017). Social cognition and interaction in stimulant use disorders. *Current Opinion in Behavioral Sciences, 13*, 55–62.

Rabinowitch, T., Cross, I., & Burnard, P. (2012). Long-term musical group interaction has a positive influence on empathy in children. *Psychology of Music, 41*(4), 484–498.

Radoje, M. (2014). Where were you born? A music therapy case study. *British Journal of Music Therapy, 28*(2), 25–35.

Ramsbotham, O., Miall, H., & Woodhouse, T. (2011). *Contemporary conflict resolution: The prevention, management and transformations of deadly conflicts* (3rd ed.). Polity Press.

Rasheed, S. (2015). Self-awareness as a therapeutic tool for nurse/ client relationship. *International Journal of Caring Sciences, 8*(1), 211–216.

Rasheed, S., Younas, A., & Sundus, A. (2019). Self-awareness in nursing: A scoping review. *Journal of Clinical Nursing, 28*(5–6), 762–774.

Ratcliffe, M. (2012). Phenomenology as a form of empathy. *Inquiry, 55*(5), 473–495.

Ratcliffe, M. (2017). Empathy without simulation. In T. Fuchs, M. Summa, & L. Vanzago (Eds.), *Imagination and social perspectives: Approaches from phenomenology and psychopathology* (pp. 199–220). Routledge.

Ratele, K. (2017). Four (African) psychologies. *Theory & Psychology, 27*(3), 313–327.

Reddekop, J. (2014). *Thinking across worlds: Indigenous thought, relational ontology, and the politics of nature; Or, if only Nietzsche could meet a Yachaj* [Doctoral dissertation, University of Western Ontario]. Electronic Thesis and Dissertation Repository. 2082.

Reddy, V. (2008). *How infants know minds*. Harvard University Press.

Reddy, W. (2001). *The navigation of feeling: A framework for the history of emotions*. University Press.

Redmond, M. (2018). *Social decentering: A theory of other-orientation encompassing empathy and perspective taking.* De Bruyter.
Reeve, D. (2002). Negotiating psycho-emotional dimensions of disability and their influence on identity constructions. *Disability & Society, 17*(5), 493–508.
Reniers, R. L., Corcoran, R., Drake, R., Shryane, N. M., & Völlm, B. A. (2011). The QCAE: A questionnaire of cognitive and affective empathy. *Journal of Personality Assessment, 93*(1), 84–95.
Renneberg, B., Heyn, K., Gebhard, R., & Bachmann, S. (2005). Facial expression of emotions in borderline personality disorder and depression. *Journal of Behavior Therapy and Experimental Psychiatry, 36*, 183–196.
Reuther, B. (2014). *Intersubjectivity*. Springer.
Reynolds, J. (2018). The extended body: On aging, disability, and well-being. *Hastings Center Report, 48*, S31–S36.
Reynolds, V. (2011). Resisting burnout with justice-doing. *International Journal of Narrative Therapy & Community Work, 4*, 27–45.
Reynolds, V. (2012). An ethical stance for justice-doing in community work and therapy. *Journal of Systemic Therapies, 31*(4), 18–33.
Ricciardi, L., Visco-Comandini, F., Erro, R., Morgante, F., Bologna, M., Fasano, A., Ricciardi, D., Edwards, M. J., & Kilner, J. (2017). Facial emotion recognition and expression in Parkinson's disease: An emotional mirror mechanism? *PloS one, 12*(1), e0169110.
Richardson, L. (2000). Writing: A method of inquiry. In N. K. Denzin & Y. S. Lincoln (Eds.), *Handbook of qualitative research* (2nd ed., pp. 923–943). Sage.
Rickson, D. (2003). The boy with the glass flute. *Voices: A World Forum for Music Therapy, 3*(2).
Rickson, D. (2010). *The development of music therapy schools consultation protocol for students with high or very high special education needs* [Doctoral dissertation, University of Wellington].
Ridder, H. M., McDermott, O., & Orrell, M. (2017). Translation and adaptation procedures for music therapy outcome instruments. *Nordic Journal of Music Therapy, 26*(1), 62–78.
Ridley, C. R., & Lingle, D. W. (1996). Cultural empathy in multicultural counseling: A multidimensional process model. In P. B. Pedersen, J. G. Draguns, W. J. Lonner, & J. E. Trimble (Eds.), *Counseling across cultures* (pp. 21–46). Sage.
Rieffe, C., Oosterveld, P., Miers, A. C., Meerum Terwogt, M., & Ly, V. (2008). Emotion awareness and internalising symptoms in children and adolescents:

The emotion awareness questionnaire revised. *Personality and Individual Differences, 45*(8), 756–761.
Riess, H. (2018). *The empathy effect*. Sounds True.
Riessman, C. (2008). *Narrative methods for the human sciences*. Sage.
Rietveld, E., Denys, D., & Van Westen, M. (2018). Ecological-enactive cognition as engaging with a field of relevant affordances. In A. Newen, L. De Bruin, & S. Gallagher (Eds.), *The Oxford handbook of 4E cognition* (pp. 41–70). Oxford University Press.
Ringrose, J., Osgood, J., Renold, E., & Strom, K. (2019). PhEmaterialism: Response-able Research & Pedagogy. *Reconceptualizing Educational Research Methodology, 3*(2).
Ringrose, J., & Renold, E. (2014). "F** k rape!" Exploring affective intensities in a feminist research assemblage. *Qualitative Inquiry, 20*(6), 772–780.
Rival, L. (2014). The materiality of life: Revisiting the anthropology of nature in Amazonia. In G. Harvey (Ed.), *The handbook of contemporary animism* (pp. 92–100). Oxon.
Rizzolatti, G., Fadiga, L., Gallese, V., & Fogassi, L. (1996). Premotor cortex and the recognition of motor actions. *Cognitive Brain Research, 3*(2), 131–141.
Robarts, J. (1996). Music therapy for children with autism. In C. Trevarthen, K. Aitken, D. Papoudi, & J. Robarts (Eds.), *Children with autism* (pp. 134–160). Jessica Kingsley.
Robarts, J. (2003). The healing function of improvised songs in music therapy with a child survivor of early trauma and sexual abuse. In S. Hadley (Ed.), *Psychodynamic music therapy: Case studies* (pp. 141–182). Barcelona.
Robarts, J. (2006). Music therapy with sexually abused children. *Clinical Child Psychology and Psychiatry, 11*(2), 249–269.
Robarts, J. (2009). Supporting the development of mindfulness and meaning: Clinical pathways in music therapy with a sexually abused child. In S. Malloch & C. Trevarthen (Eds.), *Communicative musicality: Exploring the basis of human companionship* (pp. 377–400). Oxford University Press.
Robbins, J., & Rumsey, A. (2008). Introduction: Cultural and linguistic anthropology and the opacity of other minds. *Anthropological Quarterly, 81*(2), 407–420.
Robertson, C. (2018). Musical processes as a metaphor for conflict transformation processes. *Music and Arts in Action, 6*(3), 31–46.
Robinson, D. (2014). The role of cultural meanings and situated interaction in shaping emotions. *Emotions Review, 6*(3), 189–195.

Rochat, P. (2003). Five levels of self-awareness as they unfold early in life. *Consciousness and Cognition, 12*(4), 717–731.

Rogers, C. (1951). *Client-centered therapy: Its current practice, implications, and theory.* Houghton Mifflin Company.

Rogers, C. (1957). The necessary and sufficient conditions of therapeutic personality change. *Journal of Consulting Psychology, 21*, 95–103.

Rogers, C. (1959). A theory of therapy, personality, and interpersonal relationships as developed in the client-centered framework. In S. Koch (Ed.), *Psychology: A study of a science, Vol. 3: Formulations of the person and the social contex* (pp. 184–256). McGraw Hill.

Rogers, C. (1975). Empathic: An unappreciated way of being. *The Counseling Psychologist, 5*(2), 2–10.

Rogers, C. (2007). The necessary and sufficient conditions of therapeutic personality change. *Psychotherapy, 44*(3), 240–248.

Rogers, C., & Farson, R. (1957). *Active listening.* https://wholebeinginstitute.com/wp-content/uploads/Rogers_Farson_Active-Listening.pdf

Rolli, M. (2009). Deleuze on intensity differentials and the being of the sensible. *Deleuze Studies, 3*(1), 26–53.

Rolvsjord, R. (2004). Therapy as empowerment. *Nordic Journal of Music Therapy, 13*(2), 99–111.

Rolvsjord, R. (2006a). Whose power of music? *British Journal of Music Therapy, 20*(1), 5–12.

Rolvsjord, R. (2006b). Therapy as empowerment: Clinical and political implications of empowerment philosophy in mental health practises of music therapy. *Voices: A World Forum for Music Therapy, 6*(3).

Rolvsjord, R. (2010). *Resource-oriented music therapy in mental health care.* Barcelona Publishers.

Rolvsjord, R. (2014). The competent client and the complexity of dis-ability. *Voices: A World Forum for Music Therapy, 14*(3).

Rolvsjord, R. (2016). Five episodes of clients' contributions to the therapeutic relationship: A qualitative study in adult mental health care. *Nordic Journal of Music Therapy, 25*(2), 159–184.

Rosado, A. (2019). Adolescents' experiences of music therapy in an inpatient crisis stabilization unit. *Music Therapy Perspectives, 37*(2), 133–140.

Rosaldo, M. (1984). Toward an anthropology of self and feeling. In R. Sweder & R. LeVine (Eds.), *Culture theory: Essays on mind, self, and emotion* (pp. 137–157). Cambridge University Press.

Rosan, P. (2012). The poetics of intersubjective life: Empathy and the other. *The Humanistic Psychologist, 40*, 115–135.

Ross, L. D. (1977). The intuitive psychologist and his shortcomings: Distortions in the attribution process. In L. Berkowitz (Ed.), *Advances in experimental social psychology* (Vol. 10, pp. 174–221). Academic Press.
Rovenpor, D. R., & Isbell, L. M. (2018). Do emotional control beliefs lead people to approach positive or negative situations? Two competing effects of control beliefs on emotional situation selection. *Emotion, 18*, 313–331.
Rowe, C., & Isaac, D. (2004). *Empathic attunement: The "technique" of psychoanalytic self psychology*. Rowman & Littlefield.
Ruiz-Junco, N. (2017). Advancing the sociology of empathy: A proposal. *Symbolic Interaction, 40*(3), 414–435.
Ruud, E. (1997). Music and quality of life. *Nordic Journal of Music Therapy, 6*(2), 86–91.
Ruud, E. (1998). *Music therapy: Improvisation, communication and culture.*
Ruud, E. (2003). "Burning scripts" self psychology, affect consciousness, script theory and the BMGIM. *Nordic Journal of Music Therapy, 12*(2), 115–123.
Ruud, E. (2008). Music in therapy: Increasing possibilities for action. *Music and Arts in Action, 1*(1), 46–60.
Ruud, E. (2010). *Music therapy: A perspective from the humanities.* Barcelona Publishers.
Saarni, C., Campos, J., Camras, L., & Witherington, D. (2008). Principles of emotion and emotional competence. In W. Damon & R. Lerner (Eds.), *Child and adolescent development: An advanced course* (pp. 361–405). Wiley.
Sacipa-Rodríguez, S. S., Guerra, C. T., Villarreal, L. F. G., & Bohórquez, R. V. (2009). Psychological accompaniment: Construction of cultures of peace among a community affected by war. In C. Sonn & M. Montero (Eds.), *Psychology of liberation* (pp. 221–235). Springer.
Said, E. (1978). *Orientalism: Western concepts of the Orient.* Pantheon.
Sajnani, N. (2012a). Improvisation and art-based research. *Journal of Applied Arts & Health, 3*(1).
Sajnani, N. (2012b). Response/ability: Imagining a critical race feminist paradigm for the creative arts therapies. *The Arts in Psychotherapy, 39*(3), 186–191.
Sajnani, N., Marxen, E., & Zarate, R. (2017). Critical perspectives in the arts therapies: Response/ability across a continuum of practice. *The Arts in Psychotherapy, 54*, 28–37.
Salmon, D. (2008). Bridging music and psychoanalytic therapy. *Voices: A World Forum for Music Therapy, 8*(1).
Salzen, E. (2001). A century of emotion theories—Proliferation without progress? *History and Philosophy of Psychology, 3*, 56–75.

Sass, L. (2007). Contradictions of emotion in schizophrenia. *Cognition & Emotion, 21*(2), 351–390.
Sass, L., & Parnas, J. (2007). Explaining schizophrenia: The relevance of phenomenology. In M. Chung, K. Fulford, & G. Graham (Eds.), *Reconceiving schizophrenia* (pp. 63–95). Oxford University Press.
Sawyer, K. (2010). *Group creativity: Music, theatre, collaboration.* Routledge.
Saxe, R., Carey, S., & Kanwisher, N. (2004). Understanding other minds: Linking developmental psychology and functional neuroimaging. *Annual Review of Psychology, 55*, 87–124.
Scarantino, A. (2017). How to do things with emotional expressions: The theory of affective pragmatics. *Psychological Inquiry, 28*(2–3), 165–185.
Schafer, R. (1992). *Retelling a life.* Basic books.
Scheiby, B. (2005). An intersubjective approach to music therapy: Identification and processing of musical countertransference in a music psychotherapeutic context. *Music Therapy Perspectives, 23*(1), 8–17.
Scherer, K., & Zentner, M. (2008). Music evoked emotions are different—More often aesthetic than utilitarian. *Behavioral and Brain Sciences, 31*, 595–596.
Schertz, M. (2007). Empathy as intersubjectivity: Resolving Hume and Smith's divide. *Studies in Philosophy and Education, 26*(2), 165–178.
Schiavio, A. (2015). Action, enaction, inter (en) action. *Empirical Musicology Review, 9*(3–4), 254–262.
Schiavio, A., & De Jaegher, H. (2017). Participatory sense-making in joint musical practices. In M. Lesaffre, M. Leman, & P. Maes (Eds.), *The Routledge companion to embodied music interaction* (pp. 31–39). Routledge.
Schiavio, A., van der Schyff, D., Cespedes-Guevara, J., & Reybrouck, M. (2017). Enacting musical emotions. Sense-making, dynamic systems, and the embodied mind. *Phenomenology and the Cognitive Sciences, 16*(5), 785–809.
Schimpf, M. (2021). Cultural humility in clinical music therapy supervision. In M. Belgrave & S. Kim (Eds.), *Music therapy in a multicultural context: A handbook for music therapy students and professionals* (pp. 157–184). Jessica Kingsley Publishers.
Schlegel, R. J., Vess, M., & Arndt, J. (2012). To discover or to create: Metaphors and the true self. *Journal of Personality, 80*(4), 969–993.
Schmid, W., & Rolvsjord, R. (2020). Becoming a reflexive practitioner: Exploring music therapy students' learning experiences with participatory role-play in a Norwegian context. *Scandinavian Journal of Educational Research*, 1–13.

Schnitker, S. A., Houltberg, B., Dyrness, W., & Redmond, N. (2017). The virtue of patience, spirituality, and suffering: Integrating lessons from positive psychology, psychology of religion, and Christian theology. *Psychology of Religion and Spirituality, 9*(3), 264.

Schoebi, D., & Randall, A. K. (2015). Emotional dynamics in intimate relationships. *Emotion Review, 7*(4), 342–348.

Schultze, U. (2017). What kind of world do we want to help make with our theories? *Information and Organization, 27*(1), 60–66.

Schulz, K. (2011). *Being wrong: Adventures in the margin of error.* Granta Books.

Schwaber, E. (1984). Empathy: A mode of analytic listening. In J. Lichtenberg, M. Bornstein, & D. Silver (Eds.), *Empathy II* (pp. 143–172). Routledge.

Scott, C. (2011). *Becoming dialogue: Martin Buber's concept of turning to the other as educational praxis* [Doctoral Thesis, University of British Colombia].

Scrine, E. (2016). Enhancing social connectedness or stabilising oppression: Is participation in music free from gendered subjectivity? *Voices: A World Forum for Music Therapy, 16*(2).

Scrine, E. (2018). *Music therapy as an anti-oppressive practice: Critically exploring gender and power with young people in school* [Doctoral dissertation, University of Melbourne].

Scrine, E., & McFerran, K. (2018). The role of a music therapist exploring gender and power with young people: Articulating an emerging anti-oppressive practice. *The Arts in Psychotherapy, 59*, 54–64.

Seehawer, M. K. (2018). Decolonising research in a Sub-Saharan African context: Exploring Ubuntu as a foundation for research methodology, ethics and agenda. *International Journal of Social Research Methodology, 21*(4), 453–466.

Seeley, C., & Reason, R. (2008). Expressions of energy: An epistemology of presentational knowing. In P. Liamputtong & J. Rumbold (Eds.), *Knowing differently: Arts-based and collaborative research methods* (pp. 25–46). Nova Science Publishers.

Segal, E. (2018). *Social empathy: The art of understanding others.* Columbia University Press.

Selg, P., & Ventsel, A. (2020). The "relational turn" in the social sciences. In P. Selg & A. Ventsel (Eds.), *Introducing relational political analysis* (pp. 15–40). Palgrave Macmillan.

Sels, L., Ickes, W., Hinnekens, C., Ceulemans, E., & Verhofstadt, L. (2021). Expressing thoughts and feelings leads to greater empathic accuracy during relationship conflict. *Journal of Family Psychology, 35*(8), 1199–1205.

Semenya, B., & Mokwena, M. (2012). African cosmology, psychology and community. In M. Visser & A. Moleko (Eds.), *Community psychology in South Africa* (pp. 71–84). Van Schaik Publishers.

Sena Moore, K. (2017). Understanding the influence of music on emotions: A historical review. *Music Therapy Perspectives, 35*(2), 131–143.

Seri, N., & Gilboa, A. (2017). When music therapists adopt an ethnographic approach: Discovering the music of ultra-religious boys in Israel. *Approaches: An Interdisciplinary Journal of Music Therapy*, First View, 1–15.

Seu, I. B., & Cameron, L. (2013). Empathic mutual positioning in conflict transformation and reconciliation. *Peace and Conflict: Journal of Peace Psychology, 19*(3), 266.

Seyfert, R. (2012). Beyond personal feelings and collective emotions: Toward a theory of social affect. *Theory, Culture & Society, 29*(6), 27–46.

Shaddock, D. (2000). *Contexts and connections: An intersubjective systems approach to couples therapy.* Basic Books.

Shannon, D. B., & Truman, S. E. (2020). Problematizing sound methods through music research-creation: Oblique curiosities. *International Journal of Qualitative Methods, 19*, 1609406920903224.

Shapiro, L. (2010). *Embodied cognition.* Routledge.

Shapiro, N. (2005). Sounds in the world: Multicultural influences in music therapy in clinical practice and training. *Music Therapy Perspectives, 23*(1), 29–35.

Shaver, P. R., Collins, N., & Clark, C. L. (1996). Attachment styles and internal working models of self and relationship partners. In G. Fletcher & J. Fitness (Eds.), *Knowledge structures in close relationships: A social psychological approach* (pp. 25–62). Lawrence Erlbaum.

Shaviro, S. (2009). *Without criteria: Kant, Whitehead, Deleuze and aesthetics.* MIT Press.

Shaw, W. S., Herman, R. D. K., & Dobbs, G. R. (2006). Encountering indigeneity: Re-imagining and decolonizing geography. *Human Geography, 88*(3), 267–276.

Shay, J. (1994). *Achilles in Vietnam: Combat trauma and the undoing of character.* Atheneum.

Shea, S. (1998). *Psychiatric interviewing: The art of understanding* (2nd ed.). Sanders.

Sherman, N. (2014). Recovering lost goodness: Shame, guilt, and self-empathy. *Psychoanalytic Psychology, 31*(2), 217–235.

Shields, S. A. (2000). Thinking about gender, thinking about theory: Gender and emotional experience. In A. Fischer (Ed.), *Gender and emotion: Social psychological perspectives* (pp. 3–22). Cambridge University Press.

Shimizu, H. (2000). Japanese cultural psychology and empathic understanding: Implications for academic and cultural psychology. *Ethos, 28*(2), 224–247.

Shintel, H., & Keysar, B. (2009). Less is more: A minimalist account of joint action in communication. *Topics in Cognitive Science, 1*(2), 260–273.

Shlien, J. (1997). Empathy in psychotherapy: A vital mechanism? Yes. Therapist's conceit? All too often. By itself enough? No. In A. C. Bohart & L. S. Greenberg (Eds.), *Empathy reconsidered: New directions in psychotherapy* (pp. 63–80). American Psychological Association.

Shoemaker, S. (1996). *The first-person perspective and other essays*. Cambridge University Press.

Short, H. (2017a). It feels like Armageddon: Identification with a female personality-disordered offender at a time of cultural, political and personal attack. *Nordic Journal of Music Therapy, 26*(3), 272–285.

Short, H. (2017b). "Big up West London Crew": One man's journey within a community rap/music therapy group. *Music Therapy Perspectives, 35*(2), 151–159.

Shotter, J. (2013). From inter-subjectivity, via inter-objectivity, to intra-objectivity: From a determinate world of separate things to an indeterminate world of inseparable flowing processes. In G. Sammut, P. Daanen, & F. Moghaddam (Eds.), *Understanding the self and others: Explorations in intersubjectivity and interobjectivity* (pp. 83–125). Routledge.

Siebert, D., Siebert, C., & Taylor-McLaughlin, A. (2007). Susceptibility to emotional contagion: Its measurement and importance to social work. *Journal of Social Service Research, 33*(3), 47–56.

Siegel, A. (1996). *Heinz Kohut and the psychology of the self*. Routledge.

Simionato, G., Simpson, S., & Reid, C. (2019). Burnout as an ethical issue in psychotherapy. *Psychotherapy, 56*(4), 470–482.

Simons, G., Pasqualini, M. C. S., Reddy, V., & Wood, J. (2004). Emotional and nonemotional facial expressions in people with Parkinson's disease. *Journal of the International Neuropsychological Society, 10*(4), 521–535.

Singer, T. (2006). The neuronal basis and ontogeny of empathy and mind reading: Review of literature and implications for future research. *Neuroscience & Biobehavioral Reviews, 30*(6), 855–863.

Singer, T., & Lamm, C. (2009). The social neuroscience of empathy. *Annals of the New York Academy of Sciences, 1156*(1), 81–96.

Singer, T., Seymour, B., O'Doherty, J. P., Stephan, K. E., Dolan, R. J., & Frith, C. D. (2006). Empathic neural responses are modulated by the perceived fairness of others. *Nature, 439*(7075), 466–469.

Skewes, K. (2001). *The experience of group music therapy for six bereaved adolescents* [Doctoral dissertation, University of Melbourne].

Slade, M. (2009). *100 ways to support recovery.* Rethink.

Slife, B. D. (2004). Taking practice seriously: Toward a relational ontology. *Journal of Theoretical and Philosophical Psychology, 24*(2), 157.

Sloan, M. M. (2004). The effects of occupational characteristics on the experience and expression of anger in the workplace. *Work and Occupations, 31*(1), 38–72.

Small, C. (1998). *Musicking: The meanings of performing and listening.* Wesleyan University Press.

Smith, A. (2010). *The theory of moral sentiments.* Penguin.

Smith, J. (2017). What is empathy for? *Synthese, 194*(3), 709–722.

So, H. (2017). US-trained music therapists from East Asian countries found personal therapy during training helpful but when cultural disconnects occur these can be problematic: A qualitative phenomenological study. *The Arts in Psychotherapy, 55*, 54–63.

Soenens, B., Vansteenkiste, M., Goossens, L., Duriez, B., & Niemiec, P. (2008). The intervening role of relational aggression between psychological control and friendship quality. *Social Development, 17*, 661–681.

Soley, G. (2019). The social meaning of shared musical experiences in infancy and early childhood. In S. Young & B. Ilari (Eds.), *Music in early childhood* (pp. 73–85). Springer.

Solomon, R. C. (1995). The cross-cultural comparison of emotions. In J. Marks & R. T. Ames (Eds.), *Emotions in Asian thought: A dialogue in comparative philosophy* (pp. 253–308). SUNY Press.

Sorel, S. (2010). Presenting Carly and Elliot: Exploring roles and relationships in a mother-son dyad in Nordoff-Robbins music therapy. *Qualitative Inquiries in Music Therapy, 5.*

Spandler, H. (2014). Letting madness breathe?: Critical challenges facing mental health social work today. In J. Weinstein (Ed.), *Mental health: Critical and radical debates in social work* (pp. 29–38). Policy Press.

Spandler, H., & Stickley, T. (2011). No hope without compassion: The importance of compassion in recovery-focused mental health services. *Journal of Mental Health, 20*(6), 555–566.

Spaulding, S. (2015). On direct social perception. *Consciousness and Cognition, 36*, 472–482.

Spelman, E. (1997). *Fruits of sorrow: Framing our attention to suffering.* Beacon Press.

Sperry, L. (2012). *Spirituality in clinical practice: Theory and practice of spiritually oriented psychotherapy* (2nd ed.). Routledge.

Spinrad, T., & Eisenberg, N. (2009). Empathy, prosocial behaviour, and positive development in schools. In R. Gilman, E. Huebner, & M. Furlong (Eds.), *Handbook of positive psychology in schools* (pp. 119–129). Routledge.

Spinrad, T., & Gal, D. (2018). Fostering prosocial behavior and empathy in young children. *Current Opinion in Psychology, 20*, 40–44.

Spiro, N., Schofield, M., & Himberg, T. (2013). *Empathy in musical interaction.* The 3rd International Conference on Music & Emotion, Jyväskylä, Finland, June 11–15, 2013. University of Jyväskylä, Department of Music.

Spivak, G. (1988). Can the subaltern speak? In C. Nelson & L. Grossberg (Eds.), *Marxism and the interpretation of culture* (pp. 271–313). University of Illinois Press.

Spivak, G. C. (1985). The Rani of Sirmur: An essay in reading the archives. *History and Theory, 24*(3), 247–272.

St. Pierre, E. A. (2017). Haecceity: Laying out a plane for post qualitative inquiry. *Qualitative Inquiry, 23*(9), 686–698.

Stadler, J. (2017). Empathy in film. In H. Maibom (Ed.), *Routledge handbook of philosophy of empathy* (pp. 317–326). Routledge.

Stanghellini, G. (2000). Vulnerability to schizophrenia and lack of common sense. *Schizophrenia Bulletin, 26*(4), 775–787.

Stanghellini, G. (2004). *Disembodied spirits and deanimated bodies: The psychopathology of common sense.* Oxford University Press.

Stanghellini, G., & Rosfort, R. (2013). Empathy as a sense of autonomy. *Psychopathology, 46*(5), 337–344.

Stapel, J. C., Hunnius, S., & Bekkering, H. (2012). Online prediction of others' actions: The contribution of the target object, action context and movement kinematics. *Psychological Research Psychologische Forschung, 76*(4), 434–445.

Starcevic, V., & Piontek, C. M. (1997). Empathic understanding revisited: Conceptualization, controversies, and limitations. *American Journal of Psychotherapy, 51*(3), 317–328.

Stavropoulos, D. N. (2001). *Oxford Greek-English dictionary* (12th ed.). Oxford University Press.

Stein, E. (1989). *On the problem of empathy* (W. Stein, Trans.). ICS Publishers.

Steinhardt, T., & Ghetti, C. (2020). Resonance between theory and practice: development of a theory-supported documentation tool for music therapy

as procedural support within a biopsychosocial frame. In L. Ole bonde & K. Johansson (Eds.), *Music in paediatric hospitals: Nordic perspectives* (pp. 109–140). Norwegian Academy of Music.
Stensæth, K. (2018). Music therapy and interactive musical media in the future: Reflections on the subject-object interaction. *Nordic Journal of Music Therapy, 27*(4), 312–327.
Stephens, G., Silbert, L., & Hasson, U. (2010). Speaker-listener neural coupling underlies successful communication. *Proceedings of the National Academy of Sciences of the United States of America, 107*(32), 14425–14430.
Stepien, K. A., & Baernstein, A. (2006). Educating for empathy. *Journal of General Internal Medicine, 21*(5), 524–530.
Stern, D. (2005). Intersubjectivity. In E. Person, A. Cooper, & G. Gabbard (Eds.), *Textbook of psychoanalysis* (pp. 77–92). American Psychiatric Publishing.
Stern, D. (2010a). The issue of vitality. *Nordic Journal of Music Therapy, 19*(2), 88–102, 570.
Stern, D. (2010b) *Forms of vitality: Exploring dynamic experience in psychology, the arts, psychotherapy, and development.* Oxford University Press.
Stern, J. A., Borelli, J. L., & Smiley, P. A. (2015). Assessing parental empathy: A role for empathy in child attachment. *Attachment & Human Development, 17*(1), 1–22.
Stern, S. (2009). The dialectic of empathy and freedom. *International Journal of Psychoanalytic Self Psychology, 4*, 132–164.
Stewart-Williams, S. (2007). Altruism among kin vs. nonkin: Effects of cost of help and reciprocal exchange. *Evolution and Human Behavior, 28*, 193–198.
Stige, B. (1998). Perspectives on meaning in music therapy. *British Journal of Music Therapy, 12*(1), 20–27.
Stige, B. (2002). *Culture-centered music therapy.*
Stige, B. (2003). *Elaborations towards a notion of community music therapy.* Unipub Forlog.
Stige, B. (2006). On a notion of participation in music therapy. *Nordic Journal of Music Therapy, 15*(2), 121–138.
Stige, B. (2012). Health musicking: A perspective on music and health as action and performance. In R. MacDonald, G. Kreutz, & L. Mitchell (Eds.), *Music, health, and wellbeing* (pp. 183–195). Oxford University Press.
Stige, B. (2016). *Culture-centered music therapy.* Oxford University Press.
Stige, B., & Aarø, L. E. (2011). *Invitation to community music therapy.* Routledge.
Stock, J. (2018). Violence. *Music & Arts in Action, 6*(2), 91–104.

Stolorow, R. (2014). Undergoing the situation: Emotional dwelling is more than empathic understanding. *International Journal of Psychoanalytic Self Psychology, 9*(1), 80–83.

Stolorow, R. & Atwood, G. (1996). The intersubjective perspective. *Psychoanalytic Review, 83*(2).

Stolorow, B., Atwood, G., & Orange, D. (2002). *Worlds of experience: Interweaving philosophical and clinical dimensions in psychoanalysis.* Basic.

Stone, E., & Priestley, M. (1996). Parasites, pawns and partners: Disability research and the role of non-disabled researchers. *British Journal of Sociology, 47*(4), 699–716.

Storsve, V., Westbye, I. A., & Ruud, E. (2010). Hope and recognition: A music project among youth in a Palestinian refugee camp. *Voices: A World Forum for Music Therapy, 10*(1).

Stover, C. (2017). Affect and improvising bodies. *Perspectives of New Music, 55*(2), 5–66.

Strehlow, G., & Hannibal, N. (2019). Mentalizing in improvisational music therapy. *Nordic Journal of Music Therapy, 28*(4), 333–346.

Strong, T., Gaete, J., Sametband, I. N., French, J., & Eeson, J. (2012). Counsellors respond to the DSM-Iv-tR. *Canadian Journal of Counseling and Psychotherapy, 46*(2), 85–106.

Subiantoro, M. (2018). The role of music therapy in promoting communication and social skills in children with autism spectrum disorder: A Pilot Study. *Advances in Social Science, Education and Humanities Research, 133*, 252–257.

Sue, D. W. & Sue, D. (2016). *Counseling the culturally diverse: Theory and practice* (7th ed.). Wiley.

Sundberg, J. (2013). Decolonizing posthumanist geographies. *Cultural Geographies, 21*(1), 33–47.

Sundberg, J. (2014). Decolonizing posthumanist geographies. *Cultural Geographies, 21*(1), 33–47.

Susino, M., & Schubert, E. (2017). Cross-cultural anger communication in music: Towards a stereotype theory of emotion in music. *Musicae Scientiae, 21*(1), 60–74.

Sutton, J. (2005). Hidden music-an exploration of silences in music and in music therapy. *Guidelines Article Formatting, 6*, 375.

Statssa. (2019). *Statistical release: General household survey.* http://www.statssa.gov.za/publications/P0318/P03182019.pdf

Summers-Effler, E. (2006). Ritual theory. In J. Stets & J. Turner (Eds.), *Handbook of the sociology of emotions* (pp. 135–154). Springer.

Sutton, J. (2002). The pause that follows. *Nordic Journal of Music Therapy, 11*(1), 27–38.
Swaney, M. (2020). Four relational experiences in music therapy with adults with severe and profound intellectual disability. *Music Therapy Perspectives, 38*(1), 69–79.
Swanick, R. (2019). What are the factors of effective therapy? Encouraging a positive experience for families in music therapy. *Approaches: An Interdisciplinary Journal of Music Therapy.* https://approaches.gr/wp-content/uploads/2019/11/Approaches_FirstView_a20191109-swanick-1.pdf
Swanson, D. (2012). Ubuntu, African epistemology, and development. Contributions, contradictions, tensions, and possibilities. In H. K. Wright & A. A. Abdi (Eds.), *The dialectics of African education and Western discourses: Counterhegemonic perspectives* (pp. 27–52). Peter Lang.
Swift, B. (2012). Becoming-sound: Affect and assemblage in improvisational digital music making. In *Proceedings of the SIGCHI conference on human factors in computing systems* (pp. 1815–1824). https://dl.acm.org/doi/abs/10.1145/2207676.2208315?casa_token=NDBtM6R50j8AAAAA:qWIPtHWq7WjKjq8oHhSBHhGI2kY5Mls5XcNptb8vTrQvWCqVOU-hNBMarp3T045jE5khCoBl5QqA3A
Tangen, J. L. (2017). Attending to nuanced emotions: Fostering supervisees' emotional awareness and complexity. *Counselor Education and Supervision, 56*, 65–78.
Tanyi, R. A. (2002). Towards clarification of the meaning of spirituality. *Journal of Advanced Nursing, 39*(5), 500–509.
Taringa, N. T. (2020). The potential of Ubuntu values for a sustainable ethic of the environment and development. In E. Chitando & M. R. Gunda (Eds.), *Religion and development in Africa* (pp. 387–399). University of Bamberg Press.
Taylor, S., & Statler, M. (2014). Material matters: Increasing emotional engagement in learning. *Journal of Management Education, 38*(4), 586–607.
Thamm, R. (2004). Towards a universal power and status theory of emotion. *Advances in Group Processes, 21*, 189–222.
Thomas, A., & Sham, F. (2014). "Hidden rules": A duo-ethnographical approach to explore the impact of culture on clinical practice. *Australian Journal of Music Therapy, 25*, 81–91.
Thomas, N. (2020). Community-based referential music making with limited-resource adolescents: A pilot study. *Music Therapy Perspectives, 38*(2), 112–118.

Thomas, N., & Norris, M. (2021). "Who you mean 'we?'" Confronting professional notions of "belonging" in music therapy. *Journal of Music Therapy, 58*(1), 5–11.

Thompson, G. (2012). Family-centered music therapy in the home environment: Promoting interpersonal engagement between children with autism spectrum disorder and their parents. *Music Therapy Perspectives, 30*(2), 109–116.

Thompson, R. (1994). Emotion regulation: A theme in search of definition. *Monographs of the Society for Research in Child Development, 59*, 25–52.

Thomsen, P. (2000). Obsessions: The impact and treatment of obsessive-compulsive disorder in children and adolescents. *Journal of Psychopharmacology, 14*(2 Suppl 1), S31–S37.

Thornicroft, G., Rose, D., & Kassam, A. (2007). Discrimination in health care against people with mental illness. *International Review of Psychiatry, 19*(2), 113–122.

Thrift, N. (2008). *Non-representational theory: Space/politics/affect.* Routledge.

Throop, C. J. (2016). Pain and otherness, the otherness of pain. In B. Leistle (Ed.), *Anthropology and alterity* (pp. 195–216). Routledge.

Tirch, D., Schoendorff, B., & Silberstein, L. (2014). *The ACT practitioner's guide to the science of compassion: Tools for fostering psychological flexibility.* New Harbinger Publications.

Todd, Z. (2016). An indigenous feminist's take on the ontological turn: 'Ontology' is just another word for colonialism. *Journal of Historical Sociology, 29*(1), 4–22.

Tolbert, E. (1990). Magico-religious power and gender in the Karelian lament. In M. Herndon & S. Zigler (Eds.), *Music, gender, and culture* (pp. 41–56). Institute for Comparative Music Studies.

Totton, N. (2017). Power in the therapeutic relationship. In R. Tweedy (Ed.), *The political self* (pp. 29–42). Taylor & Francis.

Tramonti, F. (2019). Steps to an ecology of psychotherapy: The legacy of Gregory Bateson. *Systems Research and Behavioral Science, 36*(1), 128–139.

Trentini, C., Tambelli, R., Maiorani, S., & Lauriola, M. (2021). Gender differences in empathy during adolescence: Does emotional self-awareness matter? *Psychological Reports*, 0033294120976631.

Trevarthen, C. (1994). The self born in intersubjectivity: The psychology of an infant communicating. In U. Neisser (Ed.), *The perceived self: Ecological and interpersonal sources of self-knowledge* (pp. 121–173). Cambridge University Press.

Trevarthen, C. (2002). Origins of musical identity: Evidence from infancy for musical social awareness. In R. A. R. MacDonald, D. J. Hargreaves, & D. Miell (Eds.), *Musical identities* (pp. 21–38). Oxford UP.

Trevarthen, C., & Fresquez, C. (2015). Sharing human movement for well-being: Research on communication in infancy and applications in dance movement psychotherapy. *Body, Movement and Dance in Psychotherapy, 10*(4), 194–210.

Trevarthen, C., & Malloch, S. (2000). The dance of wellbeing: Defining the musical therapeutic effect. *Norwegian Journal of Music Therapy, 9*(2), 3–17.

Trolldalen, G. (1997). Music therapy and interplay: A music therapy project with mothers and children elucidated through the concept of "appreciative recognition." *Nordic Journal of Music Therapy, 6*(1), 14–27.

Trondalen, G. (2016). *Relational music therapy: An intersubjective perspective.* Barcelona.

Trondalen, G. (2019). Musical intersubjectivity. *The Arts in Psychotherapy, 65*, 101589.

Trondalen, G. (2003). "Self listening" in music therapy with a young woman suffering from Anorexia Nervosa. *Nordic Journal of Music Therapy, 12*(1), 3–17.

Trzeciak, S., Mazzarelli, A., & Booker, C. (2019). *Compassionomics: The revolutionary scientific evidence that caring makes a difference* (pp. 287–319). Studer Group.

Tuastad, L., & Stige, B. (2018). Music as a way out: How musicking helped a collaborative rock band of ex-inmates. *British Journal of Music Therapy, 32*(1), 27–37.

Tuck, E., & Yang, K. W. (2014). R-words: Refusing research. In D. Paris & M. T. Winn (Eds.), *Humanizing research: Decolonizing qualitative inquiry with youth and communities* (pp. 223–247). Sage.

Tucker, I. (2010). Mental health service user territories: Enacting 'safe spaces' in the community. *Health, 14*(4), 434–448.

Tuckman, B. W., & Jensen, M. A. C. (2010). Stages of small-group development revisited. *Group Facilitation: A Research & Applications Journal, 10*, 43–48.

Turner, J., & Stets, J. (2006). Sociological theories of human emotions. *Annual Review of Sociology, 32*, 25–52.

Turry, A. (1998). Nordoff-Robbins: Transference and countertransference. In K. E. Bruscia (Ed.), *The dynamics of music psychotherapy* (pp. 161–212). Barcelona Publishers.

Turry, A. (2009). Integrating musical and psychological thinking: The relationship between music and words in clinically improvised songs. *Music and Medicine, 1*(2), 106–116.

Turry, A. (2011). *Between music and psychology.* Lambert Academic Publishing.

Turry, A., & Marcus, D. (2003). Using the Nordoff-Robbins approach to music therapy with adults diagnosed with autism. In D. Wiener & L. Oxford (Eds.), *Action therapy with families and groups* (pp. 197–228). American Psychological Association.

Turski, G. (1991). Experience and expression: The moral linguistic constitution of emotions. *Journal for the Theory of Social Behaviour, 21*(4), 373–392.

Uchida, Y., Townsend, S. S., Rose Markus, H., & Bergsieker, H. B. (2009). Emotions as within or between people? Cultural variation in lay theories of emotion expression and inference. *Personality and Social Psychology Bulletin, 35*(11), 1427–1439.

Ullman, C. (2006). Bearing witness: Across the barriers in society and in the clinic. *Psychoanalytic Dialogues, 16*(2), 181–198.

Urbain, O. (2008). *Music and conflict transformation: Harmonies and dissonances in geopolitics.* I.B. Tauris.

Ursin, G., & Lotherington, T. A. (2018). Citizenship as distributed achievement: Shaping new conditions for an everyday life with dementia. *Scandinavian Journal of Disability Research, 20*(1), 62–71.

Valdesolo, P., & DeSteno, D. (2011). Synchrony and the social tuning of compassion. *Emotion, 11*(2), 262–266.

Valente, J., & Boldt, G. (2015). The rhizome of the deaf child. *Qualitative Inquiry, 21*(6), 562–574.

Valentino, R. (2006). Attitudes towards cross-cultural empathy in music therapy. *Music Therapy Perspectives, 24*, 108–114.

Valiente, C., Eisenberg, N., Fabes, R. A., Shepard, S. A., Cumberland, A., & Losoya, S. H. (2004). Prediction of children's empathy-related responding from their effortful control and parents' expressivity. *Developmental Psychology, 40*, 911–926.

Van Baaren, R., Decety, J., Dijksterhuis, A., van der Leij, A., & Leeuwen, M. (2009). Being imitated: Consequences of nonconsciously showing empathy. In J. Decety & W. Ickes (Eds.), *The social neuroscience of empathy* (pp. 31–42). The MIT Press.

van der Schyff, D. (2013). Emotion, embodied mind and the therapeutic aspects of musical experience in everyday life. *Approaches: Music Therapy and Special Music Education, 5*(1).

Van der Schyff, D., & Krueger, J. (2019). *Musical empathy, from simulation to 4E interaction. Music, sound, and mind.* Brazilian Association of Music Cognition.
Van der Schyff, D., & Schiavio, A. (2017). The future of musical emotions. *Frontiers in Psychology, 8*, 988.
Vega, M., & Ward, J. (2016). The social neuroscience of power and its links with empathy, cooperation and cognition. In P. Garrard & G. Robinson (Eds.), *The intoxication of power: Interdisciplinary insights* (pp. 155–174). Palgrave Macmillan.
Veissière, S. P., Constant, A., Ramstead, M. J., Friston, K. J., & Kirmayer, L. J. (2020). Thinking through other minds: A variational approach to cognition and culture. *Behavioral and Brain Sciences, 43.*
Veltre, V. J., & Hadley, S. (2012). It's bigger than hip-hop: A hip-hop feminist approach to music therapy with adolescent females. In S. Hadley & G. Yancy (Eds.), *Therapeutic uses of rap and hip-hop* (pp. 79–98). Routledge.
Verduyn, P., Van Mechelen, I., & Tuerlinckx, F. (2011). The relation between event processing and the duration of emotional experience. *Emotion, 11*, 20–28.
Verney, R., & Ansdell, G. (2010). *Conversations on Nordoff-Robbins music therapy* (Vol. 5). Barcelona Publishers.
Viega, M. (2014). Listening in the ambient mode: Implications for music therapy practice and theory. *Voices: A World Forum for Music Therapy, 14*(2).
Viega, M. (2016). Exploring the discourse in hip hop and implications for music therapy practice. *Music Therapy Perspectives, 34*(2), 138–146.
Viega, M. (2018). A humanistic understanding of the use of digital technology in therapeutic songwriting. *Music Therapy Perspectives, 36*(2), 152–160.
Viega, M. (2019). Globalizing adolescence: Digital music cultures and music therapy. In K. McFerran, P. Derrington, & S. Saarikallio (Eds.), *Handbook of music, adolescents, and wellbeing* (pp. 217–224). Oxford University Press.
Vischer, R. (1873). On the optical sense of form: A contribution to aesthetics. In H. Mallgrave & E. Ikonomou (Eds.), *Empathy, form and space: Problems in German aesthetics* (pp. 89–124). Getty Centre for the History of Art and the Humanities.
Vishkin, A. (2021). Variation and consistency in the links between religion and emotion regulation. *Current Opinion in Psychology, 40*, 6–9.
Vishkin, A., Schwartz, S., Ben-Nun Bloom, P., Solak, N., & Tamir, M. (2020). Religiosity and desired emotions: Belief maintenance or prosocial facilitation? *Personality and Social Psychology Bulletin, 46*(7), 1090–1106.

Vuilleumier, P., & Trost, W. (2015). Music and emotions: From enchantment to entrainment. *Annals of the New York Academy of Sciences, 1337*(1), 212–222.

Wahman, J. (2008). Sharing meanings about embodied meaning. *The Journal of Speculative Philosophy, 22*(3), 170–179.

Waite, R., & McKinney, N. S. (2016). Capital we must develop: Emotional competence educating pre-licensure nursing students. *Nursing Education Perspectives, 37*(2), 101–103.

Waite, S., & Rees, S. (2014). Practising empathy: Enacting alternative perspectives through imaginative play. *Cambridge Journal of Education, 44*(1), 1–18.

Walker, K. L. (1997). Do you ever listen?: Discovering the theoretical underpinnings of empathic listening. *International Journal of Listening, 11*(1), 127–137.

Walsh, P. (2014). Empathy, embodiment, and the unity of expression. *Topio, 33*, 215–226.

Ward, M. (2015). Art in noise: An embodied simulation account of cinematic sound design. In M. Coëgnarts & P. Kravanja (Eds.), *Embodied cognition and cinema* (pp. 155–186). Leuven University Press.

Warnock, T. (2012). Vocal connections: How voicework in music therapy helped a young girl with severe learning disabilities and autism to engage in her learning. *Approaches: Music Therapy & Special Music Education, 4*(2), 85–92.

Warren, J. A., & Nash, A. (2019). Using expressive arts in online education to identify feelings. *Journal of Creativity in Mental Health, 14*(1), 94–104.

Warren, J. T. (2008). Performing difference: Repetition in context. *Journal of International and Intercultural Communication, 1*(4), 290–308.

Warrenburg, L. (2020). Comparing musical and psychological emotion theories. *Psychomusicology: Music, Mind, and Brain, 30*(1), 1.

Waterfall, B., Smoke, D., & Smoke, M. (2017). Reclaiming grassroots traditional indigenous healing ways and practices within urban indigenous community contexts. In S. L. Stewart, R. Moodley, & A. Hyatt (Eds.), *Indigenous cultures and mental health counselling: Four directions for integration with counselling psychology* (pp. 3–16). Routledge.

Waterhouse, M. (2011). *Experiences of multiple literacies and peace: A rhizoanalysis of becoming in immigrant language classrooms* [Doctoral dissertation, University of Ottawa, ON].

Watkins, M. (2015). Psychosocial accompaniment. *Journal of Social and Political Psychology, 3*(1), 324–341.

Watson, A., & Huntington, O. (2008). They're here—I can feel them: The epistemic apaces of Indigenous and Western knowledges. *Social and Cultural Geography, 9*, 257–281.

Watt Smith, T. (2016). *The book of human emotions.* Little Brown Spark.

Watts, V. (2013). Indigenous place-thought and agency amongst humans and non-humans (First woman and sky woman go on a European tour!). *DIES: Decolonization, Indigeneity, Education and Society, 2*(1), 20–34

Wei, L., Wu, G. R., Bi, M., & Baeken, C. (2020). Effective connectivity predicts cognitive empathy in cocaine addiction: a spectral dynamic causal modeling study. *Brain Imaging and Behavior,* 1–9.

Weine, S. M. (1996). The witnessing imagination: Social trauma, creative artists, and witnessing professionals. *Literature and Medicine, 15*(2), 167–182.

Weisz, E., & Zaki, J. (2018). Motivated empathy: A social neuroscience perspective. *Current Opinion in Psychology, 24*, 67–71.

West, T., & Kenny, C. (2017). The cultures of native Americans/first peoples: The voices of two indigenous woman scholars. In A. Whitehead-Pleaux & X. Tan (Eds.), *Cultural intersections in music therapy: Music, health, and the person* (pp. 125–136). Barcelona.

Westerman, M. (2013). Making sense of relational processes and other psychological phenomena: The participatory perspective as a post-Cartesian alternative to Gergen's relational approach. *Review of General Psychology, 17*(4), 358–373.

Westerman, M. A., & Steen, E. M. (2007). Going beyond the internal-external dichotomy in clinical psychology: The theory of interpersonal defense as an example of a participatory model. *Theory & Psychology, 17*(2), 323–351.

Westvall, M. (2021). *Participatory music-making in diasporic contexts.* Danish Musicology Online.

Wetherell, M. (2012). *Affect and emotion: A new social science understanding.* Sage.

Wheeler, B., & Thaut, M. (2011). Music therapy. In P. Juslin & J. Sloboda (Eds.), *Handbook of music and emotion* (pp. 819–848). Oxford University Press.

Whitehead-Pleaux, A., Brink, S., & Tan, X. (2017). Culturally competent music therapy assessments. In A. Whitehead-Pleaux & X. Tan (Eds.), *Cultural intersections in music therapy: Music, health, and the person* (pp. 271–283). Barcelona Publishers.

Whitman, K. (2002). *Playthroughs.* Kranky.

Whitman, W. (1904). *Song of myself.* Roycrofters.

Wigram, T. (2004). *Improvisation: Methods and techniques for music therapy clinicians, educators and students.* Jessica Kingsley.

Wigram, T., & Elefant, C. (2009). Therapeutic dialogue in music: Nurturing musicality of communication in children with autistic spectrum disorder and Rett syndrome. In S. Malloch & C. Trevathen (Eds.), *Communicative musicality* (pp. 423–445). Oxford University Press.

Wilce, J. M., & Wilce, J. M. (2009). *Language and emotion (No. 25).* Cambridge University Press.

Wilde, P., & Evans, A. (2019). Empathy at play: Embodying posthuman subjectivities in gaming. *Convergence, 25*(5–6), 791–806.

Wildman, W. J. (2010). An introduction to relational ontology. In J. Polkinghorne (Ed.), *The Trinity and an entangled world: Relationality in physical science and theology* (pp. 55–73). Wm. B. Eerdmans Publishing.

Wilkerson, A., Dimaio, L., & Sato, Y. (2017). Countertransference in end-of-life music therapy. *Music Therapy Perspectives, 35*(1), 13–22.

Wilkinson, H., Whittington, R., Perry, L., & Eames, C. (2017). Examining the relationship between burnout and empathy in healthcare professionals: A systematic review. *Burnout Research, 6*, 18–29.

Williams, A. (2017). *Therapeutic landscapes.* Routledge.

Williams, G., Gerardi, M., Gill, S., Soucy, M., & Taliaferro, D. (2009). Reflective journaling: Innovative strategy for self-awareness for graduate nursing students. *International Journal for Human Caring, 13*(3), 36–43.

Williams, J. (2005). *Understanding poststructuralism.* Acumen.

Williams, L., & Bargh, J. (2008). Experiencing physical warmth promotes interpersonal warmth. *Science, 322*(5901), 606–607.

Willink, K. G., & Shukri, S. T. (2018). Performative interviewing: Affective attunement and reflective affective analysis in interviewing. *Text and Performance Quarterly, 38*(4), 187–207.

Wilson, J., Ward, C., & Fischer, R. (2013). Beyond culture learning theory: What can personality tell us about cultural competence? *Journal of Cross-Cultural Psychology, 44*(6), 900–927.

Wilson, S. (2008). *Research is ceremony: Indigenous research methods.* Fernwood Publishing.

Winczewski, L. A., Bowen, J. D., & Collins, N. L. (2016). Is empathic accuracy enough to facilitate responsive behavior in dyadic interaction? Distinguishing ability from motivation. *Psychological Science, 27*(3), 394–404.

Winnicott, D. (1965). *The maturational processes and the facilitating environment.* International Universities Press.

Winslade, J. (2009). Tracing lines of flight: Implications of the work of Gilles Deleuze for narrative practice. *Family Process, 48*(3), 332–346.

Winter, P. (2013). *Effects of experiential music therapy education on student's reported empathy and self-esteem: A mixed methods study* [Doctoral dissertation, Temple University].

Winters, D., Wu, W., & Fukui, S. (2020). Longitudinal effects of cognitive and affective empathy on adolescent substance use. *Substance Use & Misuse, 55*(6), 983–989.

Wise, J. (2000). Home: Territory and identity. *Cultural Studies, 14*(2), 295–310.

Wood, J. (2016a). *Interpersonal communication: Everyday encounters* (8th ed.). Cengage Learning.

Woods, P. (2020). Reimagining collaboration through the lens of the posthuman: Uncovering embodied learning in noise music. *Journal of Curriculum and Pedagogy,* 1–21.

Wood, R., & Williams, C. (2008). Inability to empathize following traumatic brain injury. *Journal of the International Neuropsychological Society, 14*, 289–296.

Wood, S. (2006, November). "The Matrix": A model of community music therapy processes. *Voices: A World Forum for Music Therapy, 6*(3).

Wood, S. (2016b). *A matrix for community music therapy practice.* Barcelona Publishers.

Wright, S. (2019). *Therapist cultural humility and the working alliance: Exploring empathy, congruence, and positive regard as mediators.* SUNY.

Wu, J., Eaton, P. W., Robinson-Morris, D. W., Wallace, M. F., & Han, S. (2018). Perturbing possibilities in the postqualitative turn: Lessons from Taoism (道) and Ubuntu. *International Journal of Qualitative Studies in Education, 31*(6), 504–519.

Wyl, A. (2014). Mentalization and theory of mind. *Praxis Der Kinderpsychologie Und Kinderpsychiatrie, 63*(9), 730–737.

Xu, X., Zuo, X., Wang, X., & Han, S. (2009). Do you feel my pain? Racial group membership modulates empathic neural responses. *Journal of Neuroscience, 29*(26), 8525–8529.

Yehuda, N. (2013). 'I am not at home with my client's music… I felt guilty about disliking it': On 'musical authenticity' in music therapy. *Nordic Journal of Music Therapy, 22*(2), 149–170.

Yinger, O. S. (2016). Music therapy as procedural support for young children undergoing immunizations: A randomized controlled study. *Journal of Music Therapy, 53*(4), 336–363.

Yontef, G. M. (1993). *Awareness, dialogue & process: Essays on Gestalt therapy.* The Gestalt Journal Press.

Young, M., & Schlie, E. (2011). The rhythm of the deal: Negotiation as a dance. *Negotiation Journal, 27*(2), 191–203.

Zahavi, D. (2005). *Subjectivity and selfhood: Investigating the first-person perspective.* MIT Press.

Zahavi, D. (2008). Simulation, projection and empathy. *Consciousness and Cognition, 17*, 514–522.

Zahavi, D., & Gallagher, S. (2008). The (in) visibility of others: A reply to Herschbach. *Philosophical Explorations, 11*(3), 237–244.

Zahavi, D. (2010). Empathy, embodiment and interpersonal understanding: From Lipps to Schutz. *Inquiry, 53*(3), 285–306.

Zahavi, D. (2017). Phenomenology, empathy, and mindreading. In H. Maibom (Ed.), *The Routledge handbook of philosophy of empathy* (pp. 33–42). Routledge.

Zaki, J. (2014). Empathy: A motivated account. *Psychological Bulletin, 140*(6), 1608.

Zaki, J., & Ochsner, K. N. (2012). The neuroscience of empathy: Progress, pitfalls and promise. *Nature Neuroscience, 15*(5), 675–680.

Zapata, A., Kuby, C., & Thiel, J. (2018). Encounters with writing: Becoming-with posthumanist ethics. *Journal of Literacy Research, 50*(4), 478–501.

Zarate, R. (2020). *Critical social aesthetics and clinical listening—Cultural listening as a method for arts-based research* [Paper presentation]. World Congress of Music Therapy, Pretoria, South Africa.

Zembylas, M. (2018). Reinventing critical pedagogy as decolonizing pedagogy: The education of empathy. *Review of Education, Pedagogy, and Cultural Studies, 40*(5), 404–421.

Zembylas, M. (2007). Mobilizing anger for social justice: The politicization of the emotions in education. *Teaching Education, 18*(1), 15–28.

Zharinova-Sanderson, O. (2004). Promoting integration and socio-cultural change: Community music therapy with traumatised refugees in Berlin. In M. Pavlicevic & G. Ansdell (Eds.), *Community music therapy* (pp. 233–248). Jessica Kingsley.

Zhou, Q., Eisenberg, N., Losoya, S. H., Fabes, R. A., Reiser, M., Guthrie, I. K., et al. (2002). The relations of parental warmth and positive expressiveness to children's empathy-related responding and social functioning: A longitudinal study. *Child Development, 73*, 893–915.

Zondi, S. (2021). A fragmented humanity and monologues: Towards a diversal humanism. In M. Steyn & W. Mpofu (Eds.), *Decolonising the human: Reflections from Africa on difference and oppression* (pp. 224–242). Wits University Press.

Zorza, J. P., Marino, J., & Acosta Mesas, A. (2019). Predictive influence of executive functions, effortful control, empathy, and social behavior on the academic performance in early adolescents. *The Journal of Early Adolescence, 39*(2), 253–279.

Zurek, P. P., & Scheithauer, H. (2017). Towards a more precise conceptualization of empathy: An integrative review of literature on definitions, associated functions, and developmental trajectories. *International Journal of Developmental Science, 11*(3–4), 57–68.

Index

A. dos Santos, *Empathy Pathways*,
https://doi.org/10.1007/978-3-031-08556-7

S

Printed by Printforce, the Netherlands